HYDRODYNAMICS OF
COASTAL ZONES

Elsevier Oceanography Series, 48

HYDRODYNAMICS OF COASTAL ZONES

STANISLAW R. MASSEL
Institute of Hydroengineering, Gdansk, Poland

ELSEVIER

Amsterdam — Oxford — New York — Tokyo   1989

ELSEVIER SCIENCE PUBLISHERS B.V.
Sara Burgerhartstraat 25
P.O. Box 211, 1000 AE Amsterdam, The Netherlands

Distributors for the United States and Canada:

ELSEVIER SCIENCE PUBLISHING COMPANY INC.
655, Avenue of the Americas
New York, NY 10010, U.S.A.

ISBN 0-444-87375-9 (Vol. 48)
ISBN 0-444-41623-4 (Series)

Printed in The Netherlands

To my family

Contents

PREFACE

Coastal waters are receiving greatly increasing attention by the scientific and engineering community. The scientific interest is motivated mainly by the fact that in the coastal region the interaction between atmosphere, ocean and land can be the most clearly observed. Crests of wind waves or swell can be seen arriving at the beach at intervals of some seconds and appreciable variations in the water level with about 12 or 24 hours between occurences of maximum height are often detected. The proximity of a coast line leads to a piling–up of water and the development of surface slopes. The surface slopes give rise to horizontal pressure gradients in the water and these in turn generate currents which are superimposed on the initial wind drift. On the other hand, the river water, which often gives coastal regions their relatively low salinity, has frequently been modified by its passage through estuaries.

Coastal waters also play a special role from economic and environmental points of view. The expansion of world trade requires construction of large numbers of harbours and terminals. A striking example of the increased use of coastal waters is the exploration for oil and gas resources and their exploitation. At the same time, the recreational use of coastal zone is a very important factor in many areas.

This book does not cover all the problems mentioned above. It is intended to discuss the selected theoretical topics of coastal hydrodynamics, including basic principles and applications in coastal oceanography and coastal engineering. However, the book is not a handbook; the emphasis is placed on presentation of a number of basic problems, rather than giving detailed instructions for their application.

The bulk of the material deals with surface waves. In the author's opinion there is still a strong need for a book on wave phenomena in the coastal waters, as general textbooks on sea surface dynamics focus most of their attention on the deep ocean. The book is intended to cover this need by concentrating on the phenomena typical for the coastal zone.

The way of handling in the book is mainly based on courses of lectures given in the Institute of Hydroengineering, Polish Academy of Sciences in Gdansk. The approach throughout is a combination of the theoretical and observational. A certain amount of mathematics must play its part as the contributions by mathematicians have always been prominent in this field. The necessary mathematical background is a basic knowledge of ordinary and partial differential equations, as well as the statistical and spectral analysis of the time series. The reader should also be familiar with fundamental hydrodynamic concepts.

The book comprises nine chapters. The governing equations and conservation laws are treated in Chapter One, using the variational principles. The theory of regular surface waves is covered in Chapters Two to Four. The nonlinear effect of wave train modulation and their breaking on beaches is examined in Chapter Five. Chapters Six and Seven focus on the statistical and spectral treatment of waves induced by wind. The current generation and the circulation pattern are the subject of Chapter Eight and the sea level variations are examined in Chapter Nine. References for further reading, including both general references and some related to specific sections, can be found at the end of each chapter.

I owe a lot to many people over many years for forming the ideas expressed in the book. In particular, Professors Cz.Druet, and J.Onoszko are those who introduced the author to the field of marine hydromechanics. I wish to acknowledge the stimulating discussions provided by many colleagues and members of the staff at the Institute of Hydroengineering of the Polish Academy of Sciences in Gdansk. I am also very grateful to my wife, Barbara, for her accurate and patient typing of the manuscript and for keeping me up during the writing.

November 1988   S.R. Massel

Chapter 1

INTRODUCTION

1.1 Basic ideas and assumptions

1.1.1 Distinctive features of coastal waters

The term *coastal waters* will be taken for the purpose of this book to define the areas enclosed, towards the ocean by the continental shelf, and towards land by the upper limit for the direct action of the sea, but not extending into estuaries (Fig. 1.1). Coastal waters have features which are sufficiently distinctive from the deep oceans. The presence of the bottom at a relatively shallow depth forms a constraint on water movement, tending to divert currents so that they flow nearly parallel to it. Wind–driven currents are also strongly affected by the presence of the coastline and the bottom. These mechanisms rise in some areas the storm surges or produce the upwelling and the coastal jets.

The shallow–water waves have many specific properties which distinguish them from deep–water waves. As waves travel into shallower water, their dynamics is progressively more nonlinear and dissipative. Energy is transferred away from the peak of the spectrum to higher and lower frequencies. This is mainly due to water depth changes. In the shallow water a variety of processes composes very complicated picture of the sea surface and the wave spectrum. Among them there are nonlinear interactions between spectral components, wave transfer due to shoaling water, wave breaking, etc. However, at present we have not the theory which is able to explain upon experimental results. This is the reason that due to complexity of the wave motion in the coastal zone, our present knowledge is based mainly on the experiments and the application of statistical and spectral analysis.

The influx of fresh water run–off from the land, often passing through estuaries, has the effect of reducing the salinity, and hence also the density of coastal water. As a result of these effects, coastal waters are usually areas of relatively large horizontal gradients of density often associated with changes

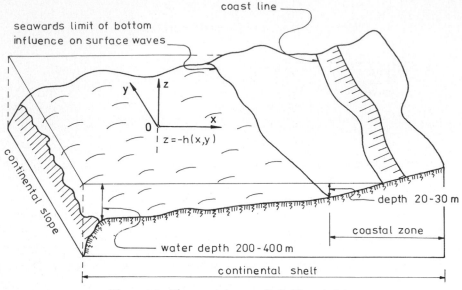

Figure 1.1: The coastal zone. Definition sketch

in currents. The coastal waters are, of course, interconnected and dependent on the adjacent ocean and land. Therefore, they cannot be considered in isolation.

1.1.2 Classification of coastal water oscillations

The periodic rise and fall of the sea surface are experienced by anyone who lives near at the coastline. As many oscillations known in physics text books, all surface variations can be classified. It is quite natural to use the period of time scale as a measure of the distinction. Example of such crude classification is given in Table 1.1, which is mainly due to Svendsen and Jonsson (1982). Of course, such classification has rather descriptive than analytical character but it reflects the variety of oscillation periods observed in the coastal zone.

Prior to introducing the basic principles of hydrodynamics, we will adopt the rectangular coordinates system $O(x, y, z)$ or $O(x_i, z)$; $i = 1, 2$. The origin of the system is at the mean sea surface (Fig. 1.1). The axes x and y are horizontal and the x axis is directed towards coastline while the y axis is parallel to the coastline. The z axis is directed opposite to the force of gravity. For some purposes, the coastal region, defined above, is quite large and diverse from the hydrodynamical point of view. Particularly, the most dynamic part of the coastal waters is the vicinity of the shore line. Thus, it will be useful to define, within the coastal waters, a region called the *coastal zone*. It is enclosed towards the ocean by the water depth contour $h = h_g$, and towards

Table 1.1: Waves, physical mechanisms, and periods

Wave type	Physical mechanism	Periods
Wind waves	Wind shear	<15s
Swell	Wind waves	<30s
Surf beat	Wave groups	1–5min
Seiche	Wind variation	2–40min
Harbour resonance	Surf beat	2–40min
Tsunami	Earthquake	10min–2h
Tides	Gravitational action of the moon and sun, earth rotation	12–24h
Storm surges	Wind stresses and atmospheric pressure variation	1–30days

the land by an upper limit for the direct action of the sea. In the coastal zone, which water depth $h \le h_g$, the water motion (especially wave motion) is strongly influenced by the bottom configuration. From the linear wave theory (see Chapter 2) we find that the bottom influence on the surface waves is observed when:

$$h \le h_g = \frac{L}{2},$$ (1.1)

where: L - wave length.
Using the classical dispersion relation (eq. 2.50), eq. (1.1) gives:

$$h_g \approx \frac{\pi g}{\omega^2},$$ (1.2)

in which: ω – angular frequency of wave motion.
For the wind–induced waves, the frequency ω should be identified with the peak frequency ω_p of wave energy spectrum (see Chapter 7). For example, in the small, semi–enclosed seas, the typical peak frequency ω_p is of order $\sim 1rd/s$. Therefore:

$$h_g \approx \pi g \approx 30.0m,$$ (1.3)

If the typical bottom slope in the coastal zone is estimated as $0.01 \div 0.015$, the average width (D) of the coastal zone is:

$$D \approx 2000 \div 3000m. \tag{1.4}$$

1.1.3 Continuous fluid and water particle concepts

One of the most important physical principles of hydrodynamics is that the fluid is *continuous*. By continuous fluid in continuous motion, we mean that the velocity $\vec{u}$ is everywhere finite and continuous while its space derivatives of the first order are finite (but not necessarily continuous). Thus, any closed surface S which moves with the fluid, permanently and completely separates the fluid matter inside S from that outside. Sometimes the following definition of continuous fluid is used, i.e., the fluid can be treated as continuous when the flow past an obstacle of the dimension A which is much larger than the average free path of the molecule l_0 (for water $l_0 \approx 3 \cdot 10^{-10}m$). The rate (l_0/A) is known as Knudsen number (**Kn**). If:

$$\mathbf{Kn} = \frac{l_0}{A} < 0.01, \tag{1.5}$$

the fluid can be treated as continuous (Puzyrewski and Sawicki, 1987). In the continuous fluid, we can define *a fluid particle* as consisting of the fluid contained within an infinitesimal volume, that is to say, a volume whose size may be considered so small that for the particular purpose in hand its linear dimensions are negligible. We can then treat a fluid particle as a geometrical point.

In general, the equation of motion for the fluid particle depends on the physical properties of the fluid and motion itself. In order to render the subject amenable to exact mathematical treatment, we make simplifying assumptions on the fluid and motion, i.e.:

a) *water is an inviscid or perfect fluid.* An inviscid fluid is a continuous fluid which can exert no shearing stress. However, the real fluids do have viscosity which creates the stresses and additional dissipations within the fluid. Thus:

$$\vec{\tau} = \mu \frac{d\vec{u}}{d\vec{n}}, \tag{1.6}$$

where: $\vec{\tau}$ - tangential stress, $\vec{u} = (u, v, w)$ - fluid velocity vector, $\vec{n}$ - vector, normal to vector $\vec{u}$, μ - coefficient of viscosity of the fluid.
Sometimes for convenience, we will represent the vector $\vec{u}$ as $\vec{u} = (u_i, w; i = 1, 2)$; therefore $u_1 = u, u_2 = v$. Thus, for the ideal fluid should be:

$$\vec{\tau} = 0 \quad \text{or} \quad \mu = 0 \quad \text{when} \quad \vec{u} \neq \vec{0}, \tag{1.7}$$

in which: $\vec{0} = (0, 0, 0)$.

To justify the identity (1.7), we adopt the differential length scale L in which the velocity varies in magnitude by U. The ratio $\mathbf{Re} = \rho U L / \mu$ (Reynolds number) represents the relative magnitudes of the inertial and viscous terms; in many oceanic motions, the Reynolds number is very large. Thus, the viscous influence is often quite negligible over most of the field of motion. The viscous forces are important only in narrow regions of the flow, where the local inertial and viscous forces are comparable. In the ocean the interfacial layer between the air and the water, as well as the bottom boundary layer, illustrates such regions quite clearly. The thickness δ of the surface boundary layer is of order $\delta \approx (2\nu/\omega)^{1/2}$, where ν - kinematic coefficient of viscosity (for water $\nu \approx 1.2 \cdot 10^{-6} m^2/s$). For the typical frequencies, the thickness $\delta \sim 0.001$m. For the boundary layer near the natural sea bottom, the eddy viscosity is much higher ($\sim 100\nu$); the thickness of the boundary layer is then about 0.1 m, which is still quite small. Therefore, the boundary layer regions are but a very small fraction of a fluid volume, and the influence of the viscosity on the wave motion can be neglected.

b) *water is an incompressible fluid*. The compressibility of water is rather small and the Young modulus is of order $E \approx 3.05 \cdot 10^8 N/m^2$ (Dera, 1983). As the typical velocity of sea water is much smaller than the sound speed, the very small water compressibility has not influence on the water motion. Therefore, the equation of continuity for the homogeneous incompressible fluid becomes (Milne–Thomson, 1974):

$$div \ \vec{u} = \nabla \cdot \vec{u} = \frac{\partial u}{\partial x} + \frac{\partial v}{\partial y} + \frac{\partial w}{\partial z}, \tag{1.8}$$

where:

$$\nabla = \frac{\partial}{\partial x}\vec{i} + \frac{\partial}{\partial y}\vec{j} + \frac{\partial}{\partial z}\vec{k}. \tag{1.9}$$

In general, vector $\vec{u}$ represents the sum of the current and wave velocities.

c) *motion is irrotational*. It means that the individual elementary particles of the fluid do not rotate. The mathematical expression of this is:

$$rot \ \vec{u} = curl \ \vec{u} = \nabla \times \vec{u} = \vec{0}, \tag{1.10}$$

When the vorticity is different from zero, the motion is defined as *rotational*. It was indicated above that in many oceanic motions, the influence of the viscous terms are quite negligible. In this event, the Lagrange theorem (Kochin et al., 1963) indicates that if, at some initial instant, the vorticity vanishes everywhere in the field of flow, the motion is irrotational. In the absence of viscous effects, it remains so. The consequence of eq. (1.10) is that the velocity $\vec{u}$ can be represented as the gradient of a scalar function, the *velocity potential* Φ:

$$\vec{u} = \nabla \Phi; \tag{1.11}$$

then in virtue of the continuity equation (1.8), the potential Φ obeys Laplace equation:

$$div(\nabla \Phi) \equiv \nabla^2 \Phi = \frac{\partial^2 \Phi}{\partial x^2} + \frac{\partial^2 \Phi}{\partial y^2} + \frac{\partial^2 \Phi}{\partial z^2}. \tag{1.12}$$

Additionally, we assume that the current velocity in the coastal zone is a slowly varying function of the horizontal coordinates (x_1, x_2) and time t. The characteristic distance $(\hat{L})$ and time $(\hat{T})$ scales of the current are much greater than those of the waves (Mei, 1983):

$$(\omega \hat{T})^{-1} \sim (k\hat{L})^{-1} \sim \frac{h}{\hat{L}} = O\left(\frac{\nabla_h h}{kh}\right) = O(\mu) \ll 1, \tag{1.13}$$

in which: ω – characteristic frequency of wave motion, k - corresponding wave number and:

$$\nabla_h = \frac{\partial}{\partial x}\vec{i} + \frac{\partial}{\partial y}\vec{j}. \tag{1.14}$$

We shall also assume, that the characteristic water depth h varies slowly in horizontal coordinates $x_i (i = 1, 2)$. The horizontal velocity components $U_i (i = 1, 2)$ usually are $O(\sqrt{gh})$. Because of (1.13), the vertical current velocity is small, $W = O(\mu)$. Then, the horizontal components of the vorticity vector take the form:

$$\left[\frac{\partial U_1}{\partial z} - \frac{\partial W}{\partial x_1}\right]\vec{i} \quad \text{and} \quad \left[\frac{\partial U_2}{\partial z} - \frac{\partial W}{\partial x_2}\right]\vec{j}, \tag{1.15}$$

where: $\vec{U} = (U_j, W), j = 1, 2$.
As the term $\partial W / \partial x_j$ is $O(\mu^2)$, the horizontal components are:

$$\frac{\partial U_j}{\partial z} = O(\mu^2). \tag{1.16}$$

1.2 Kinematics of water particle

In the coastal zone, many fluctuating motions can be identified as waves. However, the space and time scales of these oscillations are small compared with the ambient medium varying. Thus, any local property of the wave train (i.e. pressure, velocity or surface displacement) can be specified by $\eta = a\exp(i\chi)$, where a is the local amplitude and $\chi(\vec{x}, t, \vec{k}, \omega)$ is the phase. The propagation of points of constant phase is given by $\chi(\vec{x}, t) = const$. In general, for the varying bathymetry and nonuniform current, the phase $\chi(\vec{x}, t, \vec{k}, \omega)$ is also slowly varying function of coordinates and time.

Let us now define the local wave number $\vec{k}$ and frequency ω by:

$$\vec{k} = \nabla_h \chi \qquad \text{and} \qquad \omega = -\frac{\partial \chi}{\partial t}, \tag{1.17}$$

From the first of these it follows immediately that $\nabla \times \vec{k} = 0$. Thus, the local wave–number vector is irrotational. It is more convenient to eliminate phase χ to give:

$$\frac{\partial \vec{k}}{\partial t} + \nabla_h \omega = 0 \qquad \text{and} \qquad \frac{\partial k_i}{\partial x_j} - \frac{\partial k_j}{\partial x_i} = 0. \tag{1.18}$$

In order to make the physical interpretation of eq. (1.18), we restrict ourselves to the one–dimensional version, i.e.:

$$\frac{\partial k}{\partial t} + \frac{\partial \omega}{\partial x} = 0. \tag{1.19}$$

By definition, k is the number of the lines of equal phase per unit distance (or k is the *density* of phase line). On the other hand, ω is the number of phase lines passing a fixed point (or ω is the *flux* of phase lines). The rate of out–flux of phase lines is $\frac{\partial \omega}{\partial x}dx$. At the same time, the rate of decrease of phase lines is $-\frac{\partial k}{\partial t}dx$. Therefore, eq. (1.19) is a kinematical conservation equation for the density of waves.

In each point of the wave field, the frequency ω and the wave number k are related by the dispersion relation, which depends not only on the local wave number, but also on a local property as the water depth, current velocity or the ambient density gradient. In general:

$$\omega = \Omega[\vec{k}, f(\vec{x}, h, t, \rho, \ldots)]. \tag{1.20}$$

For example, in Section 2.3 will be shown that in the medium, where $\vec{U} \equiv 0$

and $h = const$, the dispersion relation takes the form:

$$\omega^2 = \sigma^2 = gk \tanh(kh),$$ (1.21)

where: $k = | \vec{k} |$.
If the medium itself is moving with velocity $\vec{U}$, the frequency of waves passing a fixed point is:

$$\omega = \Omega(\vec{k}, \vec{x}) = \sigma + \vec{k} \cdot \vec{U} = [gk \tanh kh]^{1/2} + \vec{k} \cdot \vec{U}.$$ (1.22)

The quantity ω is called the observed or apparent frequency and σ is the intrinsic frequency whose functional dependence on $\vec{k}$ is known as the dispersion relation (eq. 1.21). Substituting eq. (1.20) into eq. (1.18), we get:

$$\frac{\partial k_i}{\partial t} + \frac{\partial \Omega}{\partial k_j} \frac{\partial k_j}{\partial x_i} = -\frac{\partial \Omega}{\partial x_i}, \qquad i, j = 1, 2$$ (1.23)

or:

$$\frac{\partial k_i}{\partial t} + C_{g_j}(\vec{x}, \vec{k}, t) \frac{\partial k_j}{\partial x_i} = -\frac{\partial \Omega}{\partial x_i},$$ (1.24)

in which:

$$C_{g_j} = \frac{\partial \Omega}{\partial k_j}, \qquad i, j = 1, 2.$$ (1.25)

The velocity C_g is called *group velocity*. Eqs. (1.24) can be rewritten in the canonical form as:

$$\frac{dk_i}{dt} = -\frac{\partial \Omega}{\partial x_j} \qquad \text{and} \qquad \frac{dx_i}{dt} = \frac{\partial \Omega}{\partial k_i}.$$ (1.26)

Equations (1.26) are identical with Hamilton equations in mechanics. The vectors $\vec{x}$ should be interpreted as coordinates and vectors $\vec{k}$ represent the momenta. Then, the frequency $\Omega(\vec{k}, \vec{x}, t)$ is taken to be the Hamiltonian (Whitham, 1974; Gelfand and Fomin, 1975). Moreover, eqs. (1.26) are the canonical equations for the rays (Synge, 1963). If we are interested in the evolution of the phase χ, after substitution of eq. (1.17) in eq. (1.20) we obtain:

$$\frac{\partial \chi}{\partial t} + \Omega[\nabla_h \chi, f(\vec{x}, h, t, \rho, \ldots)] = 0,$$ (1.27)

which is the Hamilton–Jacobi equation, when the phase χ is the action.

1.3 Dynamics of water particle

The traditonal approach to hydromechanics is based on the simultaneous application of the infinitezimal and integral laws of mechanics. The first ones yields the differential motion equations for the fluid particle. From the integral laws, usually related to the variational calculus, the conservation laws result. They describe the water motion within the finite interval of time. Moreover, the corresponding integrals should reach some extremal values, usually minimal ones.

In order to complete the hydrodynamical boundary value problem, the equation of motion should be complemented by the boundary conditions at the sea surface and sea bottom. In this Section we present the basic equation of motion. We use the equivalence of infinitezimal and integral laws extensively to derive the appropriate equations. Particularly, Hamilton variational principle of the minimum action was found to be most applicable to our problem.

The wave motion at the sea surface composes many forms and scales in time and space. However, the similarity between the wave phenomena in the sea and wave motion in other continuum is quite easily observed. The investigations in the last years indicate that the attention to the Hamiltonian structure of the complete nonlinear problem and the use of methods based on infinitesimal – transformation theory provide a systematic account of symmetrics that is applicable to all wave problems mentioned above (Seliger and Whitham, 1968; Zakharov, 1974; Benjamin and Olver, 1982). The equations of motion can be easily derived from Hamilton principle when Lagrangian coordinates are used. This is because the similarity with a system of discrete particle is preserved. However, when a Eulerian description is adopted, this close similarity with a system of particles is lost. In fluid mechanics, the Eulerian description is usually preferable and the variational principle should be deduced by special methods for the fluid flow within this description.

Let us consider the principle of minimum action to our problem. This principle postulates that within the limited interval of time (t_1, t_2), the following functional J should be stationary to small variations:

$$\delta J = \delta \int_{t_1}^{t_2} \int_{x_1}^{x_2} \int_{y_1}^{y_2} \mathcal{L} \, d\vec{x} \, dt = 0. \qquad (1.28)$$

The integral eq. (1.28) has the dimension of action, i.e. energy times time and $\mathcal{L}$ is the Lagrangian density. Deriving the functional J for the perfect fluid, the main difficulty is to find the appropriate form of the Lagrangian density. For a mechanical system whose energy is completely known it is possible to use Hamilton principle under the assumption that the Lagrangian function is presented as the difference between the kinetic E_k and potential E_p energies:

$$\mathcal{L} = E_k - E_p. \tag{1.29}$$

For the water waves in the Eulerian description, the variational principle was found relatively recently by Luke (1967). Under the assumption that the wave train propagates on the slowly varying current, restricted by eq. (1.13), we get:

$$\mathcal{L} = -\rho \int_{-h(\vec{x})}^{\zeta(\vec{x},t)} \left\{ \frac{\partial \Phi^*}{\partial t} + \frac{1}{2}(\nabla_h \Phi^*)^2 + \frac{1}{2}\left(\frac{\partial \Phi^*}{\partial z}\right)^2 + gz \right\} dz, \tag{1.30}$$

where: $\zeta(\vec{x}, t)$ - free surface ordinate, $z = -h(\vec{x})$ - bottom surface, $\Phi^*(\vec{x}, z, t)$ - velocity potential.
The uniform periodic solution of the water wave equations takes the form:

$$\Phi^*(\vec{x}, z, t) = (U_i x_i + W z) + \Phi(\vec{x}, z, t), \qquad i = 1, 2, \tag{1.31}$$

in which: $\Phi(\vec{x}, z, t)$ - velocity potential corresponding to wave motion.
For later convenience, the dependence of current on time will be omitted. This study is concerned with the influence of a current, which is slowly varying in space, on waves; the opposite effect of waves on the current will be neglected. Substitution (1.31) into (1.30) gives:

$$\mathcal{L} = -\rho \int_{-h(\vec{x})}^{\zeta(\vec{x},t)} \left\{ \frac{1}{2}\left[\left(U_i + \frac{\partial \Phi}{\partial x_i}\right)^2 + \left(W + \frac{\partial \Phi}{\partial z}\right)^2\right] + \frac{\partial \Phi}{\partial t} + gz \right\} dz. \tag{1.32}$$

Please note that, the Lagrangian density is just the pressure!!! Variation of the functional (1.28) yields now:

$$\delta J = - \rho \int_{t_1}^{t_2} \int_{a_1}^{a_2} \int_{b_1}^{b_2} \left\{ \left[\frac{1}{2} \left(U_i + \frac{\partial \Phi}{\partial x_i} \right)^2 + \frac{1}{2} \left(W + \frac{\partial \Phi}{\partial z} \right)^2 + \frac{\partial \Phi}{\partial t} + gz \right]_{z=\zeta} \cdot \delta \zeta + \right.$$

$$+ \left[\frac{1}{2} \left(U_i + \frac{\partial \Phi}{\partial x_i} \right)^2 + \frac{1}{2} \left(W + \frac{\partial \Phi}{\partial z} \right)^2 + \frac{\partial \Phi}{\partial t} + gz \right]_{z=-h} \cdot \delta h +$$

$$+ \int_{-h(\bar{x})}^{\zeta(\bar{x},t)} \left[\left(U_i + \frac{\partial \Phi}{\partial x_i} \right) \delta \left(\frac{\partial \Phi}{\partial x_i} \right) + \left(W + \frac{\partial \Phi}{\partial z} \right) \delta \left(\frac{\partial \Phi}{\partial z} \right) \right] dz +$$

$$+ \left. \int_{-h(\bar{x})}^{\zeta(\bar{x},t)} \delta \left(\frac{\partial \Phi}{\partial t} \right) dz \right\} d\bar{x} dt. \tag{1.33}$$

Using the fact that (Smirnov, 1961):

$$\delta \left(\frac{\partial \Phi}{\partial x_i} \right) = \frac{\partial}{\partial x_i} \delta \Phi \qquad \delta \left(\frac{\partial \Phi}{\partial z} \right) = \frac{\partial}{\partial z} \delta \Phi \qquad \delta \left(\frac{\partial \Phi}{\partial t} \right) = \frac{\partial}{\partial t} \delta \Phi \tag{1.34}$$

and integrating by parts in (1.33) gives:

$$\delta J = - \rho \int_{t_1}^{t_2} \int_{a_1}^{a_2} \int_{b_1}^{b_2} \left\{ \left[\frac{1}{2} \left(U_i + \frac{\partial \Phi}{\partial x_i} \right)^2 + \frac{1}{2} \left(W + \frac{\partial \Phi}{\partial z} \right)^2 + \frac{\partial \Phi}{\partial t} + gz \right]_{z=\zeta} \cdot \delta \zeta + \right.$$

$$+ \left[\frac{1}{2} \left(U_i + \frac{\partial \Phi}{\partial x_i} \right)^2 + \frac{1}{2} \left(W + \frac{\partial \Phi}{\partial z} \right)^2 + \frac{\partial \Phi}{\partial t} + gz \right]_{z=-h} \cdot \delta h +$$

$$- \left[\left(U_i + \frac{\partial \Phi}{\partial x_i} \right) \frac{\partial \zeta}{\partial x_i} - \left(W + \frac{\partial \Phi}{\partial z} \right) + \frac{\partial \zeta}{\partial t} \right]_{z=\zeta} \cdot \delta \Phi +$$

$$+ \left[- \left(U_i + \frac{\partial \Phi}{\partial x_i} \right) \frac{\partial h}{\partial x_i} - \left(W + \frac{\partial \Phi}{\partial z} \right) \right]_{z=-h} \cdot \delta \Phi +$$

$$- \left. \int_{-h(\bar{x})}^{\zeta(\bar{x},t)} \left[\left(\frac{\partial U_i}{\partial x_i} + \frac{\partial W}{\partial z} \right) + \left(\frac{\partial^2 \Phi}{\partial x_i{}^2} + \frac{\partial^2 \Phi}{\partial z^2} \right) \right] \delta \Phi dz \right\} d\bar{x} \, dt = 0. \tag{1.35}$$

By choosing $\delta \zeta = 0$, $\delta h = 0$ and $\delta \Phi = 0$ on $z = \zeta$ and $z = -h$, and applying the usual variational argument, we obtain the Euler equations corresponding to the Lagrangian $\mathcal{L}$ in the form:

$$\left(\frac{\partial U_i}{\partial x_i} + \frac{\partial W}{\partial z}\right) + \left(\frac{\partial^2 \Phi}{\partial x_i{}^2} + \frac{\partial^2 \Phi}{\partial z^2}\right) = \nabla^2 \Phi + \nabla \cdot \vec{U} = 0. \tag{1.36}$$

Similarly, the appropriate choice of arguments for other variations gives:

$$\frac{1}{2}\left(U_i + \frac{\partial \Phi}{\partial x_i}\right)^2 + \frac{1}{2}\left(W + \frac{\partial \Phi}{\partial z}\right)^2 + \frac{\partial \Phi}{\partial t} + gz = 0 \qquad \text{at} \qquad z = \zeta, \tag{1.37}$$

$$-\left(U_i + \frac{\partial \Phi}{\partial x_i}\right)\frac{\partial \zeta}{\partial x_i} + \left(W + \frac{\partial \Phi}{\partial z}\right) - \frac{\partial \zeta}{\partial t} = 0 \qquad \text{at} \qquad z = \zeta, \tag{1.38}$$

$$\left(U_i + \frac{\partial \Phi}{\partial x_i}\right)\frac{\partial h}{\partial x_i} + \left(W + \frac{\partial \Phi}{\partial z}\right) = 0 \qquad \text{at} \qquad z = -h. \tag{1.39}$$

System of eqs. (1.36 - 1.39) represents the boundary value problem for the surface waves propagated over the slowly varying current. If the current is neglected, eqs. (1.36 - 1.39) may be written as:

$$\nabla^2 \Phi = 0, \tag{1.40}$$

$$\frac{1}{2}\left(\frac{\partial \Phi}{\partial x_i}\right)^2 + \frac{1}{2}\left(\frac{\partial \Phi}{\partial z}\right)^2 + \frac{\partial \Phi}{\partial t} + gz = 0 \qquad \text{at} \qquad z = \zeta, \tag{1.41}$$

$$-\frac{\partial \Phi}{\partial x_i}\frac{\partial \zeta}{\partial x_i} + \frac{\partial \Phi}{\partial z} - \frac{\partial \zeta}{\partial t} = 0 \qquad \text{at} \qquad z = \zeta \qquad i = 1, 2 \tag{1.42}$$

$$\frac{\partial \Phi}{\partial x_i}\frac{\partial h}{\partial x_i} + \frac{\partial \Phi}{\partial z} = 0, \qquad \text{at} \qquad z = -h \qquad i = 1, 2. \tag{1.43}$$

Let the equation of the rigid bottom be given by:

$$G(x, y, z) = z + h(x, y) = 0. \tag{1.44}$$

Then, in the case of an inviscid fluid, the condition to be satisfied on $G = 0$ is the same as the condition (1.43). This is a kinematic condition, i.e., the component of velocity of the fluid normal to the surface bottom must equal zero.

We suppose that the atmosphere and water environment has a common boundary surface, $S(t)$. As this surface moves, the velocity of a point (x, y, z) on it, in the direction of the normal to the surface, is equal to the normal component velocity of the particle of fluid at the same point of the surface.

For $S(t)$ to be a bounding surface means, that there can be no transfer of matter across the surface. Consequently, the mathematical formulation of the above statements is given by eq. (1.42). Let us represent the surface $S(t)$ in the form of the summation of two components:

$$z = \zeta(\vec{x}, t) = \bar{\zeta}(\vec{x}) + \eta(\vec{x}, t), \tag{1.45}$$

where: $\bar{\zeta}(\vec{x})$ - is the average ordinate of the free surface induced mainly by the storm surges, currents and long–period oscillations, $\eta(\vec{x}, t)$ - ordinate caused by the surface waves, when the $\bar{\zeta}$ is taken as the reference level. Substituting (1.45) into (1.38), we get:

$$\frac{\partial}{\partial t}(\bar{\zeta} + \eta) + \left(U_i + \frac{\partial \Phi}{\partial x_i}\right) \frac{\partial}{\partial x_i}(\bar{\zeta} + \eta) = W + \frac{\partial \Phi}{\partial z}. \tag{1.46}$$

When we linearize and use the fact that the current velocity is a slowly varying quantity, eq. (1.46) may be approximated by:

$$\frac{\partial \zeta}{\partial t} + U_i \frac{\partial \zeta}{\partial x_i} + \frac{\partial \Phi}{\partial x_i} \frac{\partial \bar{\zeta}}{\partial x_i} - \frac{\partial \Phi}{\partial z} = 0 \qquad i = 1, 2. \tag{1.47}$$

It should be mentioned that for the waves in the water of intermediate depth, the horizontal and vertical components of the orbital velocity are of the same order. Therefore, the following estimation should be satisfied:

$$\left| \frac{\partial \Phi}{\partial x_i} \frac{\partial \bar{\zeta}}{\partial x_i} \right| \cdot \left| \frac{\partial \Phi}{\partial z} \right|^{-1} \simeq \left| \frac{\partial \bar{\zeta}}{\partial x_i} \right| \ll 1.0. \tag{1.48}$$

Therefore the term $\left[(\partial \Phi / \partial x_i)(\partial \bar{\zeta} / \partial x_i)\right]$ is very small, especially compared to the term $U_i(\partial \zeta / \partial x_i)$, because the gradient operator was applied to the quantity $\bar{\zeta}$, which is varying in the scale $\hat{L}$ (eq. 1.13). Thus, eq. (1.47) takes the form:

$$\frac{\partial \zeta}{\partial t} + U_i \frac{\partial \zeta}{\partial x_i} - \frac{\partial \Phi}{\partial z} = 0 \qquad \text{at} \qquad z = \bar{\zeta}. \tag{1.49}$$

However, within the zone of very small water depths, eq. (1.47) should be used. The remaining equations of the boundary value problem, i.e., eqs. (1.36) and (1.37) or eqs. (1.40) and (1.41), are related to some conservation laws which will be discussed in the next Section.

1.4   Conservation laws

Usually the conservation principles are presented in the integral form as the physical meaning is related rather to some volume of fluid than to the separated point. Therefore, we can write:

$$\int_{t_1}^{t_2} \int_{\Omega} \left(\frac{\partial \vec{p}}{\partial t} + div\vec{A} + \vec{G} \right) d\vec{x}\,dt = 0, \tag{1.50}$$

where: $\vec{p}$ and $\vec{G}$ - vector fields and $\vec{A}$ - matrix field.
This equation can be satisfied for the arbitrary space $\Omega \times (t_1, t_2)$ only if:

$$\frac{\partial \vec{p}}{\partial t} + div\vec{A} + \vec{G} = 0, \tag{1.51}$$

for each point $(\vec{x}, t)$; (t_1, t_2) - time interval, Ω - arbitrary volume of space. Equation (1.51) represents the general differential form of the conservation laws.

The formulation of the conservation laws in the water–wave problem is simplified considerably by the application of the Hamiltonian formalism and the Noether's theorem. This theorem states that every one–parameter group of symmetries for a variational problem determines a conservation law which is satisfied by solutions of the corresponding Euler–Lagrange equations (Gelfand and Fomin, 1975). Many such conservation laws linked to these symmetries were derived by Benjamin and Olver (1982) and Longuet–Higgins (1983). In particular, when the motion is two–dimensional, they showed the existence of as many as "8" conserved "densities", each with a corresponding flux. Some of these, for example, the densities of mass, momentum and energy are well known; others, such as the angular momentum, and another similar–looking quantities, are less familar. The derivation of these conservation laws were given in a rather general form, based on the integral invariants of evolution equations under the transformations of a Lie group. However, for our purpose it will be useful to give a more simple formulation.

Therefore let us consider slowly varying wave trains. The fact that a wave train is slowly varying may be due to non–uniformity of the medium (e.g., a sloping bottom) or to the boundary and/or initial conditions imposed on the wave train. The governing equations for irrotational waves follow from the variational principle (1.28) when the Lagrangian is defined by eq. (1.30). Because Φ^* is a potential variable, the most general form for a periodic wave train is given by:

$$\Phi^* = \Phi_p(\vec{x}, t) + \Phi(\chi, z), \qquad \zeta = \zeta(\chi), \tag{1.52}$$

where:

$$\left. \begin{array}{rcll} \Phi_p &=& U_i x_i - \gamma_p t & \text{for} \quad i = 1, 2, 3 \\[2mm] \chi &=& \chi(\vec{k}, \omega, \vec{x}, t) & \end{array} \right\}. \tag{1.53}$$

Physically U_i are the components of the mean current velocity while γ_p contributes to the mean pressure; therefore, it is related to the mean water level. For slowly varying wave trains, U_i and γ_p are defined as:

$$U_i = \frac{\partial \Phi_p}{\partial x_i} \qquad \gamma_p = -\frac{\partial \Phi_p}{\partial t}. \tag{1.54}$$

In the expressions for ζ and Φ there are two additional parameters taken to be the amplitude a and the mean value $\bar{\zeta}$ of the height ζ. Therefore, the solution depends upon two triads of parameters (k, ω, a) and $(U, \gamma_p, \bar{\zeta})$. The equations for the governing functions can be derived from an averaged form of the variational principle. The average Lagrangian is found by substituting the periodic solution (1.52) into the expression (1.30), i.e.:

$$\bar{\mathcal{L}}(k, \omega, a; U, \gamma_p, \bar{\zeta}) = \frac{1}{2\pi} \int_0^{2\pi} \mathcal{L} d\chi. \tag{1.55}$$

The average variational principle:

$$\delta \int \int \int \bar{\mathcal{L}} d\vec{x} dt = 0 \tag{1.56}$$

is to be used for variations in δE, $\delta \bar{\zeta}$, $\delta \Phi_p$; thus the corresponding Euler equations are:

$$\delta E : \frac{\partial \bar{\mathcal{L}}}{\partial E} = 0, \tag{1.57}$$

$$\delta \bar{\zeta} : \frac{\partial \bar{\mathcal{L}}}{\partial \bar{\zeta}} = 0, \tag{1.58}$$

$$\delta \chi : \frac{\partial}{\partial t} \left(\frac{\partial \bar{\mathcal{L}}}{\partial \omega} \right) - \frac{\partial}{\partial x_i} \left(\frac{\partial \bar{\mathcal{L}}}{\partial k_i} \right) = 0 \qquad i = 1, 2, \tag{1.59}$$

$$\delta\Phi_p : \frac{\partial}{\partial t}\left(\frac{\partial\bar{\mathcal{L}}}{\partial\gamma_p}\right) - \frac{\partial}{\partial x_i}\left(\frac{\partial\bar{\mathcal{L}}}{\partial U_i}\right) = 0 \qquad i = 1, 2, \tag{1.60}$$

The system is completed by eliminating χ and Φ_p in eqs. (1.52) - (1.54); thus:

$$\frac{\partial k_i}{\partial t} + \frac{\partial\omega}{\partial x_i} = 0 \qquad \frac{\partial U_i}{\partial t} + \frac{\partial\gamma_p}{\partial x_i} = 0 \qquad i = 1, 2. \tag{1.61}$$

Note that the variation principle (1.28) and Lagrangian (1.55) yield another important relations. As the Lagrangian $\bar{\mathcal{L}}$ is invariant with respect to shifts in time and in space, the Noether's theorem gives (Whitham, 1967b):

$$\frac{\partial}{\partial t}\left(\omega\frac{\partial\bar{\mathcal{L}}}{\partial\omega} + \gamma_p\frac{\partial\bar{\mathcal{L}}}{\partial\gamma_p} - \bar{\mathcal{L}}\right) - \frac{\partial}{\partial x_i}\left(\omega\frac{\partial\bar{\mathcal{L}}}{\partial k_i} + \gamma_p\frac{\partial\bar{\mathcal{L}}}{\partial U_i}\right) = 0, \tag{1.62}$$

and

$$\frac{\partial}{\partial t}\left(k_i\frac{\partial\bar{\mathcal{L}}}{\partial\omega} + U_i\frac{\partial\bar{\mathcal{L}}}{\partial\gamma_p}\right) - \frac{\partial}{\partial x_j}\left(k_i\frac{\partial\bar{\mathcal{L}}}{\partial k_j} + U_i\frac{\partial\bar{\mathcal{L}}}{\partial U_j} - \bar{\mathcal{L}}\delta_{ij}\right) = 0. \tag{1.63}$$

Since exact expressions for $\Phi(\chi, z)$ and $\zeta(\chi)$ are not known, further progress is made by using the expansions of Stokes wave theory (Whitham, 1967a). For the case of an horizontal bottom we get:

$$\Phi(\chi, z) = \sum_{n=1}^{\infty}\frac{A_n}{n}\cosh[nk(z + h)]\sin(n\chi), \tag{1.64}$$

and

$$\zeta(\chi) = \bar{\zeta} + \sum_{n=1}^{\infty}a_n\cos(n\chi). \tag{1.65}$$

More information on the Stokes waves is given in Sections 2.3 and 2.4.

After substituting of eqs. (1.52), (1.53), (1.64) and (1.65) into eq. (1.55) and carrying out the integration with respect to phase χ, we obtain:

$$\bar{\mathcal{L}} = \rho\left(\gamma_p - \frac{1}{2}|\vec{U}|^2\right)(h + \bar{\zeta}) + \frac{1}{2}\rho g\left(h^2 - \bar{\zeta}^2\right) +$$

$$+ \frac{1}{2}E\left\{\frac{(\omega - U_j k_j)^2}{gk\tanh k(h + \bar{\zeta})} - 1\right\} + \frac{1}{2}\frac{k^2 E^2}{\rho g}\left\{\frac{9T^4 - 10T^2 + 9}{8T^4}\right\} +$$

$$+ O(E^3), \tag{1.66}$$

where: $T = \tanh[k(h + \bar{\zeta})]$ and $E = \frac{1}{2}\rho g a^2$.

Equations (1.57) - (1.60) can now be worked out using the $\bar{\mathcal{L}}$ from eq. (1.66). It is then readily seen that the dispersion relation follows from eq. (1.57) as:

$$\frac{(\omega - U_j k_j)^2}{gk \tanh[k(h + \bar{\zeta})]} = 1 + (ka)^2 \frac{9T^4 - 10T^2 + 9}{8T^4} + O(E^2), \tag{1.67}$$

which is indeed the dispersion relation for the second–order Stokes waves on water of finite depth. When the higher terms are omitted, and $\bar{\zeta} = 0$, the expression (1.67) yields the linear dispersion relation (2.50).

Conservation of mass. The conservation of mass can be derived by the variation $\delta \Phi_p$ (eq. 1.60), i.e.:

$$\frac{\partial}{\partial t}[\rho(h + \bar{\zeta})] + \frac{\partial}{\partial x_j}\left[\rho(h + \bar{\zeta})U_j + E\frac{k_j}{\sigma}\right] = 0. \tag{1.68}$$

In the Section 2.3.2 it was shown that:

$$E\frac{k_j}{\sigma} = \frac{E}{C_j} \equiv \hat{M}_i, \tag{1.69}$$

where:

$$\hat{M}_i \equiv \overline{\int_{-h}^{\eta + \bar{\zeta}} \rho u_i dz} \tag{1.70}$$

is the mass transport of the wave motion. The term $\rho(h + \bar{\zeta})U_j$ in eq. (1.68) represents the mass flux per unit width of the *mean flow*, i.e.:

$$M_i = \overline{\int_{-h}^{\bar{\zeta}} \rho U_i dz} = \rho U_i(h + \bar{\zeta}). \tag{1.71}$$

Thus eq. (1.68) gives for the conservation of total mass for unit area:

$$\frac{\partial}{\partial t}\left[\rho(h + \bar{\zeta})\right] + \frac{\partial}{\partial x_j}\left[M_j + \hat{M}_j\right] = 0, \qquad j = 1, 2. \tag{1.72}$$

Note, that for stationary case, eq. (1.72) can be obtained by integrating of continuity equation (1.8) against z.

Conservation of wave action. Equation (1.59) can be recognized as being a conservation equation for the quantity $\left(\partial \bar{\mathcal{L}}/\partial \omega\right)$ with $\left(\partial \bar{\mathcal{L}}/\partial k_j\right)$ as a flux (see also general form (1.51)). It is now possible to define the wave action density A as:

$$A = \frac{\partial \bar{\mathcal{L}}}{\partial \omega}. \tag{1.73}$$

Note that A is the rate of change of $\bar{\mathcal{L}}$ with ω, keeping k_j constant. It was also mentioned above that in the mechanical systems, the action is defined as the integral of the Lagrangian function $\mathcal{L}$ (eq. 1.28) between two instants; the Euler equation follows then from principle of the least action. Substituting eq. (1.66) into eq. (1.59) yields (Bretherton and Garrett, 1969):

$$\frac{\partial}{\partial t}\left(\frac{E}{\sigma}\right) + \nabla_h \left[\frac{E}{\sigma}\left(\vec{U} + \vec{C}_g\right)\right] = 0. \tag{1.74}$$

The equation (1.74) states that the local rate of change of wave action is balanced by the convergence of the flux of action. This quantity flows relative to the moving medium with the group velocity $\vec{C}_g$. Generally the conservation principle for the wave action is applied for any non–dissipative disturbances in a moving medium. In the coastal zone, the small amplitude waves propagate on the surface of water with slowly changing depth and current. If the mean velocity $\vec{U}$ is not uniform, the intrinsic frequency σ and the wave number $\vec{k}$ will vary in space and time. Thus, although the wave action $A = E/\sigma$ is conserved, the wave energy density E is not.

Conservation of energy. The equation (1.20) states that for the water waves, the local property, f, may be identified with the water depth, i.e., $f = h + \bar{\zeta}$ and $\sigma^2 = gk \tanh[k(h + \bar{\zeta})]$. Therefore, the action conservation principle takes the form:

$$\frac{\partial E}{\partial t} + \nabla_h \cdot \left[E\left(\vec{U} + \vec{C}_g\right)\right] + \frac{E}{\sigma}\left\{\vec{k}\cdot\left(\vec{C}_g\cdot\nabla_h\right)\vec{U} \; + \right.$$

$$\left. - \frac{\partial\sigma}{\partial(h + \bar{\zeta})}\left(\frac{\partial}{\partial t} + \vec{U}\cdot\nabla_h\right)(h + \bar{\zeta})\right\} = 0. \tag{1.75}$$

Taking into account that:

$$\frac{h + \bar{\zeta}}{\sigma}\frac{\partial\sigma}{\partial(h + \bar{\zeta})} = \frac{1}{2}\left(\frac{2C_g}{C} - 1\right), \tag{1.76}$$

and use of eqs. (1.71) and (1.72) for $\hat{M}_i = 0$, i.e.:

$$\left(\frac{\partial}{\partial t} + \vec{U} \cdot \nabla_h\right)(h + \bar{\zeta}) = -(h + \bar{\zeta})\nabla_h \cdot \vec{U},$$ (1.77)

we obtain:

$$\frac{\partial E}{\partial t} + \nabla_h \cdot \left[E\left(\vec{U} + \vec{C}_g\right)\right] + S_{ij}\frac{\partial U_j}{\partial x_i} = 0,$$ (1.78)

where:

$$S_{ij} = E\left\{l_i l_j \frac{C_g}{C} + \frac{1}{2}\left(\frac{2C_g}{C} - 1\right)\delta_{ij}\right\},$$ (1.79)

in which: δ_{ij} – Kronecker delta, $\vec{l} = \vec{k}/(|\vec{k}|)$ - unit vector in the direction of wave propagation, $C = \omega/k$ - phase velocity.
The equation (1.78) becomes the energy balance of the fluctuating motion. In fact, the wave energy density for the not uniform velocity $\vec{U}$ is not conserved! Only if $\frac{\partial U_j}{\partial x_i} = 0$, the equation (1.78) takes the conservation principle form (eq. 1.51). The last term on the left of the equation (1.78) represents the rate of working by the fluctuating motion against the mean rate of shear. If we consider the wave motion only, S_{ij} represents the excess momentum flux associated with this motion. In this context it was called the "radiation stress" by Longuet–Higgins and Stewart (1962). The more information on the radiation stress and its application to the coastal hydromechanics can be found in Chapters 8 and 9.

Moreover, the energy balance for the mean stream takes the form (Phillips, 1977):

$$\frac{\partial}{\partial t}\left\{\frac{1}{2}\rho(h + \bar{\zeta})\tilde{U}_i^2 + \rho g\bar{\zeta}(h + \bar{\zeta}) - \frac{1}{2}\rho g(h + \bar{\zeta})^2\right\} +$$

$$+ \frac{\partial}{\partial x_i}\left\{\rho(h + \bar{\zeta})\tilde{U}_i\left(\frac{1}{2}\tilde{U}_j^2 + g\bar{\zeta}\right)\right\} + U_j\frac{\partial S_{ij}}{\partial x_i} = 0,$$ (1.80)

where:

$$\tilde{U}_i = U_i + \frac{\hat{M}_i}{\rho(h + \bar{\zeta})}.$$ (1.81)

We note that above conservation principles for wave energy can be derived from the equation (1.62). But the proof of that lies outside the scope of this book.

Conservation of momentum. This principle for the arbitrary space occupied by fluid is expressed by the Navier–Stokes equation (Booij, 1981; Mei, 1983):

$$\rho\left[\frac{\partial^2\Phi}{\partial x_i\partial t} + \left(U_j + \frac{\partial\Phi}{\partial x_j}\right)\frac{\partial}{\partial x_j}\left(U_i + \frac{\partial\Phi}{\partial x_i}\right)\right] +$$

$$+ \ \rho\left[\left(W + \frac{\partial\Phi}{\partial z}\right)\frac{\partial}{\partial z}\left(U_i + \frac{\partial\Phi}{\partial x_i}\right)\right] + \frac{\partial}{\partial x_i}(p + \rho g z) = 0, \qquad i,j = 1,2 \quad (1.82)$$

and

$$\rho\left[\frac{\partial^2\Phi}{\partial z\partial t} + \left(U_j + \frac{\partial\Phi}{\partial x_j}\right)\frac{\partial}{\partial x_j}\left(W + \frac{\partial\Phi}{\partial z}\right)\right] \ +$$

$$+ \rho\left[\left(W + \frac{\partial\Phi}{\partial z}\right)\frac{\partial}{\partial z}\left(W + \frac{\partial\Phi}{\partial z}\right)\right] \ +$$

$$+ \frac{\partial}{\partial z}(p + \rho g z) \ = \ 0, \qquad i,j = 1,2. \qquad (1.83)$$

After neglecting the steady terms as well the higher order terms we get:

$$\rho\left[\frac{\partial^2\Phi}{\partial x_i\partial t} + U_j\frac{\partial^2\Phi}{\partial x_i\partial x_j} + W\frac{\partial^2\Phi}{\partial x_i\partial z} + \frac{\partial\Phi}{\partial x_j}\frac{\partial U_i}{\partial x_j}\right] \ +$$

$$+ \frac{\partial}{\partial x_i}\left[p + \rho g\left(z - \bar\zeta\right)\right] \ = \ 0, \qquad (1.84)$$

and

$$\rho\left[\frac{\partial^2\Phi}{\partial z\partial t} + U_j\frac{\partial^2\Phi}{\partial x_j\partial z} + W\frac{\partial^2\Phi}{\partial z^2} + \frac{\partial\Phi}{\partial z}\frac{\partial W}{\partial z}\right] \ +$$

$$+ \frac{\partial}{\partial z}\left[p + \rho g\left(z - \bar\zeta\right)\right] \ = \ 0. \qquad (1.85)$$

Equations (1.84) and (1.85) hold everywhere in the fluid, and also on the mean free surface $z = \bar\zeta(\vec{x}, t)$. Therefore:

$$\frac{\partial\Phi}{\partial t} + \frac{1}{2}\left[\left(U_j + \frac{\partial\Phi}{\partial x_j}\right)^2 + \left(W + \frac{\partial\Phi}{\partial z}\right)^2\right] + \frac{p}{\rho} + gz = 0 \qquad \text{at} \qquad z = \bar\zeta, \quad (1.86)$$

or

$$\frac{\partial \Phi}{\partial t} + \frac{1}{2}|\vec{u}_t|^2 + \frac{p}{\rho} + gz = 0,$$ (1.87)

where:

$$\vec{u}_t = \vec{U}(\vec{x}, z) + \nabla\Phi(\vec{x}, z, t) = \vec{U}(\vec{x}, z) + \vec{u}(\vec{x}, z, t).$$ (1.88)

Under the assumption that the pressure p at the mean free surface is constant and equal to atmospheric pressure, eq. (1.86) can be rewritten in the form:

$$\frac{\partial \Phi}{\partial t} + \frac{1}{2}\left(U_j + \frac{\partial \Phi}{\partial x_j}\right)^2 + \frac{1}{2}\left(W + \frac{\partial \Phi}{\partial z}\right)^2 + gz = 0 \qquad \text{at} \qquad z = \zeta.$$ (1.89)

Note that eq. (1.89) is identical with eq. (1.37) derived from the variational principle. Moreover, the conservation of momentum can be expressed conveniently in terms of the vertically integrated mean value, i.e.:

$$\frac{\partial}{\partial t}\left[\rho(h+\zeta)\tilde{U}_i\right] + \frac{\partial}{\partial x_j}\left[\rho(h+\zeta)\tilde{U}_i\tilde{U}_j + S_{ij}\right] = -\rho g(h+\zeta)\frac{\partial \bar{\zeta}}{\partial x_i},$$ (1.90)

in which: $\tilde{U}_i$ is given by eq. (1.81).
Note that the conservation law for the total momentum follows also from variational principle (1.63).

Conservation of angular momentum. The particles in the wave motion describe roughly circular (deep water) or elliptic (finite water depth) orbits - see Section 2.3.2. Therefore, it would be intuitively clear to expect that the wave train posses, on the average, a positive angular momentum, when considered from a Lagrangian point of view. Longuet–Higgins (1980) has shown that the mean orbital angular momentum of a wave train of speed C and amplitude a is indeed positive and equal:

$$\bar{A}_L = \frac{1}{4}a^2 C = \frac{EC}{2g}, \qquad E = \frac{1}{2}ga^2,$$ (1.91)

about any point in the mean level. It should be noted that the dimensions of angular momentum are the same as those of wave action. For example, for deep water we have:

$$\sigma = \frac{g}{C}, \qquad \bar{A}_L = \frac{1}{2}\frac{E}{\sigma}. \tag{1.92}$$

Above we have shown that wave action is a conservation quantity. Thus, although the angular momentum and wave action are quite distinct, the known conservation of wave action in weakly nonlinear wave interactions implies also the conservation of Lagrangian angular momentum. Indead, Longuet–Higgins (1983) has obtained in the simple way that for the irrotational flow:

$$\frac{dA_L}{dt} = -\int_C p(x\,dx + y\,dy) - Mg\bar{x}, \tag{1.93}$$

where: C - any simple closed contour bounding a domain D of the fluid, p - pressure as given by Bernoulli equation, M - mass of water enclosed in a domain D, $\bar{x}$ - coordinate of the centre of mass.

1.5   References

Benjamin, T.B. and Olver, P., 1983. Hamiltonian structure, symmetries and conservation laws for water waves. *Jour. Fluid Mech.*, 125: 137–185.

Booij, N., 1981. Gravity waves on water with non-uniform depth and current. Rep. 81–1, Dept. Civil Eng., Delft Univ. of Tech., 130 pp.

Bretherton, F.P. and Garrett, C.J.R., 1969. Wavetrains in inhomogeneous moving media. *Proc. Roy. Soc. London*, A 302: 529–554.

Dera, J., 1983. *Fizyka morza.* Panstwowe Wydawnictwo Naukowe, Warszawa, 432 pp (in Polish).

Gelfand, I.M. and Fomin, S.V., 1975. *Rachunek wariacyjny.* Panstwowe Wydawnictwo Naukowe, Warszawa, 268 pp (in Polish).

Kochin, N.J., Kibel, I.A. and Roze, N.V., 1963. *Teoreticheskaya gidromekhanika.* Chast.I. Gos. Izdat. Fiz. Mat. Liter., Moskva, 583 pp (in Russian).

Longuet-Higgins, M.S., 1980. Spin and angular momentum in gravity waves. *Jour. Fluid Mech.*, 97: 1–25.

Longuet-Higgins, M.S., 1983. On integrals and invariants for inviscid, irrotational flow under gravity. *Jour. Fluid Mech.*, 134: 155–159.

Longuet-Higgins, M.S. and Stewart, R.W., 1962. Radiation stress mass transport in gravity waves, with application to "surf beats". *Jour. Fluid Mech.*, 13: 481–504.

Luke, J.C., 1967. A variational principle for a fluid with a free surface. *Jour. Fluid Mech.*, 27: 395–397.

Mei, C.C., 1983. *The applied dynamics of ocean surface waves.* A Wiley Interscience Publication, New York, 734 pp.

Milne-Thomson, L.M., 1974. *Theoretical hydrodynamics.* Macmillan Press Ltd., London, 743 pp.

Phillips, O.M., 1977. *The dynamics of the upper ocean.* Cambridge Univ. Press, 336 pp.

Puzyrewski, R. and Sawicki, J., 1987. *Podstawy mechaniki plynow i hydrauliki.* Panstwowe Wydawnictwo Naukowe, Warszawa, 332 pp (in Polish).

Seliger, R.L. and Whitham, G.B., 1968. Variational principles in continuum mechanics. *Proc. Roy. Soc. London,* A 305: 1–25.

Smirnov, V.I., 1961. *Matematyka wyzsza.,* T.IV, Panstwowe Wydawnictwo Naukowe, Warszawa, 320 pp (in Polish).

Svendsen, Ib.A. and Jonsson, I.G., 1982. *Hydrodynamics of coastal waters.* Technical Univ. of Denmark, 285 pp.

Synge, J.L., 1963. The hamiltonian method and its application to water waves. *Proc. Roy. Irish Acad.* 63A: 1–33.

Whitham, G.B., 1967a. Non–linear dispersion of water waves. *Jour. Fluid Mech.,* 27: 399–412.

Whitham, G.B., 1967b. Variational methods and applications to water waves. *Proc. Roy. Soc. London,* A 299: 6–25.

Whitham, G.B., 1974. *Linear and nonlinear waves.* A Wiley Interscience Publication, New York, 636 pp.

Zakharov, W.E., 1974. Gamiltonovskiy formalizm dla voln v nelineynykh sredakh s dispersey. *Izv. Vyz. Uchebn. Zav., "Radiofizika",* XVII: 431-453 (in Russian).

Chapter 2

SHORT SURFACE WAVES

2.1   General characteristics

The Table 1.1 shows a large variety of the periodic oscillations which are observed in the coastal zone. This is due to action of many physical mechanisms and their interactions. In the effect, the ocean surface is composed of a large variety of waves moving in different directions and with different frequencies, phases, and amplitudes. The almost always random wind stress is obviously the primary cause of surface waves. In order to bring some degree of order out of chaos, it is necessary to treat an actual wave field by the statistical and spectral methods. The simpliest approach is first to define the average value of wave height, period and wave length, and treat the waves train as monochromatic (i.e., they have only one frequency). The regular waves can be treated not only as some theoretical idea. They are, in fact, more ubiquitous than one would expect. The waves produced by a paddle in the wave flume and the swell propagating on the sea surface can be determined resonably well from the theory of the regular waves. Then, the frequency of wave motion corresponds usually to some characteristic frequency (for example, the peak frequency of wave spectrum).

If we consider the wave motion on a certain (constant) water depth h, the specification of H (wave height), k (i.e. L), C (phase velocity) and ω (i.e. T) is needed. However, these parameters are not independent. The dispersion relation (2.50) links L and T together and the formula $L = C\,T$ form another constraint. Hence, in principle, the wave motion will be specified, provided h and H plus one of the three quantities L, C or T are given. Although, the frequency structure of wave motion was substantially simplified, the boundary value problem for the wave motion (eqs. (1.40) - (1.43)) is still very difficult to solve. It is mainly due to nonlinear boundary conditions at the free surface and sea bottom. Therefore, the rigorous general solutions to the wave boundary value problem, which are valid in the whole frequency range, i.e. from short waves to the long ones, are not existing.

Prior to discussing the theoretical solutions to the specific wave motion, we will introduce the more detailed classification of the regular waves. If the wave height H is much smaller then wave length L, i.e.:

$$\frac{H}{L} \ll 1.0, \tag{2.1}$$

the names *small waves* or *small amplitude waves* are often used. Moreover, the wave solution, which results, is called *linear waves* ("first order Stokes waves") or *sinusoidal waves*. The last is referring to the function which turns out to describe the phase variations. For waves occurring in nature, the wave steepness H/L is usually at most $0.05 - 0.08$. Thus, the wave height is allowed to be small but finite. These waves are known as the *waves of finite height*. Sometimes, if $H = O(h)$, the waves are called the *steep waves*.

To define the classification uniquely we relate the water depth to wave length. It turns out that three different cases are important:

$$\left.\begin{array}{lll}
\text{shallow water waves} & \text{when} & \frac{h}{L} < \frac{1}{10}\left(\frac{1}{20}\right) \\[2mm]
\text{waves on intermediate water depth} & \text{when} & \frac{1}{10}\left(\frac{1}{20}\right) < \frac{h}{L} < \frac{1}{2} \\[2mm]
\text{deep} - \text{water waves} & \text{when} & \frac{h}{L} > \frac{1}{2}
\end{array}\right\} . \tag{2.2}$$

The first condition yields that the water length is much longer than water depth and the waves are called the *long waves*. Goda (1983) has shown that the parameters mentioned above can be summarized in the form of one non-dimensional parameter Π, i.e.:

$$\Pi = \frac{H}{L} \coth^3\left(\frac{2\pi h}{L}\right) = \frac{H}{L} \coth^3(kh). \tag{2.3}$$

If $h/L > 1/2$ (deep-water waves), from eq. (2.3) we get:

$$\Pi = \frac{H}{L}. \tag{2.4}$$

Note that for deep–water waves, the parameter Π is identical with the wave steepness. On the other hand, if $h/L \leq 1/10$ the parameter Π takes the form:

$$\Pi = \left(\frac{1}{2\pi}\right)^3 \left(\frac{H}{h}\right) \left(\frac{L}{h}\right)^2. \tag{2.5}$$

Now, the Π parameter is proportional to so called Ursell parameter U, i.e.:

$$\Pi = (2\pi)^{-3} U, \qquad \text{where} \qquad U = \left(\frac{H}{h}\right)\left(\frac{L}{h}\right)^2. \tag{2.6}$$

When:

$$U = O(1) \qquad \text{or} \qquad \Pi = O\left((2\pi)^{-3}\right), \tag{2.7}$$

the linear theory is useful in many respects, even when the requirements of linear theory, small $kH/2$, is violated (Section 2.3). However, when Π parameter is going to be larger, the more sofisticated nonlinear theories (for example Stokes theory) should be applied.

In many applications, in particular in coastal regions, the wave length L, is much larger than the water depth h, and the wave height H is an appreciable fraction of h so that the Π parameter becomes too large even for the application of the Stokes theory. The wave theories, corresponding to the long waves, satisfy other assumptions than Stokes waves. Moreover, they need other methods of solution (Chapter 4). To define the problem uniquely, we define two subranges for the Π (or U) parameter:

- short waves when $\Pi \leq 0.30$ or $U \leq 75$,
- long waves when $\Pi \geq 0.30$ or $U \geq 75$.

The limit value should be taken as approximation only. The more precise analysis will be given in the following Sections.

2.2   Problem formulation for short waves ($\Pi \leq 0.30$ or $U \leq 75$)

Consider two–dimensional, periodic, surface waves of wave length L propagating over a horizontal bottom (water depth is equal h) with the speed $\vec{C}$. For simplicity, the current velocity $\vec{U}$ will be omitted. Choose rectangular coordinates (x, z) such that the x - axis is horizontal and the z - axis is directed vertically upwards (Fig. 2.1). From equations (1.40) - (1.43) we have:

$$\nabla^2 \Phi = 0, \tag{2.8}$$

$$\frac{\partial \Phi}{\partial t} + \frac{1}{2}\left[\left(\frac{\partial \Phi}{\partial x}\right)^2 + \left(\frac{\partial \Phi}{\partial z}\right)^2\right] + gz = 0 \qquad \text{at} \qquad z = \zeta, \tag{2.9}$$

$$\frac{\partial \zeta}{\partial t} + \frac{\partial \Phi}{\partial x}\frac{\partial \zeta}{\partial x} = \frac{\partial \Phi}{\partial z} \qquad \text{at} \qquad z = \zeta, \tag{2.10}$$

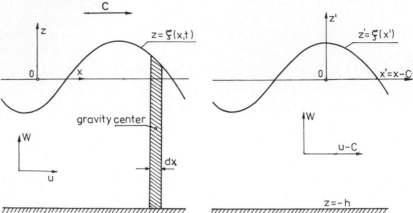

Figure 2.1: The stationary $(0, x, z)$ and moving $(0, x', z')$ coordinate systems

$$\frac{\partial \Phi}{\partial z} = 0 \qquad \text{at} \qquad z = -h. \tag{2.11}$$

Now consider a second rectangular coordinate system $(0, x', z')$ moving in the positive x - direction with the wave at the phase speed C. In this reference frame the motion is independent of time, t. Thus:

$$x' = x - Ct, \qquad z' = z, \qquad \frac{\partial}{\partial t} = -C\,\frac{\partial}{\partial x'}, \qquad \frac{\partial}{\partial x'} = \frac{\partial}{\partial x}. \tag{2.12}$$

The velocity potential, Φ', stream function, Ψ', and orbital velocity, (u', w'), in this frame are related to similar quantities in the (x, z) - frame by:

$$\left.\begin{aligned}
\Phi' &= \Phi - Cx, \qquad \Psi' = \Psi - Cz \\
u' &= u - C = \frac{\partial \Phi'}{\partial x'} = \frac{\partial \Psi'}{\partial z'} \\
w' &= w = \frac{\partial \Phi'}{\partial z'} = -\frac{\partial \Psi'}{\partial x'}
\end{aligned}\right\}. \tag{2.13}$$

Note, that the velocity potential Φ' and stream function Ψ' satisfy the known Cauchy–Riemann relations:

$$\frac{\partial \Phi'}{\partial x'} = \frac{\partial \Psi'}{\partial z'}, \qquad \frac{\partial \Phi'}{\partial z'} = -\frac{\partial \Psi'}{\partial x'} \tag{2.14}$$

Eq. (2.14) yields:

$$(\nabla \Phi') \cdot (\nabla \Psi') = 0, \tag{2.15}$$

which means isolines for Φ' are perpendicular to isolines for Ψ'.
Substitution of eqs. (2.13) and (2.14) into the continuity equation (1.8), shows that Φ' and Ψ' have to satisfy the Laplace equation as well:

$$\nabla^2 \Phi' = \nabla^2 \Psi' = 0. \tag{2.16}$$

Because of (2.14), the formulation of the wave boundary value problem in terms Φ' and Ψ' are quite equivalent. However, the boundary conditions are expressed much simplier by the stream function Ψ'. Therefore, in the following, the stream function Ψ' representation will be used. The free surface and bottom are streamlines, i.e.:

$$\Psi'(x', -h) = 0, \tag{2.17}$$

$$\Psi'[x', h + \zeta(x')] = -Q, \tag{2.18}$$

where: $\zeta(x')$ is the free surface and $(-Q)$ is the constant volume flow rate per unit width. Q is positive and this flow is in the negative x direction.
Please note, eqs. (2.17) and (2.18) represent the kinematic free surface boundary conditions (no flow through the free surface). Dynamic free surface boundary condition, representing constant atmospheric pressure on the free surface, is:

$$\left(\frac{\partial \Psi'}{\partial x'}\right)^2 + \left(\frac{\partial \Psi'}{\partial z'}\right)^2 + 2g\zeta = R \qquad \text{at} \qquad z' = \zeta(x'), \tag{2.19}$$

where: R is the Bernoulli constant.

2.3 First Stokes perturbation method

The problem which has been just formulated for two dimensional steady periodic gravity wave is the simpliest of all. Despite its relative simplicity, however, it contains the full nonlinearity of the surface boundary conditions. Recent studies have shown that a solution to this problem exists even for a wave of the greatest height (Toland, 1978). Up to now, the only known exact solution of the problem is the Gerstner trochoidal wave (Milne–Thomson, 1974). The motion is, however, not irrotational. Since the pioneering work of Stokes (1847, 1880), most applications have relied on approximations, based on the assumption that the wave slopes are small. Stokes first used a systematic perturbation technique assuming that the unknown functions Φ', Ψ', ζ, C may be represented by an infinite Fourier series. The coefficients in these series can be written as perturbation expansions in terms of a parameter ka, in which k - wave number, a - length scale which is equal to the amplitude of the wave at the lowest order. When these series are substituted into the free–surface conditions, a set of equations is obtained from which the coefficients in the power series can be found successfully. For arbitrary constant water depth, Stokes developed a solution to second order in ka. Laitone (1962) generalized Stokes first method by the use of the classical small–perturbation expansion method with the parameter $\sigma = 1/\sinh(2\pi h/L)$. For finite amplitude waves, the series expansion (up to third approximation) is found to be most suitable for wave length less than 8 times the depth. The similar approximation was obtained by Isobe and Kraus (1983), when the perturbation parameter $\epsilon = k(H/2)$. Skjelbreia and Hendrickson (1961) have developed an alternative fifth–order solution in terms of ka. As it was recently shown by Fenton (1985), the Skjelbreia and Hendrickson theory is, in fact, wrong at fifth order. It should be added that an alternative third–order of Stokes theory for the steady waves in water of constant depth was obtained by Aleshkov and Ivanova (1972) using the method proposed by Sretenskiy (1977).

All solutions mentioned above, are valid under the assumption that the phase velocity C is given by the Stokes first definition. Thus, the velocity of propagation C is defined to be the velocity with which the wave form is propagated in space, while the mean (in time) horizontal velocity at each point of space occupied by the fluid is zero, i.e.:

$$C = \frac{1}{L} \int_0^{x+L} [u(x', z') + C]dx' = 0. \tag{2.20}$$

In the laboratory conditions, Stokes 2–nd definition usually is used and the velocity of propagation C is defined to be the velocity with which the wave form is propagated in space, while the mean horizontal velocity of the mass of fluid which is comprised between two planes perpendicular to the direction of propagation of the waves is zero, i.e.:

$$\frac{\int_0^L \int_{-h}^{\zeta(x')} [u(x', z') + C]dx'\, dz'}{\int_0^L \int_{-h}^{\zeta(x')} dx'\, dz'} = 0. \tag{2.21}$$

2.3.1 Nonlinear approximation

Let us consider a fifth–order of Stokes theory for the steady waves. Instead of an unknown Fourier coefficient ka being used as expansion parameter, the dimensionless wave amplitude $\epsilon = kH/2$ will be used. As suggested by previous Stokes wave theories, the following expansion for stream function Ψ' is assumed (Fenton, 1985):

$$\frac{k\Psi'}{\bar{C}} = -k(z'+h) + \sum_{i=1}^{\infty}\sum_{j=1}^{i} \tilde{A}_{ij}\epsilon^i \sinh[jk(z'+h)]\cos(jkx'), \qquad (2.22)$$

where: $\bar{C}$ - mean fluid speed (in frame in which waves are stationary) over one wave length. This is equal to wave speed relative to frame in which there is zero current.

Eq. (2.22) satisfies the eq. (2.16) and the boundary condition on the bottom (2.17). After substitution the expansion for Ψ' into the boundary conditions at the sea surface (2.18) - (2.19) we get:

$$\frac{kQ}{\bar{C}} = -k[h+\zeta(x')] + \sum_{i=1}^{\infty}\sum_{j=1}^{i} \tilde{A}_{ij}\epsilon^i \sinh jk[h+\zeta(x')]\cos(jkx') = 0, \qquad (2.23)$$

and

$$\left\{ -1 + \sum_{i=1}^{\infty}\sum_{j=1}^{i} j\tilde{A}_{ij}\epsilon^i \cosh jk[h+\zeta(x')]\cos(jkx') \right\}^2 +$$

$$+ \left\{ \sum_{i=1}^{\infty}\sum_{j=1}^{i} j\tilde{A}_{ij}\epsilon^i \sinh jk[h+\zeta(x')]\sin(jkx') \right\}^2 +$$

$$+ \frac{2g}{\bar{C}^2}[h+\zeta(x')] - \frac{2R}{\bar{C}^2} = 0. \qquad (2.24)$$

Now we assume that $\bar{C}, Q, R$ and ζ can be represented in terms of ϵ:

$$\bar{C}\left(\frac{k}{g}\right)^{1/2} = C_o + \sum_{i=1}^{\infty} C_i\epsilon^i, \qquad (2.25)$$

$$Q\left(\frac{k^3}{g}\right)^{1/2} = \bar{C}h\left(\frac{k^3}{g}\right)^{1/2} + \sum_{i=1}^{\infty} D_i\epsilon^i, \qquad (2.26)$$

$$\frac{Rk}{g} = \frac{1}{2}C_o^2 + kh + \sum_{i=1}^{\infty} E_i\,\epsilon^i, \qquad (2.27)$$

$$k\zeta(x') = \sum_{i=1}^{\infty} \sum_{j=1}^{i} B_{ij}\epsilon^i \cos j_{kx'}. \tag{2.28}$$

Please note, that the surface elevation in (2.28) is presented as a power series with terms like $\cos j_{kx'}$, rather than as a Fourier series with terms like $\cos(jkx')$. In order to find the unknown coefficients $A_{ij}, B_{ij}, C_o, C_i, D_i$ and E_i, the expansions (2.25) - (2.28) are substituted into (2.23) - (2.24) and the hyperbolic functions in these equations are also expanded in the power series. The boundary conditions must be satisfied for all value of ϵ and all values of x. After subsequent mathematical manipulations we obtain with accuracy up to fifth–order (Fenton, 1985):

$$\Phi'(x', z') = -\bar{C} x' + C_o \left(\frac{g}{k^3}\right)^{1/2} \sum_{i=1}^{5} \epsilon^i \sum_{j=1}^{i} A_{ij} \cdot$$

$$\cdot \cosh\left[jk(z' + h)\right]\sin(jkx') + O(\epsilon^6), \tag{2.29}$$

where:

$$\bar{C}\left(\frac{k}{g}\right)^{1/2} = C_o + \epsilon^2 C_2 + \epsilon^4 C_4 + O(\epsilon^6), \tag{2.30}$$

and

$$k\zeta(x') = \epsilon \cos(kx') + \epsilon^2 B_{22} \cos(2kx') + \epsilon^3 B_{31}[\cos(kx') - \cos(3kx')] +$$

$$+ \epsilon^4[(B_{42}\cos(2kx') + B_{44}\cos(4kx')] +$$

$$+ \epsilon^5[-(B_{53} + B_{55})\cos(kx') + B_{53}\cos(3kx')] +$$

$$+ \epsilon^5 B_{55}\cos(5kx') + O(\epsilon^6), \tag{2.31}$$

and

$$Q\left(\frac{k^3}{g}\right)^{1/2} = C_o kh + \epsilon^2(C_2 kh + D_2) + \epsilon^4(C_4 kh + D_4) + O\left(\epsilon^6\right), \tag{2.32}$$

$$\frac{Rk}{g} = \frac{1}{2}C_o^2 + kh + \epsilon^2 E_2 + \epsilon^4 E_4 + O\left(\epsilon^6\right). \tag{2.33}$$

Table 2.1: Coefficients of the Fenton approximation

Coeff.	Expression
S	$\cosh^{-1}(2kh)$
SH	$\sinh(kh)$
CTH	$\coth(kh)$
TH	$\tanh(kh)$
A_{11}	SH^{-1}
A_{22}	$\dfrac{3S^2}{2(1-S)^2}$
A_{31}	$\dfrac{-4-20S+10S^2-13S^3}{8SH(1-S)^3}$
A_{33}	$\dfrac{-2S^2+11S^3}{8SH(1-S)^3}$
A_{42}	$\dfrac{12S-14S^2-264S^3-45S^4-13S^5}{24(1-S)^5}$
A_{44}	$\dfrac{10S^3-174S^4+291S^5+278S^6}{48(3+2S)(1-S)^5}$
A_{51}	$\dfrac{-1184+32S+13232S^2+21712S^3+20940S^4+12554S^5-500S^6-3341S^7-670S^8}{64SH(3+2S)(4+S)(1-S)^6}$
A_{53}	$\dfrac{4S+105S^2+198S^3-1376S^4-1302S^5-117S^6+58S^7}{32SH(3+2S)(1-S)^6}$
A_{55}	$\dfrac{-6S^3+272S^4-1552S^5+852S^6+2029S^7+430S^8}{64SH(3+2S)(4+S)(1-S)^6}$
B_{22}	$\dfrac{CTH(1+2S)}{2(1-S)}$
B_{31}	$\dfrac{-3(1+3S+3S^2+2S^3)}{8(1-S)^3}$
B_{42}	$\dfrac{CTH(6-26S-182S^2-204S^3-25S^4+26S^5)}{6(3+2S)(1-S)^4}$
B_{44}	$\dfrac{CTH(24+92S+122S^2+66S^3+67S^4+34S^5)}{24(3+2S)(1-S)^4}$
B_{53}	$\dfrac{9(132+17S-2216S^2-5897S^3-6292S^4-2687S^5+194S^6+467S^7+82S^8)}{128(3+2S)(4+S)(1-S)^6}$
B_{55}	$\dfrac{5(300+1579S+3176S^2+2949S^3+1188S^4+675S^5+1326S^6+827S^7+130S^8)}{384(3+2S)(4+S)(1-S)^6}$
C_0	$TH^{1/2}$
C_2	$\dfrac{TH^{1/2}(2+7S^2)}{4(1-S)^2}$
C_4	$\dfrac{TH^{1/2}(4+32S-116S^2-400S^3-71S^4+146S^5)}{32(1-S)^5}$
D_2	$\dfrac{-CTH^{1/2}}{2}$
D_4	$\dfrac{CTH^{1/2}(2+4S+S^2+2S^3)}{8(1-S)^3}$
E_2	$\dfrac{TH(2+2S+5S^2)}{4(1-S)^2}$
E_4	$\dfrac{TH(8+12S-152S^2-308S^3-42S^4+77S^5)}{32(1-S)^5}$

The coefficients $A_{ij}, B_{ij}, C_0, C_i, D_i$ and E_i can be presented in terms of the hyperbolic function $S = \cosh^{-1}(2kh)$, which are given in Table 2.1. In the constant reference frame $(0, x, z)$, the velocity potential $\Phi(x, z, t)$ takes the form:

$$\Phi(x, z, t) = (C - \bar{C})x + C_o\left(\frac{g}{k^3}\right)^{1/2} \cdot$$

$$\cdot \sum_{i=1}^{5} \epsilon^i \sum_{j=1}^{i} A_{ij} \cosh[jk(z + h)] \sin[jk(x - Ct)], \tag{2.34}$$

and the elevation of the surface is given by:

$$k\zeta(x, t) = \epsilon \cos k(x - Ct) + \text{ other terms in eq. (2.31)}, \tag{2.35}$$

while for the pressure we have:

$$\frac{p}{\rho} = R - g(z + h) - \frac{1}{2}\left[(u - C)^2 + w^2\right], \tag{2.36}$$

Up to now, the mean velocity $\bar{C}$ is unknown. In order to find this value, we consider three various cases (Fenton, 1985):

1) If the mean uniform current is equal zero, the velocity $\bar{C} = C$.

2) If the mean current velocity is U and it is in the wave direction, the Eulerian time mean fluid velocity is given by:

$$\bar{C} = C - U. \tag{2.37}$$

3) In the laboratory experiments, instead of mean fluid velocity, the mass flux under the waves is known or can be assumed. In a frame moving with the waves the volume flow rate underneath the waves per unit span is Q (in the negative x–direction). Therefore, the mean velocity with which the fluid is transported is equal $(-Q/h)$, and the mean fluid transport velocity in the direction of propagation of the waves is C_t, i.e.:

$$C_t = C - \frac{Q}{h}. \tag{2.38}$$

The theory, outlined above, can be directly applied when the following parameters are known: wave height H, wave length L, water depth h and current velocity U (or mean fluid transport velocity C_t). If the wave period T is known, the wave length L (or wave number k) can be calculated from the transcendental equation (see (2.30) or (2.32)):

$$U\left(\frac{k}{g}\right)^{1/2} - \frac{2\pi}{T(gk)^{1/2}} + C_o(kh) + \left(\frac{kH}{2}\right)^2 C_2(kh) + \left(\frac{kH}{2}\right)^4 C_4(kh) = O, \quad (2.39)$$

or

$$\left(\frac{k}{g}\right)^{1/2} C_t - \frac{2\pi}{T(gk)^{1/2}} + C_o(kh) + \left(\frac{kH}{2}\right)^2 \left[C_2(kh) + \frac{D_2(kh)}{kh}\right] +$$

$$+ \left(\frac{kH}{2}\right)^4 \left[C_4(kh) + \frac{D_4(kh)}{kh}\right] = 0. \quad (2.40)$$

In the case of deep and shallow water, the expressions for the particular wave parameters becomes considerably simple. When $kh \to \infty$ (deep water), we have:

$$k\zeta(x,t) = \epsilon \cos k(x - Ct) + \frac{1}{2}\epsilon^2 \cos 2k(x - Ct) +$$

$$- \frac{3}{8}\epsilon^3 [\cos k(x - Ct) - \cos 3k(x - Ct)] +$$

$$+ \frac{1}{3}\epsilon^4 [\cos 2k(x - Ct) + \cos 4k(x - Ct)] +$$

$$+ \frac{1}{384}\epsilon^5 [-422 \cos k(x - Ct) + 297 \cos 3k(x - Ct) +$$

$$+ 125 \cos 5k(x - Ct)] + O(\epsilon^6), \quad (2.41)$$

and

$$C = \left(\frac{g}{k}\right)^{1/2} \left[1 + \frac{1}{2}\epsilon^2 + \frac{1}{8}\epsilon^4\right] + O(c^6). \quad (2.42)$$

For the shallow water, $kh \to 0$. However, it is an inappropriate limit for the application of Stokes theories. The more adequate limits are those given through the parameters Π and U. If the values of these parameters are high, the perturbation Stokes solution should not be used. The Fenton estimation yields the conclusion that present fifth-order theory is quite accurate for waves shorter than 10 times the water depth (Fenton, 1985).

Let us consider the limit $kh \to 0$ in details. The inspection of the Table 2.1 shows that the ratio $\tilde{R}$ of the $n+1$ term divided by the n term in the expressions for the velocity potential and the surface ordinate is:

$$\tilde{R} \approx \epsilon(kh)^{-3} \qquad \text{when} \qquad kh \to 0. \tag{2.43}$$

This ratio must be less than 1 in order for the series to converge:

$$\tilde{R} < 1.0, \tag{2.44}$$

or:

$$U = \left(\frac{H}{h}\right)\left(\frac{L}{h}\right)^2 < 8\pi^2 \qquad \text{and} \qquad \Pi < \frac{1}{\pi}. \tag{2.45}$$

The values of the Ursell parameter or Π parameter actually should be less than indicated above. It is due to the fact that in shallow water the theoretical wave form will developed an anomalous bump in the wave trough. The secondary crest in the wave trough will occur when:

$$\frac{\partial \zeta}{\partial x'} = 0 \qquad \text{and} \qquad \frac{\partial^2 \zeta}{\partial x'^2} < 0 \qquad \text{for} \qquad x' = \pi. \tag{2.46}$$

After substitution (2.31) into (2.46) we get:

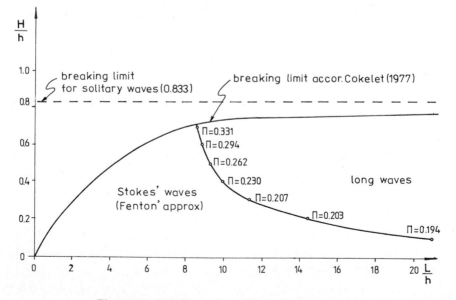

Figure 2.2: Validity region of Fenton approximation

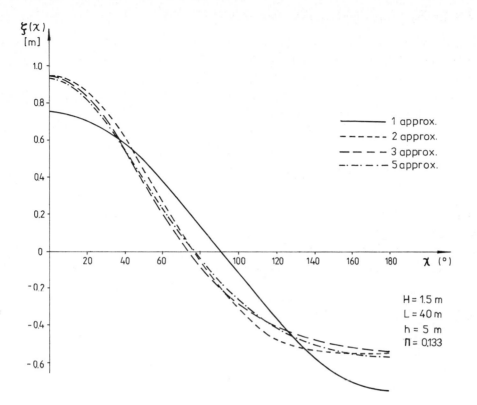

Figure 2.3: Wave profile. Comparison of various approximations

$$4B_{22}\epsilon + 8B_{31}\epsilon^2 + 4(B_{42} + 4B_{44})\epsilon^3 - (8B_{53} + 24B_{55})\epsilon^4 - 1 = 0. \qquad (2.47)$$

The condition (2.47) is presented in the plane $(L/h, H/h)$ in Fig. 2.2. In the same Figure, the breaking limit (see Chapter 5) is indicated. Additionally, in Fig. 2.3, the wave profiles corresponding to the various approximations are given.

Note that Fig. 2.2 illustrates the repartition, as suggested above, of all wave types into two groups: short waves (Stokes waves) and long waves (cnoidal and solitary waves). The dividing line corresponds to the Ursell parameter $U \approx 75$. It is in agreement with the Van Dorn (1966) idea that for the short waves parameter U should be less than $U \approx 100$. More detailed partitions can be found elsewhere (e.g. Druet, 1965).

2.3.2   Small amplitude approximation

Small amplitude wave theory is based upon further assumption that all motions are small and eqs. (2.9) - (2.10) may be linearized. In particular, this assumes that squares of the velocity components (u^2 and w^2) are negligible in comparison with remaining terms in the dynamic boundary condition. Moreover, the product of the free surface slope and orbital velocity in the kinematic boundary condition can be omitted. Under this assumption, the velocity potential (see eq. (2.34)) takes the form (current velocity is zero):

$$\Phi(x,z,t) = \frac{gH}{2\omega} \frac{\cosh k(z+h)}{\cosh kh} \sin(kx - \omega t). \qquad (2.48)$$

For other quantities we get:

free surface ordinate

$$\zeta(x,t) = \frac{H}{2} \cos(kx - \omega t), \qquad (2.49)$$

dispersion relation

$$\frac{\omega^2 h}{g} = kh \tanh(kh). \qquad (2.50)$$

In *deep water*, kh is large and $tanh(kh) \approx 1.0$; therefore:

$$\omega^2 = gk, \qquad T = \sqrt{\frac{2\pi L}{g}}, \qquad L = \frac{g}{2\pi}T^2, \qquad (2.51)$$

Dispersion relation for *shallow water* reduces in the following manner:

$$\omega = \sqrt{gh}\,k, \qquad T = \frac{L}{\sqrt{gh}}, \qquad L = \sqrt{gh}\,T. \qquad (2.52)$$

The wave speed is now determined solely by the water depth. Thus, the waves are nondispersive.

On the other hand, dispersion of waves takes place when the phase velocity is changing with the wave length or with the wave frequency, i.e. $\omega = f(k)$ (see for example eq. (2.50)). The dispersion relations of some selected wave phenomena are listed in Table 2.2 which is mainly due to Gran (1985). The solution to the dispersion relationship, (2.50), for k can be considerably simplified using the formula proposed by Hunt (1979):

Table 2.2: The dispersion relations for selected wave phenomena

Wave phenomenon	Dispersion relation	Velocity relation
Shallow water Acoustic Electromagnetic Elastic shear Streched string	$\omega = \sqrt{gh}\,k$ $\omega = \sqrt{B/\rho}\,k$ $\omega = ck$ $\omega = \sqrt{D/\rho}\,k$ $\omega = \sqrt{T/m_0}\,k$	$C_g = C$
Deep water	$\omega^2 = gk$	$C_g = \frac{1}{2}C$
Capillary	$\omega^2 = (T/\rho)k^3$	$C_g = \frac{3}{2}C$
Flexural bar Free slow particle (Schrodinger)	$\omega = (EI/m_0)k^2$ $\omega = (h/2m)k^2$	$C_g = 2C$
Relativistic particle (Klein–Gordon)	$\omega^2 = \left(\frac{mc^2}{\hbar}\right)^2 + c^2k^2$	$C_g = \frac{c^2}{C}$
Gravity waves, general	$\omega^2 = gk\tanh kh$	$C_g = \frac{1}{2}\left(\frac{2kh}{\sinh 2kh} + 1\right)C$
Surface waves, general	$\omega^2 = \left(gk + \frac{T}{\rho}k^3\right)\tanh kh$	$C_g = \frac{1}{2}\left[\frac{2kh}{\sinh 2kh} + \frac{g+3Tk^2/\rho}{g+Tk^2/\rho}\right]C$

B - bulk modulus,
D - shear elasticity coefficient,
E - Young modulus,
I - second moment of cross section area,
T - surface, string or membrane tension,
c - light speed,
$\hat{h}$ - Planck constant/2π,
m - mass of particle,
m_0 - mass per unit area or length,
ρ - mass density.

$$k = \frac{\omega}{\sqrt{gh}} [\alpha + P^{-1}(\alpha)]^{1/2}, \tag{2.53}$$

where:

$$P(\alpha) = 1.0 + 0.6522\alpha + 0.4622\alpha^2 + 0.0864\alpha^4 + 0.0675\alpha^5, \tag{2.54}$$

and $\alpha = \frac{\omega^2 h}{g}$. The formula (2.53) is accurate to 0.001 for all α.

phase velocity

$$C^2 = \frac{g}{k} \tanh(kh), \tag{2.55}$$

group velocity

$$C_g = \frac{d\omega}{dk} = mC, \tag{2.56}$$

in which:

$$m = \frac{1}{2} \left(1 + \frac{2kh}{\sinh 2kh} \right), \tag{2.57}$$

particle velocity components

$$\left. \begin{array}{rcl} u & = & \dfrac{\partial \Phi}{\partial x} = \dfrac{gkH}{2\omega} \dfrac{\cosh k(z+h)}{\cosh kh} \cos(kx - \omega t) \\[3mm] w & = & \dfrac{\partial \Phi}{\partial z} = \dfrac{gkH}{2\omega} \dfrac{\sinh k(z+h)}{\cosh kh} \sin(kx - \omega t) \end{array} \right\}, \tag{2.58}$$

particle displacements

Consider the water particle which position is denoted as (x_1, z_1). The displacements components (ξ, η) of the particle can be found by integrating the velocity components with respect to time:

$$\xi(x_1, z_1, t) = \int u(x_1, z_1, t) dt, \tag{2.59}$$

$$\eta(x_1, z_1, t) = \int w(x_1, z_1, t) dt, \tag{2.60}$$

Substituting (2.58) into (2.59) and (2.60) we get:

$$\xi = -\frac{H}{2}\frac{gk}{\omega^2}\frac{\cosh k(z_1 + h)}{\cosh kh}\sin(kx_1 - \omega t), \qquad (2.61)$$

$$\eta = \frac{H}{2}\frac{gk}{\omega^2}\frac{\sinh k(z_1 + h)}{\cosh kh}\cos(kx_1 - \omega t), \qquad (2.62)$$

or

$$\left.\begin{array}{rl} \xi = - & A \sin(kx_1 - \omega t) \\ \eta = & B \cos(kx_1 - \omega t) \end{array}\right\}, \qquad (2.63)$$

where:

$$\left.\begin{array}{rl} A & = \dfrac{H}{2}\dfrac{gk}{\omega^2}\dfrac{\cosh k(z_1 + h)}{\cosh kh} \\[2ex] B & = \dfrac{H}{2}\dfrac{gk}{\omega^2}\dfrac{\sinh k(z_1 + h)}{\cosh kh} \end{array}\right\}. \qquad (2.64)$$

Squaring and adding of eq. (2.63) we obtain:

$$\left(\frac{\xi}{A}\right)^2 + \left(\frac{\eta}{B}\right)^2 = 1. \qquad (2.65)$$

Eq. (2.65) reflects the fact that the water particle follows an ellipse trajectory with semiaxes A and B.

pressure in fluid

$$p(x, z, t) = -\rho gz + \rho g\frac{H}{2}\frac{\cosh k(z + h)}{\cosh kh}\cos(kx - \omega t), \qquad (2.66)$$

or:

$$p(x, z, t) = -\rho gz + \rho g\zeta K_p(z), \qquad (2.67)$$

where:

$$K_p(z) = \frac{\cosh k(z + h)}{\cosh kh}. \qquad (2.68)$$

The first term on the right–hand side of (2.66) represents the hydrostatic pressure, which exists without the presence of the wave motion. The second term describes the dynamic pressure. The dynamic pressure is induced mainly

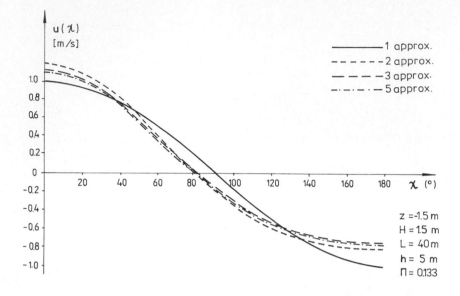

Figure 2.4: Horizontal orbital velocity. Comparison of various approximations

due to two factors. The first is the free surface displacement. If $K_p(z)$ is equal 1.0 the pressure contribution associated with the surface displacement would be purely hydrostatic. The second contributor is the vertical acceleration, which is 180° out of phase with the free surface displacement. Thus the pressure distribution is modified. Note that factor $K_p(z)$ has a maximum of unity at $z = 0$, and a minimum of $1/\cosh(kh)$ at the bottom.

The pressure distribution above the mean water level can be obtained by expanding the pressure in the Taylor series for a small positive distance $z_1 (0 < z_1 < \zeta)$, i.e.:

$$
\begin{aligned}
p(z_1) &= [-\rho g z + \rho g \zeta K_p(z)]_{z=0} + \\
&+ z_1 \frac{\partial}{\partial z} [-\rho g z + \rho g \zeta K_p(z)]_{z=0} + \ldots = \rho g (\zeta - z_1).
\end{aligned} \tag{2.69}
$$

Therefore, the pressure under the wave crest is hydrostatic down to $z = 0$. In Figs. 2.4 - 2.6 the comparison of the particle velocity components and pressure, calculated with various accuracy, is given.

Within the small amplitude approximation, some nonlinear quantities (they involve the wave height to the second power) can be obtained. In the following we will consider such quantities as: wave energy, energy flux, mass transport and mean water level. The analysis is mainly due to Dean and Dalrymple (1984).

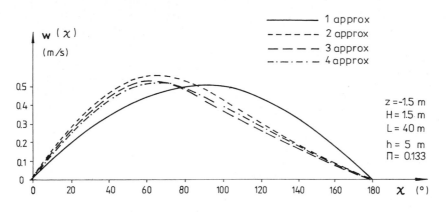

Figure 2.5: Vertical orbital velocity. Comparison of various approximations

potential energy

A displacement of the water mass from a position of equilibrum against a gravitational field results in an increase in potential energy. The potential energy of a small column of fluid with mass dm (Fig. 2.1) relative to the bottom is:

$$d\hat{E}_p = g\,\frac{h+\zeta}{2}\,dm,\tag{2.70}$$

or:

$$d\hat{E}_p = \rho g\,\frac{(h+\zeta)^2}{2}\,dx,\tag{2.71}$$

Thus, the averaged (over one wave length) potential energy takes the form:

$$\bar{\bar{E}}_p = \frac{1}{L}\int_x^{x+L} d\hat{E}_p = \frac{1}{L}\int_x^{x+L} \rho g\,\frac{(h+\zeta)^2}{2}dx.\tag{2.72}$$

If the surface elevation ζ is represented by (2.49), from (2.72) we obtain:

$$\bar{\bar{E}}_p = \rho g\,\frac{h^2}{2} + \rho g\,\frac{H^2}{16}.\tag{2.73}$$

The first term on the rigth–hand side of (2.73) is the potential energy, which exists without the presence of waves. The second term is related to the potential energy due to waves, i.e.:

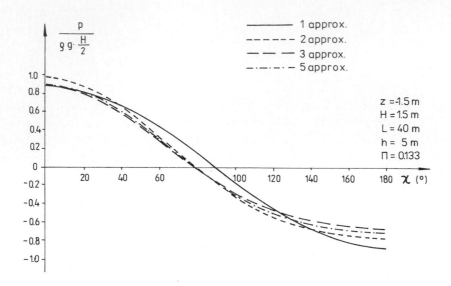

Figure 2.6: Dynamical pressure. Comparison of various approximations

$$\bar{E}_p = \rho g \frac{H^2}{16},$$ (2.74)

kinetic energy

The kinetic energy reflects the motion of the water particle due to wave. Therefore, we have:

$$dE_k = dm \frac{u^2 + w^2}{2} = \rho \frac{u^2 + w^2}{2} \, dx \, dz.$$ (2.75)

In order to find the averaged kinetic energy we integrate dE_k over depth and average the result over a wave length:

$$\bar{E}_k = \frac{1}{L} \int_x^{x+L} \int_{-h}^{\zeta} \rho \frac{u^2 + w^2}{2} \, dx \, dz.$$ (2.76)

Thus:

$$\bar{E}_k = \rho g \frac{H^2}{16}.$$ (2.77)

This is exactly equal to the potential energy. The total average energy of the wave is now the sum of the potential and kinetic energy:

$$\bar{E}_t = \bar{E}_p + \bar{E}_k = \rho g \frac{H^2}{8}. \tag{2.78}$$

energy flux

The mass transport in the small amplitude wave field is equal zero because the water particles follow the closed trajectories (2.65). However, even simple observations indicate that water waves transmit energy. The rate at which the energy is transferred is called the *energy flux* $\mathcal{P}$. It is equal to the work done by the fluid on one side of a vertical section on the fluid on the other side. This work can be written as:

$$\mathcal{P} = \int_{-h}^{\zeta}(p + \rho g z)u dz. \tag{2.79}$$

Substituting (2.58) and (2.66) into (2.79) and integrating we obtain the average energy flux over a wave period in the form:

$$\bar{\mathcal{P}} = \bar{E}_t \, Cm = \bar{E}_t \, C_g. \tag{2.80}$$

mass transport

The observed motion of the small neutrally buoyant float, induced by waves, is the indicative of the mean fluid motion. In order to find the mean mass flux in the Eulerian frame, we calculate first the average horizontal velocity at any point bellow the water surface (Dean and Dalrymple, 1984):

$$\bar{u}(x, z) = \frac{1}{T}\int_0^T u(x, z, t)dt = 0. \tag{2.81}$$

The special attention is needed to estimate the horizontal velocity in the region between the trough and the wave crest, where we have:

$$u(x, \zeta) = \left\{u(x, 0) + \zeta\frac{\partial u}{\partial z}\right\}_{z=0} =$$

$$= \frac{gkH}{2\omega}\cos(kx - \omega t) + \frac{H^2}{4}k\omega\cos^2(kx - \omega t). \tag{2.82}$$

Averaging the velocity $u(x, \zeta)$ over a wave period yields the mean transport of water:

$$\bar{u}(x, \zeta) \equiv \frac{1}{T}\int_0^T u(x, \zeta)dt = \frac{H^2 k\omega}{8} = \frac{(kH)^2 \, C}{8}. \tag{2.83}$$

Finally, the total mass flux will be done by the following integration (Dean, Dalrymple, 1984):

$$M \equiv \overline{\int_{-h}^{\zeta} \rho u \, dz} = \overline{\int_{-h}^{0} \rho u \, dz} + \overline{\rho \zeta u} \frac{E}{C}. \tag{2.84}$$

More information on the mass transport will be given in Chapter 8.

mean water level

Let us first apply the expansion of the Bernoulli equation (2.9) into Taylor series at $z = \zeta$. After averaging over a wave period we obtain:

$$\overline{\frac{1}{2}\left[\left(\frac{\partial \phi}{\partial x}\right)^2 + \left(\frac{\partial \phi}{\partial z}\right)^2\right]} + g\bar{\zeta} - \overline{\zeta \frac{\partial^2 \phi}{\partial t \partial z}} = 0. \tag{2.85}$$

Substituting for ϕ and ζ from the linear wave theory yields:

$$\bar{\zeta} = -\frac{a^2 k}{2 \sinh(2kh)}. \tag{2.86}$$

Eq. (2.86) predicts the always negative deviation of the mean water level from $z = 0$ (set down). In Chapter 9, the general analysis of the variation of mean water level will be given.

2.4   Second Stokes perturbation method

In order to overcome the problem of the unknown location of the free surface, Stokes (1880) transformed the wave problem from the $(x', z',)$ - plane to the (Φ', Ψ') - plane. In this inverse problem, where Φ' and Ψ' rather than x' and z' are the independent variables, the calculations can be greatly reduced. As the waves are assumed to be symmetrical with respect to the crests and the troughs, the transformation maps the area under a wave in the (Φ', Ψ') - plane into a rectangle (Fig. 2.7). As the free surface and the bottom are the stream lines, there the stream function Ψ' has to be constant. Therefore the bottom and the free surface are mapped in the (Φ', Ψ') - plane into straight lines where $\Psi' = $ constant. Moreover, the stream lines, under the crest and trouhgs, are horizontal and perpendicular to the isolines for Φ' due to symmetry of the wave profile. Thus, these lines are mapped in the (Φ', Ψ') - plane on vertical, straight lines where $\Phi' = $ constant. Formally, this transformation takes the form:

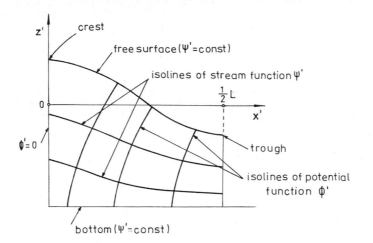

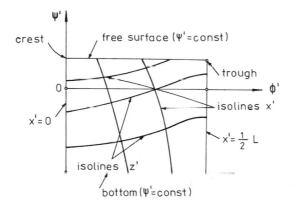

Figure 2.7: Transformation of the coordinate system

$$\hat{z} = F(\hat{w}), \tag{2.87}$$

where: $\hat{z} = x' + iz'$, $\qquad \hat{w} = \Phi' + i\Psi'$.

The governing differential equations in the (Φ', Ψ) - plane are again the Laplace equation:

$$\frac{\partial^2 x'}{\partial \Phi'^2} + \frac{\partial^2 x'}{\partial \Psi'^2} = \frac{\partial^2 z'}{\partial \Phi'^2} + \frac{\partial^2 z'}{\partial \Psi'^2} = 0, \tag{2.88}$$

Moreover, the bottom condition, and the kinematic free surface condition are transformed into the simple conditions which are expressing the constancy of the stream function there. The dynamical free surface condition (2.19), becomes:

$$z'(\Phi', \Psi') + \frac{1}{2g} V^2 = const,$$

(2.89)

where:

$$V^2 = \left[\left(\frac{\partial x'}{\partial \Phi'} \right)^2 + \left(\frac{\partial x'}{\partial \Psi} \right)^2 \right]^{-1}.$$

(2.90)

As the solution is periodic, the following condition holds:

$$F(\Phi' + \Delta\Phi' + i\Psi) - F(\Phi' + i\Psi') = L,$$

(2.91)

where: $\Delta\Phi'$ is the range of Φ' over one wave length L. Moreover, the symmetry of wave profile yields:

$$F(-\hat{w}) = -F(\hat{w}).$$

(2.92)

By a unique choice of the origin of z', the constant on the right–hand side may vanish. Then since (2.90) we have:

$$2gz' \left[\left(\frac{\partial z'}{\partial \Phi'} \right)^2 + \left(\frac{\partial z'}{\partial \Psi'} \right)^2 \right] = -1;$$

(2.93)

a condition which is cubic in z' and its derivatives.

In order to solve the above boundary value problem, Stokes (1880) assumed that the solution $x'(\Phi', \Psi)$ and $z'(\Phi', \Psi')$ can be described in term of a series with unknown coefficients, which finally has to be determined. The coefficient of the n–term of the series is of the order n in a small dimensionless parameter ($\sim$ wave height/wave length). When the coefficients have been found, the role of (Φ', Ψ') and (x', z') is reversed and then the wave characteristics in the (x', z') plane can be derived. Using this method Stokes obtained the solution to the fifth order for infinitely deep water and to the third order for finite depth of water. It should be noted that the important problem of the series convergence for the second Stokes method was solved and the solution can be found elsewhere (for example Hunt, 1953).

The Stokes results were extended by De (1955) who obtained a fifth–order solution for water of finite depth. In his solution, $\Psi' = -Q$ at the bottom and $\Psi' = 0$ at the free surface (in the reference frame in which x' - axis is located at a distance Q/C above the bottom and z' - axis is directed vertically downwards). The De solution in (Φ', Ψ') - plane takes the form (Vis and Dingemans, 1978):

$$
\left.
\begin{aligned}
x' &= -\frac{\Phi'}{C} + \sum_{n=1}^{\infty} \left[A_n e^{nk\Psi'/C} + B_n e^{-nk\Psi'/C} \right] \sin\left(n\frac{k\Phi'}{C}\right) \\
z' &= -\frac{\Psi'}{C} + \sum_{n=1}^{\infty} \left[A_n e^{nk\Psi'/C} - B_n e^{-nk\Psi'/C} \right] \cos\left(n\frac{k\Phi'}{C}\right)
\end{aligned}
\right\}.
\tag{2.94}
$$

The bottom condition yields:

$$
A_n \exp(-nkQ/C) = B_n \exp(nkQ/C),
\tag{2.95}
$$

The remaining equation for the coefficients A_n and B_n can be found from the dynamical condition at the free surface (2.89). After these coefficients have been calculated and the series (2.94) have been reversed, the final calculation formulas can be obtained. Similar to De results, Chappelear (1961) has based his solution on the second Stokes method. For practical application, however, this method has a limitation in that the theory is presented in terms of ka and depth scale Q/C. In general, this scale is not equal to the depth, and is initially unknown.

2.5 Analytical and numerical solution for steep waves

The coastal engineer frequently requires the information concerning very nonlinear ocean waves, including the limiting breaking waves. In the previous Sections the Stokes methods of calculation for steady, irrotational waves of finite amplitude on water of uniform mean depth were presented. The first method involves expansion of the complex velocity potential in powers of a small parameter ϵ, proportional to the maximum surface slope. In the second method the calculations are simplified by the inverse technique, in which Φ' and Ψ' rather than x' and z' are the independent variables. However, because of the enormous amount of work, extension to higher than fifth order is hardly possible. In the recent years, a number of wave theories have been developed which are considerably more computer–oriented. These theories are as accurate as the series results and are often easier to implement computationally. Moreover, these methods generally do not need the convergence improvement techniques of the series. Although the numerical methods are, in fact, still based on the Stokes teories, the advantage of these methods is that, the solution can be found to as high an order as required.

In the following, we describe a few such approaches. Let us consider first the boundary value problem for the function Ψ' (eqs. (2.16) - (2.19)) and assume that N^{th} order stream function takes the form (Dean, 1965):

$$\Psi'(x',z') = Cz' + \sum_{n=1}^{N} X(n)\sinh[nk(z'+h)]\cos(nkx'). \tag{2.96}$$

Contrary to Stokes wave theories, no restriction have been imposed to wave height, wave length and wave depth relations. The stream function method is applicable to all permanent waves at the constant depth, even for near–breaking waves. The only condition not satisfied by the above form is the dynamic free surface boundary condition. Therefore, the problem is to find $X(n)$ so that the dynamic free surface boundary condition is the best satisfied at I discrete points along the wave profile, each point being denoted by i. Thus:

$$R_i = \left(\frac{\partial \Psi'}{\partial x'}\right)_i^2 + \left(\frac{\partial \Psi'}{\partial z'}\right)_i^2 + 2g\zeta_i = R = const. \tag{2.97}$$

In order to calculate R_i values, the coefficients $X(n), n = 1, 2, \ldots, N$ should be known. It is assumed that an approximate solution is available and R_i are determined. The R_i are then used to obtain new $X(n)$ and so on, according to the least–squares criteria; so the error E_1 will be minimum, i.e.:

$$E_1 = \frac{2}{L}\int_0^{L/2} (R_i - R)^2 dx' = min, \tag{2.98}$$

where:

$$R = \frac{2}{L}\int_0^{L/2} R_i \, dx'. \tag{2.99}$$

For design purposes, two additional conditions should be satisfied, that is:
- $\zeta(x')$ must have a zero mean:

$$\frac{2}{L}\int_0^{L/2} \zeta(x')dx' = 0, \tag{2.100}$$

- calculated wave height should be equal to the wave height known *a priori*. Therefore, we must minimize function E_c, when the above conditions are treated as the constraints (Dean and Dalrymple, 1984):

$$E_c = E_1 + \frac{2\lambda_1}{L} \int_0^{L/2} \zeta(x')dx' + \lambda_2[\zeta(0) - \zeta(L/2) - H], \qquad (2.101)$$

where: λ_1, λ_2 - Lagrange multipliers.

In order to facilitate the solution, we expand the function E_c into truncated Taylor series:

$$E_c^{(j+1)} = E_c^{(j)} + \sum_{n=1}^{N+2} \frac{\partial E_c^{(j)}}{\partial X(n)} \Delta X^{(j)}(n). \qquad (2.102)$$

Minimizing the expanded E_c function with respect to all the $X(n)$ plus λ_1 and λ_2 yields a series of linear equation for the $\Delta X^{(j)}(n)$ for fixed j. The process is repeated for several iterations until $E_c^{(j+1)}$ is acceptably small.

The stream function wave theory has been used to tabulate the 40 cases of nonlinear waves in dimensionless form with emphasis on ready reference by the engineer and oceanographer (Dean, 1974). Extension of the theory for the irregular measured water surface is possible, and it requires determining the best–fit stream function to a given free surface shape (Dean, 1965).

There are a number of further variations on semi–numerical methods in the literature. In the first approaches, as in the Dean method, the attempts (Chappelear, 1961, Dalrymple, 1974, Chaplin, 1980) were made to over–specify the problem and solve the series of linear equations in a least–squares sense. The later mathematical formulations are essentially identical and only the small differences relate largely to the details of the numerical solution algorithms. In particular, Rienecker and Fenton (1981) introduced the normalization of eq. (2.96) on the $1/\cosh(nkh)$, necessity of which was pointed out by Huang and Hudspeth (1984).

Very important results were obtained by the inverse plane method in which Φ' and Ψ' functions are adopted as the dependent variables rather than (x', z') of the common physical plane method. Schwartz (1974) and Cokelet (1977) established the Fourier coefficients as series in terms of a perturbation parameter by computer manipulation. Let us consider the boundary value problem (2.16) - (2.19) in the complex plane $\hat{z} = x' + iz'$ (Fig. 2.8). Thus:

$$\nabla^2 \Phi' = \nabla^2 \Psi' = 0, \qquad (2.103)$$

where:

$$\hat{\omega}' = \Phi' + i\Psi', \qquad (2.104)$$

and

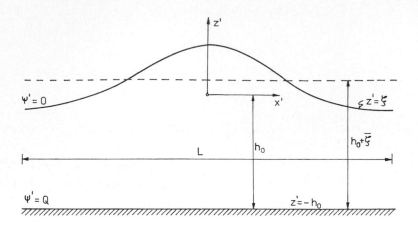

Figure 2.8: Coordinate system.

$$\Psi' = 0 \quad \text{at} \quad z' = \zeta \quad \text{and} \quad \Psi' = Q \quad \text{at} \quad z' = -h_0, \qquad (2.105)$$

$$u'^2 + w'^2 + 2g\zeta = R \quad \text{at} \quad z' = \zeta, \qquad (2.106)$$

We denote the mean elevation of the free surface as $\bar{\zeta}$. Thus, the mean water depth is $h = h_0 + \bar{\zeta}$ and it does not in general equal h_0. We take the following units of length, time and mass:

$$L^* = k^{-1}, \qquad T^* = (gk)^{-1/2}, \qquad M^* = \rho, \qquad (2.107)$$

such that the following dimensionless quantities (with subscripts "b") are:

$$\left.\begin{aligned}
x'_b &= \frac{x'}{L^*} = kx', \qquad L_b = \frac{L}{L^*} = 2\pi, \qquad k_b = 1.0 \\[2mm]
g_b &= 1.0, \qquad (kh_0)_b = (h_0)_b = kh_0 \\[2mm]
u_b &= u\left(\frac{k}{g}\right)^{1/2}, \qquad \Phi'_b = \Phi'\left(\frac{k^3}{g}\right)^{1/2}, \qquad \Psi'_b = \Psi'\left(\frac{k^3}{g}\right)^{1/2} \\[2mm]
\bar{E}_b &= E\frac{k^2}{\rho g}
\end{aligned}\right\} . \qquad (2.108)$$

In the following the subscripts will be omitted. The solution for x' and z' is assumed in the form (Cokelet, 1977):

$$
\left.
\begin{aligned}
x' &= -\frac{\Phi'}{C} - \sum_{j=1}^{\infty} \frac{a_j}{j} \left(e^{-j\Psi'/C} + e^{-2jh_0} e^{j\Psi'/C} \right) \sin\left(j\frac{\Phi'}{C}\right) \\
z' &= -\frac{\Psi'}{C} + \sum_{j=1}^{\infty} \frac{a_j}{j} \left(e^{-j\Psi'/C} - e^{-2jh_0} e^{j\Psi'/C} \right) \cos\left(j\frac{\Phi'}{C}\right)
\end{aligned}
\right\},
\tag{2.109}
$$

where the height of the origin above the bottom has been defined such that $h \equiv Q/C$.

In the eq. (2.109) the symmetry of the wave profile about the wave crest and the bottom conditions are satisfied. The real constants a_j are determined by satisfying the Bernoulli equation on the free surface (2.106) in which:

$$
u' - iw' = \frac{d\hat{w}}{d\hat{z}} = \left(\frac{d\hat{z}}{d\hat{w}} \right)^{-1},
\tag{2.110}
$$

Substitution of (2.109) and (2.110) into (2.106) gives:

$$
\left.
\begin{aligned}
C^2 + 2\sum_{l=1}^{\infty} \frac{a_l \, \delta_l \, f_l}{l} &= R \, f_0 \\
\sum_{l=1}^{\infty} \frac{a_l \, \delta_l}{l} \left(f_{|l-j|} + f_{l+j} \right) &= R \, f_j, \qquad (j = 1, 2, ...)
\end{aligned}
\right\},
\tag{2.111}
$$

in which:

$$
\left.
\begin{aligned}
f_0 &= 1 + \sum_{l=1}^{\infty} a_l^2 \sigma_{2l} \\
f_l &= a_j \sigma_j + \sum_{l=1}^{\infty} a_j a_{l+j} \sigma_{2l+j} + \\
&\quad + \frac{1}{2} \sum_{l=1}^{j-1} a_l a_{j-l} (\sigma_l - \delta_{j-l}), \qquad (j = 1, 2, ...)
\end{aligned}
\right\},
\tag{2.112}
$$

and

$$
\left.
\begin{aligned}
\sigma_j &= 1 + e^{-2jh_0} \\
\delta_j &= 1 - e^{-2jh_0}
\end{aligned}
\right\}.
\tag{2.113}
$$

Equations (2.111) and (2.112) are a set of nonlinear algebraic equations which determine the Fourier coefficients a_j completely. In agreement with Schwartz (1974), Cokelet solves these equations by means of a perturbation method, using the small parameter ϵ. Therefore:

$$
\left.
\begin{aligned}
a_j &= \sum_{k=0}^{\infty} \alpha_{jk}\epsilon^{j+2k}, & j &= 1, 2, \dots \\[2mm]
f_j &= \sum_{k=0}^{\infty} \beta_{jk}\epsilon^{j+2k}, & j &= 1, 2, \dots \\[2mm]
C^2 &= \sum_{l=0}^{\infty} \gamma_l \epsilon^{2l} \\[2mm]
R &= \sum_{l=0}^{\infty} \Delta_l \epsilon^{2l}
\end{aligned}
\right\}
\tag{2.114}
$$

Substitution of (2.114) into (2.111) and (2.112) and equating coefficients of equal powers of ϵ yield the recurrence relations for α, β, γ and Δ coefficients.

The proper choice of the expansion parameter ϵ is of great importance to this problem. The good perturbation parameter should encompass waves from the lowest to the very highest and gives rapid convergence in the perturbation series (2.111). In the classical Stokes work, this parameter was identical with the first Fourier coefficient, $\epsilon = a_1/L$ which is zero for infinitesimal waves and increases with the wave height. However, this statement is true only to the order to which Stokes carried his approximation, i.e. $O(a_1^3)$ for arbitrary depth and $O(a_1^5)$ for deep water. Schwartz (1974) has shown that for steep waves this expansion fails, because a_1 surprisingly does not increase monotonically with the wave steepness. In fact a_1 reaches a maximum and then decreases before the highest wave is reached. Therefore a_1 is not a suitable expansion parameter for waves near the highest, since a single value of a_1 can correspond to two different wave heights. In order to remove this ambiguity, Schwartz (1974) used the parameter $\epsilon = 2a/L = H/L$, where H denotes the crest - to - trough height. He obtained convergence with the aid of Pade summation and carried the expansion in (2.109) out to $O(\epsilon^{48})$ for arbitrary depths and to $O(\epsilon^{117})$ for deep water. However, the convergence of Schwartz series worsens for very high waves and for very shallow depths. Following the idea of Longuet–Higgins (1975), Cokelet (1977) defined the expansion parameter ϵ as:

$$
\epsilon^2 = 1 - \frac{u'^2_c \, u'^2_t}{C^4},
\tag{2.115}
$$

in which: u'_c - fluid speed at the wave crest (in the reference moving with the wave), u'_t - fluid speed at the wave trough.

It should be noted that ϵ has the advantage that its range, $0 \le \epsilon^2 \le 1$, is known *a priori*. For very small waves, $u'_c \sim u'_t \sim -C$, and $\epsilon^2 \approx 0$. On the other hand, as the wave steepness increases to its maximum value, $u'_t \sim 0$

Table 2.3: The limiting wave height for various fluid depth

e^{-kh}	L/h	H/L			H/h	Cokelet	
		Schwartz	Cokelet	Wiliams	H/h	U	Π
0	0.00	0.1412	0.1411	0.1411	0.0000	-	0.1410
0.1	2.73	0.1380	0.1378	0.1378	0.3760	2.7996	0.1463
0.2	3.90	0.1285	0.1285	0.1285	0.5016	7.6446	0.1634
0.3	5.22	0.1145	0.1144	0.1144	0.5971	16.2619	0.1965
0.4	6.86	0.0975	0.0974	0.0974	0.6680	31.4239	0.2566
0.5	9.07	0.0791	0.0791	0.0791	0.7171	58.9427	0.3663
0.6	12.30	0.0614	0.0609	0.0610	0.7491	113.3331	0.5844
0.7	17.62	0.0450	0.0437	0.0440	0.7698	238.8924	1.0897
0.8	28.16		0.0279	0.0283	0.7857	623.0975	2.6397
0.9	59.63		0.015	0.0136	0.8150	2898.421	12.9533
1.0	∞		0.000	0.0000	0.8332*	∞	∞

* accor. Williams (1981)

(stagnation point) and $\epsilon^2 \approx 1.0$. Cokelet applied his method to the full range of water depth and wave heights, and computed the wave profile, wave celerity and the integral properties (mean momentum, kinetic and potential energy, radiation stress, momentum flux) for the cases $e^{-kh_0} = 0, 0.1, 0.2, ..., 0.9$ and $\epsilon^2 = 0.0$ to 1.0.

Because of the slow convergence of the series, they were summed by using Pade approximation technique to the order $O(\epsilon^{110})$. An interesting results is that the wave speed, wave energy, and wave momentum all exhibit maxima at wave height slightly smaller than the breaking height. This is demonstrated in Figs. 2.9 and 2.10 for the wave speed C. Fig. 2.9 is a graph of the normalized wave height, H/h, against the normalized wave speed squared, C^2/gh, for various depths. The small difference between h_0 and h was omitted, therefore $h \approx h_0$. In fact, for the entire depth range the highest waves are not the fastest. The end points of the particular curves in Figure 2.9 correspond to the highest wave heights ($\epsilon^2 = 1.0$) for given $e^{-kh}(L/h)$. The detail of the function $H/h = f\left(C^2/gh\right)$ for $e^{-kh} = 0.4385$ is given in Figure 2.10. Moreover, the Table 2.3 gives the highest wave heights for various fluid depths and compares them with Schwartz and Williams (1981) results. Additionally the corresponding Ursell number U and Goda parameter Π values are given.

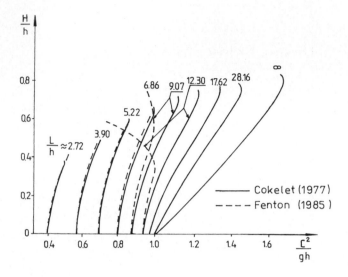

Figure 2.9: Normalized wave height (H/h) versus normalized wave squared velocity (C^2/gh). (From Cokelet, 1977)

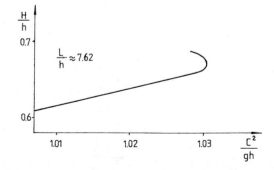

Figure 2.10: The detail of the function $\frac{H}{h} = \left(\frac{C^2}{gh}\right)$ for $e^{-kh} = 0.4385$. (From Cokelet, 1977)

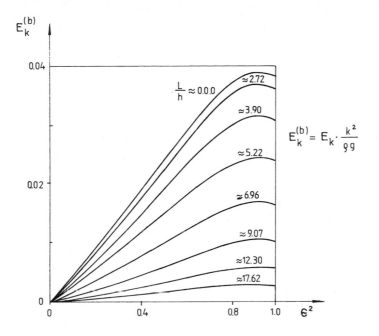

Figure 2.11: The kinetic energy versus parameter ϵ^2. (From Cokelet, 1977)

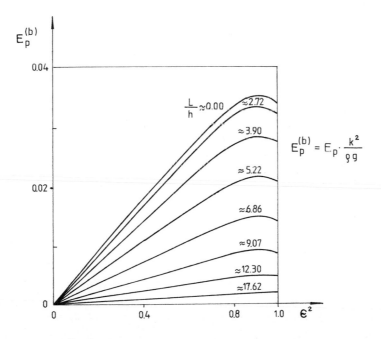

Figure 2.12: The potential energy versus parameter ϵ^2. (From Cokelet, 1977)

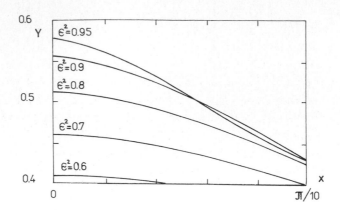

Figure 2.13: The profiles of the almost highest wave heights. (From Longuet–Higgins et al., 1976)

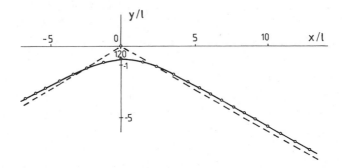

Figure 2.14: Asymptotic form of the highest wave

The analysis of these values yields the conclusion that the Fenton approximation given in Section 2.3.1, has in fact only limited application for steeper waves.

The pairs (H/h) and (L/h) form a curve which describes the breaking limit. This curve was included in Figure 2.2.

Figures 2.11 - 2.12 show the kinetic energy and potential energy plotted against ϵ^2. The most important result is that these quantities achieve maxima for waves not of the highest height. The same conclusion is also valid for other integral quantities. It should noted that rate of kinetic and potential energy in all range of e^{-kh} and ϵ^2 varies from 1.0 to ~ 1.17. The difference between both energies can also be calculated for the lower order of approximation by the other methods (see for example Scarsi and Stura, 1971).

An explanation for the maxima in the wave energy and other physical properties can be given in terms of the wave profiles (Fig. 2.13). It appears that the profile of the steepest waves, with very sharp curvature at the crest, intersect those of the not–so–steep waves, and hence lie below them over most of the wave crest, but above them in the trough (Longuet–Higgins et al., 1976). It should be noted that similar phenomenon was found previously for solitary waves by Longuet-Higgins and Fenton (1974). As the limiting wave is approaching, the curvature at the crest becomes increasingly sharp, but the crest is still rounded before the limit of a 120^0 angle is reached. Longuet–Higgins and Fox (1977) have shown that the flow takes an asymptotic form as in Figure 2.14 where an appropriate scale l for the local flow is given by:

$$l = \frac{u'^2}{2g},$$
(2.116)

where: u'^2 - sharp corner flow.

At infinity (at radial distance r large compare to l) the flow tends to the well–known Stokes, corner flow with a crest angle of 120^0 (Chapter 5). However, the free surface first intersects its plane asymptote at the distance $r = 3.32\,l$ and then approaches them gradually from above.

Williams (1985) has calculated and collected in the form of Tables of progressive gravity waves, the most comprehensive set of the displacements, velocities, accelerations and pressures at all points in the fluid. The discussion of the method and evaluation of his solution is given in Williams (1981).

Precise numerical calculations of steep waves have generally employed power series expansions with a large number of coefficients, which must first be calculated. The condition of constant pressure at the free surface yields a series of cubic relations between the coefficients (2.93). Longuet–Higgins (1978, 1988) discovered certain identities between the Fourier coefficients which are only of the second degree. Therefore, a new method of calculation which is based on these relations is essentially simpler and much more efficient. If we take $\psi' = 0$ at the free surface and $\psi' = \psi_B'$ on the bottom then the coordinates x', y' can be expressed in terms of Φ', Ψ' by the following Fourier series:

$$\left.\begin{aligned}
z' + \Psi'/C &= \frac{1}{2}A_0 + \sum_{n=1}^{\infty} A_n \sinh[n(\Psi_B' - \Psi')/C]\cos(n\Phi'/C) \\
x' + \Phi'/C &= -\sum_{n=1}^{\infty} A_n \cosh[n(\Psi_B' - \Psi')/C]\sin(n\Phi'/C)
\end{aligned}\right\},$$
(2.117)

where: A_n being real constants, to be determined.

Therefore, the equation of the surface elevation takes the parametrical form:

$$
\left.
\begin{array}{l}
\zeta \;=\; \dfrac{1}{2}a_0 + \displaystyle\sum_{1}^{\infty} a_n \cos(n\theta) \\[4mm]
x' \;=\; z'_0\theta + \displaystyle\sum_{1}^{\infty} a_n z'_n \sin(n\theta)
\end{array}
\right\}, \tag{2.118}
$$

in which:

$$
\left.
\begin{array}{ll}
a_0 \;=\; A_0, & a_n = A_n \sinh(n\psi_B/C) \\[2mm]
z'_0 \;=\; 1.0, & z'_n = \coth(n\psi_B/C)
\end{array}
\right\}, \tag{2.119}
$$

and $\theta = -\Phi/C$.

Longuet–Higgins (1988) has shown that the A_n satisfy the quadratic relations which can be expressed compactly in the form:

$$
\frac{\partial F}{\partial a_n} = 0, \qquad n = 0, 1, 2, \dots , \tag{2.120}
$$

where:

$$
F = (J + a_0 K) + \frac{1}{2}(\alpha + \hat{\alpha}) + \frac{1}{4}(a_0 + C^2 + \bar{u}_B^2)^2, \tag{2.121}
$$

$$
J = \frac{1}{2}\sum_{m=1}^{\infty} a_m^2, \tag{2.122}
$$

$$
K = \frac{1}{2}\sum_{m=1}^{\infty} m\,\gamma_m\, a_m^2, \tag{2.123}
$$

$$
\alpha = \sum_{m=2}^{\infty} a_m \sum_{n=1}^{m-1} n\,\gamma_n\, a_{m-n}\, a_n, \tag{2.124}
$$

$$
\hat{\alpha} = \sum_{m=2}^{\infty} a_m \sum_{n=1}^{m-1} n\,\gamma_m\, a_{m-n}\, a_n, \tag{2.125}
$$

$$
\gamma_n = \coth(nk\psi_B/C). \tag{2.126}
$$

u_B - horizontal velocity at the bottom.

For any given values of C^2 and h the possible values of $a_0, a_1, a_2 \dots$ are those that correspond to stationary values of the function $F(a_0, a_1, a_2, \dots)$.

The system of equations (2.120) can be solved for $a_0, a_1, a_2, \ldots$ by the method described by Longuet–Higgins (1985, 1988). The calculations show quite unexpected result that the most symmetric orbits, in the Lagrangian sense, occur not in deep water but in water of intermediate depth. For example, when $kh = 2.0$, for very nonlinear waves, the vertical displacement is highly symmetric about the midpoint between crest and trough. Due its simplicity, the Longuet–Higgins expansion method is relatively easy to use with commonly available microcomputers (Naguszewski, personal communication, 1988).

2.6 Validity of theoretical solutions

In the Sections above, the various wave theories which provide useful predictions in the ocean and coastal engineering were presented. However, the range of physical validity of each theories is not infinite. It is often possible to obtain the mathematical prediction from a theory outside its range of physical validity but in practice, some very high–order Stokes expansions provide a reference against which the accuracy of the particular solutions can be tested (see for example Cokelet, 1977 or Longuet–Higgins, 1985, 1988). In the simpliest case, the integral quantities are used as a basis for comparison.

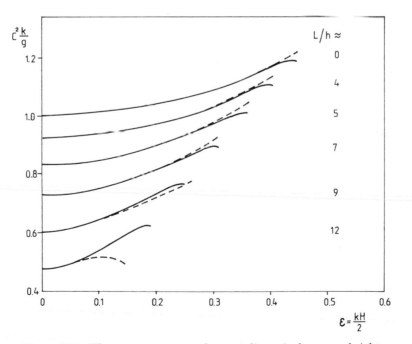

Figure 2.15: The square wave speed versus dimensionless wave height

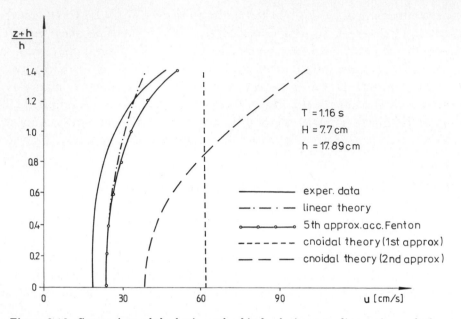

Figure 2.16: Comparison of the horizontal orbital velocity according various solutions

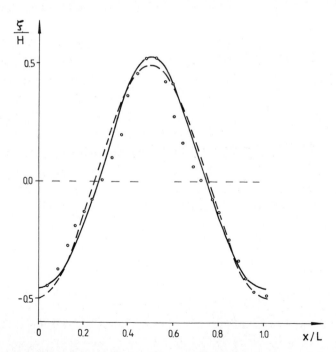

Figure 2.17: Comparison of theoretical surface profiles with experimental data. Legend as in Fig. 2.18

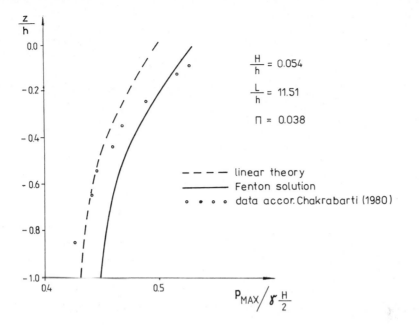

Figure 2.18: Comparison of the theoretical pressure with experimental data

In Fig. 2.15 the square of the dimensionless wave speed, $C^2 k/g$, against the dimensionless wave height, $kH/2$, for several values of the approximate wave length to water depth ratios was presented. The Fenton results (the broken lines) are compared with the Cokelet (1977) calculations. It can be seen that the Fenton theory is quite accurate for waves shorter than about 10 times the water depth. For the longer waves, the fifth–order theory should not be used. In this frequency range, the theory of long waves is more appropriate. In order to examine the validity of the Stokes theory with respect to the details of actual flow, we can choose the horizontal particle velocity under the crest, the free surface ordinates, and wave pressure. The comparison is shown in Figs. 2.16 - 2.18.

2.7 References

Aleshkov, J.Z. and Ivanova, S.V., 1972. Begushchiye svobodnye volny na poverkhnosti zhidkosti postoyannoy glubiny. *Trudy Koord. Soveshchaniy po Gidrotekhnike*, Moskva, 1: 18–23 (in Russian).

Chaplin, J.R., 1980. Developments of stream–function wave theory. *Coastal Eng.*, 3: 179–205.

Chappelear, J.E., 1961. Direct numerical calculation of wave properties. *Jour. Geoph. Res.*, 66: 501–508.

Cokelet, E.D., 1977. Steep gravity waves in water of arbitrary uniform depth. *Phil. Trans. Roy. Soc. London,* A286: 183–230.

Dalrymple, R.A., 1974. A finite amplitude wave on a linear shear current. *Jour. Geoph. Res.,* 79: 4498–4504.

De, S.C., 1955. Contributions to the theory of Stokes waves. Proc. Camb. Phil. Soc., 51: 713–736.

Dean, R., 1965. Stream function representation of nonlinear ocean waves. *Jour. Geoph. Res.,* 70: 4561–4572.

Dean, R.G., 1974. Evaluation and development of water wave theories for engineering application. Spec. Rep. 1, U.S. Army, Coastal Engineering Research Center, 655 pp.

Dean, R.G. and Dalrymple, R.A., 1984. *Water wave mechanics for engineers and scientists.* Prentice–Hall Inc., Englewood Cliffs, 353pp.

Druet, Cz., 1965. Obciazenia hydrodynamiczne falochronow portowych posadowionych w strefie transformacji falowania. *Rozpr. Hydrot.,* 17: 3–70 (in Polish).

Fenton, J.D., 1985. A fifth–order Stokes theory for steady waves. *Proc. ASCE, Jour. Waterway, Port, Coastal and Ocean Eng.,* 111: 216–234.

Goda, Y., 1983. A unified nonlinearity parameter of water waves. Rep. Port, Harb. Res. Inst., 22: 3–30.

Gran, S., 1985. Lecture in ocean engineering - waves and wave forces. Tech. Rep. University of Oslo, 251 pp.

Huang, M.C. and Hudspeth, R.T., 1984. Stream function solutions for steady water waves. *Continental Shelf Reasearch,* 3: 175–190.

Hunt, J.N., 1953. A note on gravity waves of finite amplitude. *Quart. Jour. Mech.* Appl. Math., 6: 336–343.

Hunt, J.N., 1979. Direct solution of wave dispersion equation. *Proc. ASCE, Jour. Waterway, Port, Coastal and Ocean Eng.,* 105: 457–459.

Isobe, M. and Kraus, N.C., 1983. Derivation of a third-order Stokes waves theory. Hydr. Lab., Dept. Civil Eng., Yokohama Nat. Univ., Tech. Rep. 83-1, 37 pp.

Laitone, E.V., 1962. Limiting conditions for cnoidal and Stokes waves. *Jour. Geoph. Res.*, 67: 1555–1564.

Longuet–Higgins, M.S., 1975. Integral properties of periodic gravity waves of finite amplitude. *Proc. Roy. Soc. London*, A342: 157–174.

Longuet–Higgins, M.S., 1978. Some new relations between Stokes' coefficients in the theory of gravity waves. *Jour. Inst. Math. Applics.*, 22: 261–273.

Longuet–Higgins, M.S., 1985. A new way to calculate steep gravity waves. In: Y. Toba, H. Mitsuyasu (Editors), *The ocean surface.* pp. 1–15.

Longuet–Higgins, M.S., 1988. Lagrangian moments and mass transport in Stokes waves. II. Water of finite depth. *Jour. Fluid Mech.*, 186: 321–336.

Longuet–Higgins, M.S. and Fenton, J.D., 1974. On mass, momentum, energy and circulation of a solitary wave. II. *Proc. Roy. Soc. London*, A340: 471–493.

Longuet–Higgins, M.S., Cokelet, E.D. and Fox, M.J.H., 1976. The calculation of steep gravity waves. *Proc. Int. Conf. BOSS'76*, Trondheim, I: 1–13.

Longuet–Higgins, M.S. and Fox, M.J.H., 1977. Theory of the almost–highest wave: the inner solution. *Jour. Fluid Mech.*, 80: 721–741.

Milne–Thomson, L.M., 1974. *Theoretical hydrodynamics.* Macmillan Press Ltd., 743 pp.

Rienecker, M.M. and Fenton, J.D., 1981. A Fourier representation method for steady water waves. *Jour. Fluid Mech.*, 104: 119–137.

Scarsi, G., Stura, S., 1971. Le caratteristiche energetiche delle onde cilindriche di ampiezza e ripidita finite su fondali a dolce acclivita. *L'Energia Electrica*, XLVIII: 3–12.

Schwartz, L.W., 1974. Computer extention and analytic continuation of Stokes' expansion for gravity waves. *Jour. Fluid Mech.*, 62: 553–578.

Skjelbreia, L. and Hendrickson, J., 1961. Fifth order gravity wave theory. *Proc. 7th Coastal Eng. Conf.*, 1: 184–196.

Sretenskiy, L.I., 1977. *Teoriya volnovykh dvizheniy zhidkosti.* Izd. Nauka, Moskva, 815 pp (in Russian).

Stokes, G.G., 1847. On the theory of oscillatory waves. *Trans. Camb. Phil. Soc.*, 8: 441–455.

Stokes, G.G., 1880. Considerations relative to the greatest height of oscillatory waves which can be propagated without change of form. *Mathematical and physical papers*, 1: 225–228.

Toland, J.K., 1978. On the existence of a wave of greatest height and Stokes' conjecture. *Proc. Roy. Soc. London*, A363: 469–485.

Van Dorn, W.G., 1966. Theoretical and experimental study of wave enhancement and run–up on uniformly sloping impermeable beaches. Scripps Inst. Oceanogr., Rep. SIO 66.

Vis, C. and Dingemans, M.W., 1978. An evolution of some wave theories. Delft Hydraulics Lab., Rep. R1192, 177 pp.

Williams, J.M., 1981. Limiting gravity waves in water of finite depth. *Phil. Trans. Roy. Soc. London*, A302: 139–188.

Williams, J.M., 1985. *Tables of progressive gravity waves.* Pitman Advanced Publishing Program, Boston/London/Melbourne, 640 pp.

Chapter 3

REFRACTION OF SHORT WAVES BY SLOWLY VARYING DEPTH

3.1 Geometrical optics approximation

In all previous derivations it has been assumed that waves are propagating on the water of constant depth. However, the sea bottom only very seldom can be treated as the horizontal one. In the deep ocean, the depth changing (even by few hundreds meters), does not influence the short surface waves. It is not the case for the coastal zone where the shallower water depth begins to effect the propagation speed. Since the phase velocity of water waves increases with the local water depth, the parts of a wave crest lying over deeper water travel faster than the parts of the same wave crest lying over shallower water. In the course of its propagation, such a wave front therefore turns gradually toward the shallows. This is in agreement with the common observation on beaches that the crests end up almost parallel to the shore line, even when they were approaching the coast at an oblique angle from the sea. The dependence of propagation speed on the depth may be tenuous, but natural seabed slopes are generally very small and as the waves travel over long distances, the locally small modulation due to changes in the depth often accumulates to bring about important effects. The modulation is therefore characterized by a small parameter; it characterizes the ratio of the wave length to the horizontal length scale of depth variation (Meyer, 1979):

$$\mu = O\left(\frac{\nabla_h h}{kh}\right) \ll 1.0, \tag{3.1}$$

where: $\nabla_h h$ - horizontal gradient of the depth changes.

In order to derive the equations for the wave rays pattern and wave amplitude variations under the assumption (3.1), we use the variational principles given in Chapter 1.

Let us consider the (x, y) plane to be filled with $\vec{k}$ vectors varying in magnitude and direction from point to point. At the given point we draw a curve, which is tangent to the local $\vec{k}$ vector at every point along the curve. This curve is called *wave ray*. The wave ray is always orthogonal to the local crests or phase lines, where $\chi = $ const. Moreover, we assume that function Ω (1.20) is not depending on time. Taking the square of the first eq. (1.17) we obtain a nonlinear differential equation for χ:

$$| \nabla_h \chi |^2 = k^2 \qquad \text{or} \qquad \left(\frac{\partial \chi}{\partial x} \right)^2 + \left(\frac{\partial \chi}{\partial y} \right)^2 = k^2. \tag{3.2}$$

Equation (3.2) is called the *eikonal equation*. For the later convenience we present the ray equations in the more elementary way. If an angle of incidence Θ is defined as the angle made between the beach normal (the x direction) and the wave direction, then:

$$\vec{k} = k_x \vec{i} + k_y \vec{j} = k \cos \Theta \, \vec{i} + k \sin \Theta \, \vec{j}. \tag{3.3}$$

Substituting (3.3) into second equation (1.18), we obtain:

$$\frac{\partial(k \sin \Theta)}{\partial x} - \frac{\partial(k \cos \Theta)}{\partial y} = 0. \tag{3.4}$$

For a shoreline where the alongshore variations in the y direction of all variables are zero, eq. (3.4) reduces to:

$$\frac{d(k \sin \Theta)}{dx} = 0 \qquad \text{or} \qquad k \sin \Theta = const, \tag{3.5}$$

and

$$\frac{\sin \Theta}{C} = const. \tag{3.6}$$

The constant value in eq. (3.6) is given through the boundary condition at the given initial point, i.e.:

$$\frac{\sin \Theta}{C} = \frac{\sin \Theta_0}{C_0}. \tag{3.7}$$

Eq. (3.7) expresses the known *Snell's law*. According to this law, the angle of the wave propagation Θ decreases as the wave shoals and the wave rays are

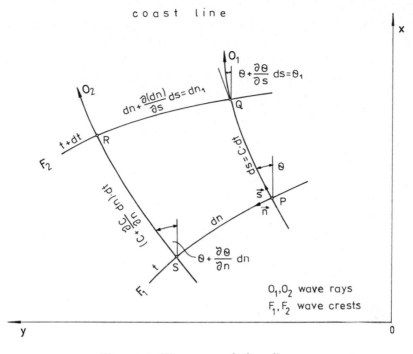

Figure 3.1: Wave rays and phase lines

going to be perpendicular to the shoreline.

In general case, when the bottom contours are irregular and varying along a coast, eq. (3.4) should be used:

$$k \cos \Theta \frac{\partial \Theta}{\partial x} + k \sin \Theta \frac{\partial \Theta}{\partial y} = \cos \Theta \frac{\partial k}{\partial y} - \sin \Theta \frac{\partial k}{\partial x}. \tag{3.8}$$

Usually eq. (3.8) is considered in the orthogonal coordinate system $(\vec{s}, \vec{n})$, where $\vec{s}$ is in the wave direction, while $\vec{n}$ is normal to it (Fig. 3.1). Thus:

$$\left. \begin{array}{l} x = |\vec{s}| \cos \Theta - |\vec{n}| \sin \Theta \\ y = |\vec{s}| \sin \Theta + |\vec{n}| \cos \Theta \end{array} \right\}. \tag{3.9}$$

Substitution of (3.9) into (3.8) yields (Dean, Dalrymple, 1984):

$$\left.\begin{aligned}
\frac{\partial \Theta}{\partial s} &= \frac{1}{k}\frac{\partial k}{\partial n} = -\frac{1}{C}\frac{\partial C}{\partial n}\\[2mm]
\frac{\partial C}{\partial n} &= -\frac{\partial C}{\partial x}\sin\Theta + \frac{\partial C}{\partial y}\cos\Theta
\end{aligned}\right\}. \tag{3.10}$$

Over an arbitrary bottom topography the refraction problem (eq. (3.10)) can be solved by graphical methods which are described elsewhere (Shore Protection Manual, 1977; Svendsen and Jonsson, 1982). However, the construction by hand of wave orthogonals and calculation of wave heights is a very time consuming task. Graphical methods are therefore not much in use today. They are replaced by numerical methods which are more suitable for digital computers programs.

It was shown in Chapter 1 that the wave action in a wave train is transported in the direction of wave propagation with the group velocity C_g. We will now consider the flux of wave action past two vertical sections SP and RQ in three–dimensional coastal waters (Fig. 3.1). The motion will be assumed stationary ($\frac{\partial}{\partial t} = 0$) and the current is neglected ($\vec{U} = 0$). From eq. (1.74) we have:

$$\nabla_h \cdot (E\vec{C_g}) = 0. \tag{3.11}$$

Let us consider two rays separated by dn and dn_1, respectively and integrate eq. (3.11) along the closed contour SPRQ. Along the rays which are perpendicular to the lines of the equal phase is:

$$\vec{n} \cdot \nabla_h \chi = 0. \tag{3.12}$$

From the Gauss theorem and the fact that C_g is tangent to the ray, it follows that the energy fluxes through SP and RQ are equal:

$$[EC_g dn]_{SP} = [EC_g dn]_{RQ} = const. \tag{3.13}$$

Therefore, the wave amplitude (a) is changed along the wave ray according to the following formula:

$$\frac{a_{RQ}}{a_{SP}} = \left\{ \frac{(C_g)_{SP}\,(dn)_{SP}}{(C_g)_{RQ}\,(dn)_{RQ}} \right\}^{1/2}, \tag{3.14}$$

or:

$$\frac{a_{RQ}}{a_{SP}} = K_T K_R.$$ (3.15)

The coefficient K_T:

$$K_T = \left\{ \frac{(C_g)_{SP}}{(C_g)_{RQ}} \right\}^{1/2} = \left\{ \frac{m_{SP}}{m_{RQ}} \frac{C_{SP}}{C_{RQ}} \right\}^{1/2},$$ (3.16)

is called the *shoaling coefficient*. The coefficient m is defined in eq. (2.57). For the very small water depth and normal wave incidence, eq. (3.16) gives:

$$\frac{a_{RQ}}{a_{SP}} = K_T = \left(\frac{h_{SP}}{h_{RQ}} \right)^{1/4} \quad \text{or} \quad H_{RQ} h_{RQ}^{1/4} = H_{SP} h_{SP}^{1/4}.$$ (3.17)

Thus, eq. (3.17) expresses the *Green's law*.

The rate:

$$\left\{ \frac{(dn)_{SP}}{(dn)_{RQ}} \right\}^{1/2} = K_R,$$ (3.18)

is the *refraction coefficient*. For the case of straight and parallel bottom, the coefficient K_R is given by Snell's law, i.e.:

$$K_R = \left(\frac{dn}{dn_1} \right)^{1/2} = \left(\frac{\cos \Theta}{\cos \Theta_1} \right)^{1/2} \ll 1.0.$$ (3.19)

In general, when the isobats are arbitrary, the orthogonal spacing rate $\alpha = dn/dn_1$ should be found from the following equations (Svendsen and Jonsson, 1982; Selezov et al., 1983; Dean and Dalrymple, 1984):

$$\frac{d^2\alpha}{ds^2} + P(s)\frac{d\alpha}{ds} + Q(s)\alpha = 0,$$ (3.20)

where:

$$
\left.\begin{aligned}
\alpha &= \frac{dn_1}{dn} = \left(\frac{1}{K_R}\right)^2 \\[2mm]
P(s) &= -\frac{1}{C}\left(\frac{\partial C}{\partial x}\cos\Theta + \frac{\partial C}{\partial y}\sin\Theta\right) \\[2mm]
Q(s) &= \frac{1}{C}\left(\frac{\partial^2 C}{\partial x^2}\sin^2\Theta - \frac{\partial^2 C}{\partial x\partial y}\sin 2\Theta + \frac{\partial^2 C}{\partial y^2}\cos^2\Theta\right) \\[2mm]
\frac{dx}{ds} &= \cos\Theta, \qquad \frac{dy}{ds} = \sin\Theta
\end{aligned}\right\}, \tag{3.21}
$$

or:

$$
\left.\begin{aligned}
\frac{d\alpha}{ds} &= \sigma, &\qquad \frac{d\sigma}{ds} &= -P(s)\sigma - Q(s)\alpha \\[2mm]
\frac{dx}{ds} &= \cos\Theta, &\qquad \frac{dy}{ds} &= \sin\Theta \\[2mm]
\frac{d\Theta}{ds} &= \frac{1}{C}\left[\frac{\partial C}{\partial x}\sin\Theta - \frac{\partial C}{\partial y}\cos\Theta\right]
\end{aligned}\right\}. \tag{3.22}
$$

In many simple cases, the differential equations ((3.21) - (3.22)) for the rays pattern may be found in the analytical way. The following example gives some idea of such approach. Let the bottom contour be parallel to the y axis so that $h = h(x)$ and $k = k(x)$. Thus, the eikonal equation (3.2) gives:

$$
\frac{d}{dx}\left[\frac{ky'}{(1 + y'^2)^{1/2}}\right] = 0, \tag{3.23}
$$

in which: $y' = \dfrac{dy}{dx} = \dfrac{\partial \chi}{\partial y}\left(\dfrac{\partial \chi}{\partial x}\right)^{-1}$.

Eq. (3.23) implies that:

$$
\frac{ky'}{(1 + y'^2)^{1/2}} = K = const. \tag{3.24}
$$

Taking into account that:

$$
\frac{y'}{(1 + y')^{1/2}} = \frac{dy}{ds} = \sin\Theta, \tag{3.25}
$$

eq. (3.24) is, in fact, the known Snell's law, i.e.:

$$k \sin \Theta = k_0 \sin \Theta_0 = K, \tag{3.26}$$

where: k_0 and Θ_0 refer to a known point (x_0, y_0) on the ray.
The solution of (3.24) gives:

$$y - y_0 = \pm \int_{x_0}^{x} \frac{K \, dx}{[k^2(x) - K^2]^{1/2}}. \tag{3.27}$$

Therefore, the rays can exist only when $k^2 > K^2$. The orthogonality of the rays and wave phase lines yields the following equations for the phase lines and their slope (Mei, 1983):

$$\mp K y = \int^{x} \left(k^2 - K\right)^{1/2} dx + const, \tag{3.28}$$

and

$$\frac{dy}{dx} = \frac{\mp (k^2 - K^2)^{1/2}}{K}. \tag{3.29}$$

3.2 Waves in neighbourhood of a straight caustic

Let us assume now that a plane incident wave approaches from the left $(x < 0)$ into deeper water (Fig. 3.2). Physically, these waves may represent the waves approaching the coast and then reflected from shore. In the profile A, the wave number and the approach angle are equal k_0 and Θ_0, respectively. First, we assume that $K = K' = k_0 \sin \Theta_0 < k(x)$ everywhere. Since $K < k(x)$, the square root $(k^2 - K^2)^{1/2}$ is always real. As h increases, wave number k decreases. Then, the Snell's law indicates that $\Theta(x) > \Theta_0$ and the wave ray turns away from the normal.

Let us now assume that $k_0 < K'' = k_0 \sin \Theta_0 < k_1$. The rays can exist only when $k^2(x) > K^2 = K''^2$. From the incident point (x_0, y_0) to the point C, the slope $dy/dx > 0$ and the angle Θ is increasing. At the point C, $k(x) = K''$ and the slope dy/dx is infinite. It means that the line $x = x_G$ is the envelope of all the rays. The possibility therefore arises that wave rays may be turned back toward the coast before they can reach the deep sea. The envelope mentioned above, is called *caustic* by analogy with optics. Figure 3.2 indicates that in the shallow water area, i.e. till water depth $h = h_C$, where waves are refracted back from a caustic barrier, we encounter cross–waves. Through each point near the caustic line, accordingly, there passes both an "incident" and a

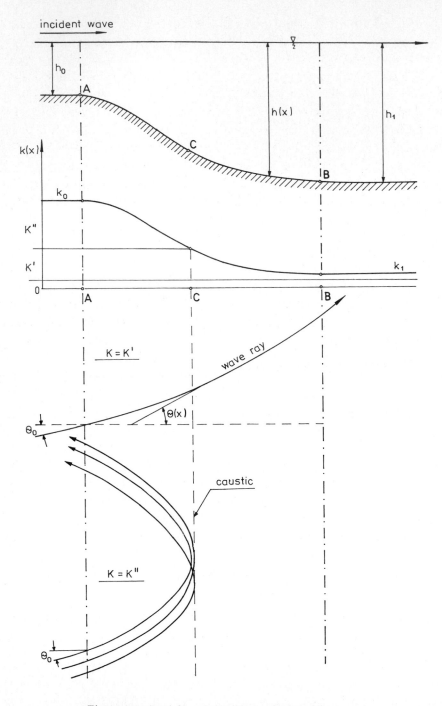

Figure 3.2: Straight caustic line in the coastal zone

"reflected" ray. On the other hand, in the deeper water area $(h > h_C)$, the wave motion is not existing. In the vicinity of the caustic line, the wave rays are approaching closely each other $(dn \to 0)$ and finally they are intersecting. Thus, the energy flux is increasing infinitely and the geometrical optics approximation is no longer applied. This obstacle was overcome by Chao (1971) by the construction of the uniformized asymptotic representation which is valid everywhere. Another approach to the caustic problem, based on the stationary phase method, was proposed by Lighthill (1978).

However, for our purpose, a more simple approximate treatment of the caustic region will be given. The following analysis is mainly to Mei (1983). For a sufficiently smooth bottom, the wave number k close to the caustic line can be expanded in a Taylor series:

$$k^2 = K^2 + (x - x_C)\frac{dk^2}{dx} + \dots \tag{3.30}$$

For the later convenience we assume the origin of the coordinate system at the point C on the caustic line. Moreover, following Mei (1983), we introduce the slow coordinates:

$$\bar{x} = \mu x, \qquad \bar{y} = \mu y, \qquad \bar{z} = z, \qquad \bar{t} = \mu t \qquad \text{with} \qquad \mu \ll 1.0. \tag{3.31}$$

Thus, the caustic line is represented by $\bar{y}$ axis. In terms of the stretched variables, eq. (3.30) takes the form:

$$k^2 = K^2 + \bar{x}\frac{dk^2}{d\bar{x}} = K^2 - \left(-2k\frac{dk}{d\bar{x}}\right) = K^2 - \gamma\bar{x}, \tag{3.32}$$

in which: $\gamma = -2k(dk/d\bar{x})$.

With Snell's law (3.26) and eq. (3.32), we get for wave number k_x:

$$k_x = (-\gamma\bar{x})^{1/2} \qquad \text{or} \qquad \int k_x d\bar{x} = \frac{2}{3}\gamma^{1/2}(-\bar{x})^{3/2}. \tag{3.33}$$

The conservation for the energy flux applied to the profiles associated with points A and C (Fig. 3.2) yields the following expression for the wave amplitude at point C:

$$a_C = a_A\left(\frac{C_g k_x}{k}\right)^{1/2}\left(\frac{C_g}{K}\right)^{-1/2}(-\gamma\bar{x})^{-1/4}, \tag{3.34}$$

or:

$$a_C = \tilde{a}_0(-\gamma\bar{x})^{-1/4} \quad \text{and} \quad \tilde{a}_0 = \left(\frac{C_g k_x}{k}\right)^{1/2}_{\bar{x}=\bar{x}_A} \left(\frac{C_g}{K}\right)^{-1/2}_{\bar{x}=\bar{x}_C} a_A. \tag{3.35}$$

The free surface to the left of the caustic consist of incident and reflected wave fields. Thus:

$$\zeta = \tilde{a}_0(-\gamma\bar{x})^{-1/4} \exp\left(\frac{iK\bar{y}}{\mu}\right) \left\{\exp\left[i\frac{\gamma^{1/2}}{\mu}\frac{2}{3}(-\bar{x})^{3/2}\right] + \right.$$

$$\left. + R\exp\left[-i\frac{\gamma^{1/2}}{\mu}\frac{2}{3}(-\bar{x})^{3/2}\right]\right\}\exp\left(\frac{i\omega\bar{t}}{\mu}\right), \tag{3.36}$$

in which: R - reflection coefficient.

Eq. (3.36) indicates that the wave amplitude increases infinitely as $\bar{x} \to 0$. In order to improve local theory near the caustic, we assume the velocity potential in the form:

$$\Phi = -\frac{ig\tilde{A}(\bar{x})}{\omega}\frac{\cosh k(z+h)}{\cosh kh}\exp\left(\frac{iK\bar{y}}{\mu} - \frac{i\omega\bar{t}}{\mu}\right). \tag{3.37}$$

Substituting (3.37) into Laplace equation (expressed in the coordinate system $0(\bar{x}, \bar{y}, \bar{z})$) and keeping the leading terms gives:

$$\mu^2\frac{d^2\tilde{A}}{d\bar{x}^2} + (k^2 - K^2)\tilde{A} = 0. \tag{3.38}$$

At the caustic line ($\bar{x} \approx 0$), eq. (3.38) takes the form:

$$\mu^2\frac{d^2\tilde{A}}{d\bar{x}^2} - \gamma\bar{x}\tilde{A} = 0. \tag{3.39}$$

If we introduce the new variable:

$$\xi = \gamma^{1/3}\bar{x}\mu^{-2/3}, \tag{3.40}$$

eq. (3.39) becomes the Airy equation:

$$\frac{d^2\tilde{A}}{d\bar{x}^2} - \xi\tilde{A} = 0. \tag{3.41}$$

The solution of eq. (3.41) for one caustic line and unbounded area of motion is expressed by Airy function:

$$\tilde{A}(\xi) = \tilde{a}_1 Ai(\xi). \tag{3.42}$$

The form of the solution $\tilde{A}(\xi)$ depends on the sign of ξ. The solution is oscillatory for $\xi < 0$ and monotonic for $\xi > 0$. It may be shown from eq. (3.42) that the sea surface ordinate ζ takes the form:

$$\zeta = \tilde{a}_1 Ai(\xi) \exp\left(\frac{ik\bar{y}}{\mu}\right) \exp\left(\frac{i\omega\bar{t}}{\mu}\right). \tag{3.43}$$

The unknown coefficients $\tilde{a}_1$ and R can be found by matching eq. (3.43) with eq. (3.36) at $\xi \to -\infty$; hence:

$$Ai(\xi) \sim \frac{1}{\sqrt{\pi}}(-\xi)^{1/4} \sin\left[\frac{2}{3}(-\xi)^{3/2} + \frac{\pi}{4}\right]. \tag{3.44}$$

Thus:

$$\tilde{a}_1 = 2\sqrt{\pi}i(\gamma\mu)^{-1/6} \exp\left(i\frac{\pi}{4}\right) \tilde{a}_0 \quad \text{and} \quad R = \exp\left(i\frac{\pi}{2}\right). \tag{3.45}$$

For the positive ξ (area behind the caustic line), the asymptotic solution may be represented in the form:

$$Ai(\xi) \sim \frac{1}{2\sqrt{\pi}} \xi^{-1/4} \exp\left(-\frac{2}{3}\xi^{3/2}\right), \qquad \xi \to \infty. \tag{3.46}$$

Let us apply the expressions, given above, to the coastal zone in which the gentle bottom slope $(1/100)$ is situated between water depth $h_A = 5\ m$ and $h_B = 100\ m$ (Fig. 3.2). The incident wave of the period $T = 10\ s$ approaches from the left at the angle $\Theta_A = 45°$. Hence, $K = k_a \sin\Theta_A = 0.06562$. At the profile C (the caustic line), the water depth $h_C \approx 10.817\ m$ and $\gamma \approx 0.023$. Substituting these values into eqs. (3.36) and (3.43), we get the surface oscillations as in Fig. 3.3, in which $\bar{y} = 0$ and $\bar{t} = 0$ was assumed. The largest amplitude is finite and occurs before the caustic is reached. The reflected wave has the same amplitude as the incident one but differs in phase by $1/2\pi$. Before the caustic line, the resulting wave is similar to the standing wave, while behind the caustic, it is exponentially decreasing with the distance $\bar{x}$ from the caustic. Fig. 3.3 indicates that the geometrical optics methods (broken line) yields the infinite oscillation amplitude at the caustic line and no waves for $\bar{x} > 0$. It should be noted that the troubles related to the caustics can be rather easily overcome using so called parabolic approximation to the refraction–diffraction equation (see Section 3.4).

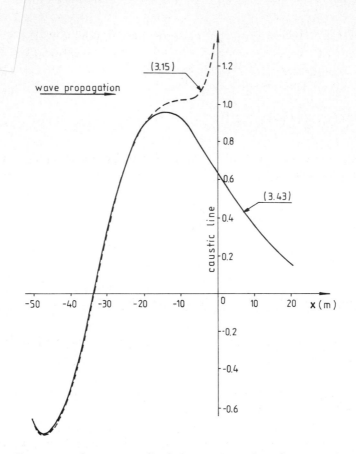

Figure 3.3: The wave amplitude in vicinity and on the caustic line

3.3   Influence of bottom friction

The variation of wave height due to variations in water depth only, in two–dimensional case, yields the simple formula (3.15). This expression is valid only when the energy dissipation between profiles SP and RQ can be omitted. However, it is not true in general case. In shoaling water, outside the surf zone, the energy dissipation will partly be due to bottom friction and percolation in the permeable bed. Laboratory experiments indicate that the former is normally the most important one. For example, in the regular progressive wave with $H = 2\ m, T = 8\ s$, and $h = 7\ m$, the relative changes in mean energy flux over one wave length due to turbulent bottom friction and percolation equal 1.7% and 0.06%, respectively (Svendsen and Jonsson, 1982). Therefore, in the following, the energy losses due to percolation in the porous bottom will be neglected. Some information on the percolation mechanisms can be found elsewhere (Svendsen and Jonsson, 1982, Longuet–Higgins 1983). In the surf zone, the dissipation due to breaking is at least an order of magnitude larger than the bottom friction. We will discuss this problem in Chapter 5.

Therefore, in the following we restrict ourselves to the dissipation due to bottom friction and rewrite eq. (3.15) in the form:

$$\frac{a_{RQ}}{a_S} = K_T K_R K_S,$$

(3.47)

in which:

$$K_S = \left[\frac{\bar{I}_{RQ}\, dn_1}{\bar{I}_{SP}\, dn}\right]^{1/2}.$$

(3.48)

and: $\bar{I}_{RQ}, \bar{I}_{SP}$ energy fluxes (see eq. (2.80)). Thus:

$$\frac{d}{ds}(\bar{I}dn) = -\bar{E}_S\, dn,$$

(3.49)

where: $\bar{E}_S$ - mean energy dissipated due to bottom friction, per unit area of the bed and per unit time.

Using (3.48) and (3.49) we eliminate the flux $(\bar{I}dn)$ and obtain:

$$\frac{d}{ds}(K_S^2) = -\frac{C\bar{E}_S\, \alpha}{\bar{I}_{SP}}$$

(3.50)

The dissipation rate $\bar{E}_S$ in the wave boundary layer is given by:

$$\bar{E}_S = \frac{2}{3\pi}\,\rho f_e u_{b,max}^3,$$

(3.51)

where: f_e - wave energy loss factor, $u_{b,max}$ - velocity amplitude at the bed according to potential theory.

After substituting (3.51) into (3.50) we find (Svendsen and Jonsson, 1982):

$$\frac{d}{ds}(K_S) = \frac{-8}{3L}\frac{dC}{dh} A_{-h} f_e K_S,$$

(3.52)

where: A_{-h} - particle amplitude at the bed according to potential theory.

The factor f_e is a function of the ratio A_{-h} and the Nikuradse roughness parameter k_N; thus, it is not a constant. Laboratory experiments have shown that f_e can be determined from the following expression:

$$\left.\begin{array}{lll} f_e = 0.30 & \text{for} & \dfrac{A_{-h}}{k_N} < 1.57 \\[3mm] \dfrac{1}{4\sqrt{f_e}} + \lg\left(\dfrac{1}{4\sqrt{f_e}}\right) = -0.08 + \lg\left(\dfrac{A_{-h}}{k_N}\right) & \text{for} & \dfrac{A_{-h}}{k_N} > 1.57 \end{array}\right\} . \tag{3.53}$$

In order to simplify the factor f_e for $(A_{-h})/k_N > 1.57$, Kamphuis (1975) suggested that:

$$f_e = 0.4\left(\frac{A_{-h}}{k_N}\right)^{0.75} \qquad \text{for} \qquad \frac{A_{-h}}{k_N} < 100. \tag{3.54}$$

In some special cases, the refraction and shoaling coeffcients can be expressed more explicitly. For example, using the linear theory, the coefficient K_T takes the form:

$$K_T = \frac{H_{PQ}}{H_{SP}} = (2m \tanh kh)^{-1/2}, \tag{3.55}$$

under the assumption that profile SP is in the deep water $(h/L > 1/2)$. Usually, the coefficient K_T is shown as a function of h/L_0 (H_0 and L_0 - wave parameters in the deep water). In Fig. 3.4, the function $H/H_0 = f(h/L_0)$ has been presented. The linear formula (3.55) is compared with the nonlinear Cokelet (1977) solution (Sakai and Battjes, 1980). Each particular curve corresponds to the selected steepness of the deep water waves. The curve with infinitezimal steepness represents the linear solution. The difference between linear and nonlinear solutions is increasing when approaching the breaking limit. This rapid variation of wave height calculated from Cokelets theory is inconsistent with the basic assumption of gradually varying wave parameters. Outside of the breaking area, the nonlinear shoaling according to Cokelet (1977) is close to the shoaling predicted by other nonlinear solutions (see for example Le Mehaute and Webb, 1964; Le Mehaute and Wang, 1980). The shoaling coefficient K_T for the long waves will be discussed in Chapter 4.

Moreover, it should be noted that hydraulic experiments yield the conclusion that the shoaling coefficient K_T based on the Cokelet's theory gives quite reasonable and practically applicable results, except near the breakpoint.

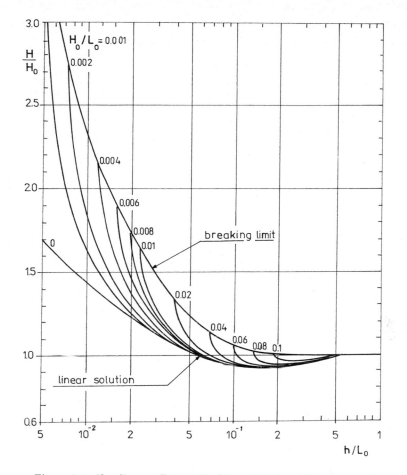

Figure 3.4: Shoaling coefficient K_T.(From Sakai and Battjes, 1980)

3.4 Refraction - diffraction processes

For the above solutions to be relevant over the whole beach, wave slopes must be very small. Moreover, under the assumption that the rate of change of water depth is small within a characteristic wave length, the wave reflection can be neglected. The ray theory further assumes that wave rays can be defined as a family of curves to which the wave number vectors are tangential and wave energy flux is conserved between two adjacent wave rays. There are, however, some difficulties in using and interpreting the results based on these assumptions. Usually on sandy beaches nearly periodic longshore sandbars can be found. The number of bars can range from 3 to 17 (Mei, 1985). The local slopes in the vicinity of the underwater bars are sometimes substantial

(Boczar-Karakiewicz et al., 1981). Within the coastal zone, the approaching channels to ports are located. For waves propagated in the coastal zone, they present obstacles with rapidly increasing water depth.

The solutions obtained by using ray theory sometimes give rays which are crossing. Simple examples occur for water of constant depth where rays are straight lines. Any initial wave front which is concave in the direction of propagation leads to crossed rays and the flux of wave energy then appears to have a singularity. The approximation described above can be used to modify these ray solutions only locally near a simple caustic, but not for complex types of caustics.

Waves usually become steeper as they propagate into shallow water because the phase and group velocities both decrease. To maintain energy flux, amplitude increases at the same time as wave length shortens, so any wave eventually becomes too steep for linear theory to hold. Since the wave ray theory also exludes wave diffraction, it is unable to predict the wave characteristics near coastal structures such as breakwaters. To include the diffraction effects, the combined refraction and diffraction problem should be formulated. There are several different ways of making suitable approximations. Two particularly significant developments of recent years are:

- the introduction of the *mild–slope equation*. It results from eliminating the vertical coordinate in cases where bottom slopes are gentle,
- the introduction of the concept of wave action (see Chapter 1).

In the following we introduce the mild–slope equation for the wave train propagating on slowly varying water depth when the uniform current $\vec{U} = (U_1, U_2, 0)$ is taken into account. In addition to this we assume that the spatial and time variation of the current velocity and the mean free surface are very small (see Section 1.1). As this study is concerned with the influence of current on waves, the opposite effect of waves on current is neglected. The mild–slope equation will be derived from the variation principle (1.28) with the Lagrangian (1.32). The linear approach to the boundary value problem involving the wave propagating over the slowly changing bottom suggests the following representation for the potential Φ and free surface ζ:

$$\left. \begin{aligned}
\Phi^*(\vec{x}, z, t) &= \Phi_p(\vec{x}, t) + \epsilon Z(z, h^*)\tilde{\varphi}(\vec{x}, t) \\
\zeta(\vec{x}, t) &= \bar{\zeta}(\vec{x}, t) + \epsilon \eta(\vec{x}, t) \\
h^* &= h(\vec{x}) + \bar{\zeta}(\vec{x}, t) \\
Z(z, h^*) &= \cosh[k(z + h^*)] / \cosh(kh^*)
\end{aligned} \right\}, \tag{3.56}$$

in which a non-linearity scale ϵ was introduced. Therefore, $\tilde{\varphi}$ is $O(1)$ with respect to ϵ. It should be added that the water depth h and the current velocity vary with respect to the modulation scale μ (3.1). After substituting (3.56) into (1.32) and neglecting terms with ∇Z, we get (Dingemans, 1985):

$$-\frac{\mathcal{L}}{\rho} = -\frac{1}{\rho}(\mathcal{L}_0 + \epsilon\mathcal{L}_1 + \epsilon^2\mathcal{L}_2), \tag{3.57}$$

in which:

$$\mathcal{L}_0 = (h + \bar{\zeta})\frac{\partial\Phi_p}{\partial t} + \frac{1}{2}(h + \bar{\zeta})(\nabla\Phi_p)^2 + \frac{1}{2}g(\bar{\zeta}^2 - h^2), \tag{3.58}$$

$$\mathcal{L}_1 = \eta\frac{\partial\Phi_p}{\partial t} + \frac{\partial\Phi}{\partial t}\int_{-h}^{\bar{\zeta}} Z\,dz + \frac{1}{2}(h + \bar{\zeta})(\nabla\Phi_p)^2 +$$

$$+ \nabla\Phi_p \cdot \nabla\tilde{\varphi}\int_{-h}^{\bar{\zeta}} Z\,dz + g(h + \bar{\zeta})\eta, \tag{3.59}$$

$$\mathcal{L}_2 = \frac{1}{\epsilon}\frac{\partial\tilde{\varphi}}{\partial t}\int_{\bar{\zeta}}^{\epsilon\eta} Z\,dz + \frac{1}{\epsilon}\nabla\Phi_p \cdot \nabla\varphi\int_{\bar{\zeta}}^{\epsilon\eta} Z\,dz +$$

$$+ \frac{1}{2}(\nabla\tilde{\varphi})^2\int_{-h}^{\bar{\zeta}} Z^2\,dz + \frac{1}{2}\tilde{\varphi}^2\int_{-h}^{\bar{\zeta}}\left(\frac{\partial Z}{\partial z}\right)^2 dz + \frac{1}{2}g\eta^2. \tag{3.60}$$

The variation of $\mathcal{L}_0$ with respect to Φ_p and $\bar{\zeta}$ yields the following equations for current velocity $\vec{U}(\vec{U} = \nabla\Phi_p)$:

$$\left.\begin{array}{l} \dfrac{\partial\bar{\zeta}}{\partial t} + \nabla_h \cdot [(h + \bar{\zeta})\vec{U}] = 0 \\[2ex] \dfrac{\partial\vec{U}}{\partial t} + (\vec{U} \cdot \nabla_h)\vec{U} + g\nabla_h\bar{\zeta} = 0 \end{array}\right\}. \tag{3.61}$$

In the same way, the variation of $\mathcal{L}_2$ with respect to η and $\tilde{\varphi}$ gives:

$$\frac{D\tilde{\varphi}}{Dt} + g\eta = 0, \tag{3.62}$$

and:

$$g\frac{D\eta}{Dt} + g\eta\nabla_h \cdot \vec{U} + \nabla_h \cdot (CC_g\nabla\tilde{\varphi}) - (\sigma^2 - kCC_g)\tilde{\varphi} = 0, \tag{3.63}$$

$$\frac{D}{Dt} = \left(\frac{\partial}{\partial t} + \vec{U} \cdot \nabla_h\right).$$

(3.64)

Eliminating the ordinate η from eqs. (3.62) and (3.63), we get time–dependent form of the mild–slope equation:

$$\frac{D^2\tilde{\varphi}}{Dt^2} + (\nabla_h \cdot \vec{U})\frac{D\tilde{\varphi}}{Dt} - \nabla_h \cdot (CC_g\nabla_h\,\tilde{\varphi}) \; +$$

$$+ (\sigma^2 - k^2CC_g)\tilde{\varphi} \; = \; 0.$$

(3.65)

For the time–harmonic motion, in which:

$$\tilde{\varphi}(\vec{x}, t) = \mathcal{R}e\left[\varphi(\vec{x})\exp(-i\omega t)\right],$$

(3.66)

eq. (3.65) may be written as (Dingemans, 1985):

$$\frac{\partial}{\partial x_i}\left[U_iU_j\frac{\partial\varphi}{\partial x_j} - CC_g\frac{\partial\varphi}{\partial x_i}\right] - 2i\omega\vec{U} \cdot \nabla_h\varphi \; +$$

$$+ (\sigma^2 - \omega^2 - k^2CC_g - i\omega\nabla_h \cdot \vec{U})\varphi \; = \; 0.$$

(3.67)

Equation (3.67) is the time–independent mild–slope equation with the current. For $\vec{U} = 0$, eq. (3.67) is simplyfing considerably:

$$\nabla_h \cdot (CC_g\nabla_h\varphi) + k^2CC_g\varphi = 0.$$

(3.68)

Equation (3.68) was first introduced by Berkhoff (1972, 1976). By demanding conservation of wave energy, two different functionals for the finite element solution of the mild–slope equation were recently constructed by Behrendt and Jonsson (1984). Both functionals were constructed in a straight–forward way that leads to a better physical understanding of the functionals.

Comparisons with solutions for the full linear equations by Booij (1983) give confidence in the use of the mild–slope equation for quite large slopes in suitable circumstances. The mild–slope equation (3.68), like the full linear wave equation are elliptic in the (x, y) plane and hence requires solution methods dealing with the whole region of interest in (x, y) space. However, for mild–slope equation with current (3.67), the classification depends on the sign of the expression $P = CC_g(|\vec{U}|^2 - CC_g)$. Only if $P < 0$, the equation (3.67) is elliptic. For the currents, large with respect to $(CC_g)^{1/2}$, the mild–slope equation becomes hyperbolic ($P > 0$).

In the special case of arbitrary constant kh, eq. (3.68) reduces to the Helmholtz equation:

$$\nabla_h^2 \varphi + k^2 \varphi = 0,$$

(3.69)

while in the shallow water $(kh \ll 1)$ it is:

$$g\nabla_h \cdot (h\nabla_h \varphi) + \omega^2 \varphi = 0,$$

(3.70)

which is valid for long waves even if $\mu = O(1)$. Thus, eq. (3.68), while approximate in intermediate depth, is exact in both deep and shallow water.

In order to clarify the relation between mild–slope equation and the geometrical optics approximation we consider wave–like structure:

$$\varphi(x) = b(\vec{x}) \exp[iS(\vec{x})],$$

(3.71)

in which:

$$\chi(\vec{x}, t) = S(\vec{x}) - \omega t,$$

(3.72)

and phase function $\chi(\vec{x}, t)$ is given by (3.2). Substituting of eq. (3.71) into (3.68) yields:

$$(\nabla_h S)^2 = k^2 + \frac{\nabla_h(CC_g)}{CC_g} \frac{\nabla_h b}{b} + \frac{\nabla_h^2 b}{b},$$

(3.73)

and

$$\frac{\partial}{\partial x_i} \left[\left(CC_g \frac{\partial S}{\partial x_i} \right) b^2 \right] = 0.$$

(3.74)

Taking into account that the wave energy is given by $E = (1/2)\rho g a^2$ (a - amplitude of the free surface elevation), the potential amplitude b takes the form:

$$b = \frac{ga}{\omega}.$$

(3.75)

Therefore, eqs. (3.73) and (3.74) become as:

$$(\nabla S)^2 = k^2 + \frac{\nabla_h(CC_g)}{CC_g} \frac{\nabla_h a}{a} + \frac{\nabla_h^2 a}{a},$$

(3.76)

$$\nabla_h \cdot (CC_g a^2 \nabla S) = 0. \tag{3.77}$$

On the other hand, the geometrical optics approximation assumes that (see eq. (3.2)): $(\nabla S)^2 = k^2$. Therefore, for the geometrical optics approximation to be valid, the following conditions should be held:

$$\left. \begin{aligned} \frac{\nabla_h CC_g}{CC_g} \frac{\nabla_h a}{a} &\ll k^2 \\ \frac{\nabla_h^2 a}{a} &\ll k^2 \end{aligned} \right\}. \tag{3.78}$$

The term $(\nabla_h^2 a)/a$ describes the effect of diffraction. This fact becomes clear after substituting $\zeta = a \exp(iS)$ into the Helmholtz equation $\nabla^2 \zeta + k^2 \zeta = 0$. Thus, the real and imaginary part gives:

$$(\nabla S)^2 = k^2 + \frac{\nabla_h^2 a}{a}, \qquad \nabla_h \cdot (a^2 \nabla S) = 0. \tag{3.79}$$

The first term of eq. (3.78) can be rewritten in the form (Dingemans, 1985):

$$\frac{\nabla_h CC_g}{CC_g} = G \frac{\nabla_h h}{h}, \tag{3.80}$$

where:

$$G = 1 - \frac{1}{2} \frac{C}{C_g} \left[1 - \frac{gh}{CC_g} + 2kh \tanh(kh) \right]. \tag{3.81}$$

Because $max(G) = 1$, eq. (3.80) yields:

$$\frac{\nabla_h CC_g}{CC_g} < \frac{\nabla_h h}{h}. \tag{3.82}$$

Therefore, the first relation in (3.78) becomes:

$$\frac{\nabla_h h}{kh} \frac{\nabla_h a}{ka} \ll 1. \tag{3.83}$$

Thus, the geometrical optics approximation is valid only if the bottom slope is small with respect to kh and the wave amplitude variation in space is also small with respect to see surface slope ka. It should be noted that the energy transport equation in the mild–slope equation representation (3.77) and in the geometrical optics approximation (3.79) is the same when the respective expressions for ∇S are used.

The discussion given above can be extended on the case of current. Particularly for the energy transport equation, we have (Kirby, 1984; Dingemans, 1985):

$$\nabla_h \cdot \left[\frac{E}{\sigma} \left(\vec{U} + \vec{C}_g \right) \right] = 0, \tag{3.84}$$

which is the wave action conservation equation.

Eq. (3.84) is identical with eq. (1.74) (for steady motion) obtained from the variational principle. Due to elliptic nature of the mild–slope equation (3.68), it is difficult to implement for typical wave–propagation problems extending over many wave length. There are also boundaries, such as beaches, where waves break, where no sensible boundary conditions have been devised. In Fig. 3.5 the comparison of the reflection coefficients from the plane slope is illustrated. The coefficients were calculated by the mild–slope equation and by the full two–dimensional model. For the normal approach to beach slope, eq. (3.68) takes the form:

$$\frac{d}{dx} \left(CC_g \frac{d\varphi}{dx} \right) + k^2 CC_g \varphi = 0. \tag{3.85}$$

Thus:

$$\Phi(x, z, t) = \mathcal{R}e \left\{ Z(z) \varphi(x) \exp(-i\omega t) \right\}. \tag{3.86}$$

On the other hand, the full two–dimensional model yields:

$$\Phi(x, z, t) = \mathcal{R}e \left\{ \hat{\varphi}(x, z)(-i\omega t) \right\}, \tag{3.87}$$

in which:

$$\frac{\partial^2 \hat{\varphi}}{\partial x^2} + \frac{\partial^2 \hat{\varphi}}{\partial z^2} = 0. \tag{3.88}$$

The solution of the eqs. (3.85) and (3.88) were obtained by the finite element method. Fig. 3.5 indicates that the approximate mild–slope equation provides quite good accuracy for the slope up to $\tan \alpha \approx 1/3$. For more complicated conditions, the usual method of dealing with this problem is to introduce the approximations which give hyperbolic (Copeland, 1985; Madsen and Larsen, 1987) or parabolic equations. The parabolic approximation

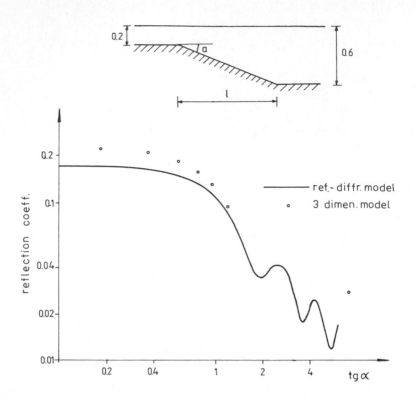

Figure 3.5: The comparison of the reflection coefficient calculated according to mild–slope equation and three–dimensional model. (From Booij, 1983)

has proved of value for radio–wave propagation, underwater accoustics and in seismic waves (Radder, 1979). Recently, the theory based on parabolic approximation has been developed for water waves and has emerged as a powerful tool for studying combined wave refraction and diffraction phenomena in coastal region (Booij, 1981; Kirby and Dalrymple, 1983b; Kirby, 1984; Liu and Tsay, 1984, 1985). The practical advantages over more accurate elliptic equations are that numerical integration can be obtained by marching from a deep water region toward the shoreline. The parabolic approximation modifies the ray theory and allows the wave energy to flow perpendicular to the direction of wave motion so as to take wave diffraction into account. The parabolic approximation to be obtained may be regarded as being an approximation to the mild–slope equation of Berkhoff (1972). For the simplest case of constant water depth, mild–slope equation reduces to the Helmholtz equation (3.69) which permits wave propagation in all directions. The propagation in the positive x–direction can be described assuming:

$$\varphi(x, y) = A(x, y)e^{ikx}, \tag{3.89}$$

where $A(x, y)$ is slowly varying in the sense that:

$$|\nabla A| \ll kA. \tag{3.90}$$

Substituting of (3.89) into (3.69) gives:

$$\frac{\partial}{\partial x}\left(2ik + \frac{\partial}{\partial x}\right)A + \frac{\partial^2 A}{\partial y^2} = 0, \tag{3.91}$$

and by virtue of the assumption (3.90) the second x derivative is neglected to give:

$$2ik\frac{\partial A}{\partial x} + \frac{\partial^2 A}{\partial y^2} = 0. \tag{3.92}$$

Eq. (3.92) represents the linearized, diffraction limit of the lowest–order parabolic approximation; this is in fact the one–dimensional Schrodinger equation. Eq. (3.89) indicates that the parabolic approximation is applicable when the wave field does not vary greatly from a undirectional wave train. Therefore, no returning reflections are permitted, though forward ones are. Moreover, in a moving medium, due to current, the amplitude function becomes $A(x, y)$ though only small changes in frequency can be allowed.

In order to obtain the parabolic approximation for the general form of the mild–slope equation, we consider the time–dependent form (3.65) in which h and $\vec{U}$ are arbitrary, slowly varying functions of x and y. The potential $\tilde{\varphi}(\vec{x}, t)$ is assumed to take on a form given by:

$$\tilde{\varphi}(\vec{x}, t) = -ig\hat{\varphi}(\vec{x})\exp(i\psi), \tag{3.93}$$

in which:

$$\hat{\varphi}(\vec{x}) = \frac{A(x, y)}{\sigma}, \qquad \psi = \int^x k\,dx - \omega t. \tag{3.94}$$

Substituting (3.93) into (3.65) yields (Kirby, 1984):

$$i\left\{\frac{\partial \sigma}{\partial t}\hat{\varphi} + 2\sigma\frac{\partial \hat{\varphi}}{\partial t} + 2\sigma\vec{U}\cdot\nabla_h\hat{\varphi} + \nabla_h\cdot(\sigma\vec{U})\hat{\varphi}\right\} +$$

$$+i\left\{\frac{\partial(\sigma C_g)}{\partial x}\hat{\varphi} + 2\sigma C_g\left(\frac{\partial\hat{\varphi}}{\partial x}\right)\right\} - \frac{\mathcal{D}^2\hat{\varphi}}{\mathcal{D}t^2} +$$

$$-(\nabla_h\cdot\vec{U})\frac{\mathcal{D}\hat{\varphi}}{\mathcal{D}t} + \nabla_h\cdot(CC_g\nabla_h\hat{\varphi}) = 0. \tag{3.95}$$

If we neglect time dependence and assume additionally that $\partial/\partial x \sim \mu(\partial/\partial y)$, we obtain the parabolic approximation for linear waves in two dimensions in the form:

$$2ik\frac{\partial A}{\partial x} + 2ik\frac{U_2}{C_g+U_1}\frac{\partial A}{\partial y} + i\frac{k\sigma}{(C_g+U_1)} \cdot$$

$$\cdot\left\{\frac{\partial}{\partial x}\left(\frac{C_g+U_1}{\sigma}\right) + \frac{\partial}{\partial y}\left(\frac{U_2}{\sigma}\right)\right\}A +$$

$$+\frac{k}{\sigma(C_g+U_1)}\frac{\partial}{\partial y}\left(CC_g\frac{\partial A}{\partial y}\right) = 0, \tag{3.96}$$

where it is assumed that $O(|\vec{U}|^2) \ll CC_g$.

Extending the multiple scale perturbation expansion of Yue and Mei (1980) to the case non–constant depth, Kirby and Dalrymple (1983b) obtained a parabolic equation for the propagation of weakly nonlinear Stokes waves. For the complex amplitude $A(x,y)$, this equation takes the form:

$$2ikCC_g\frac{\partial A}{\partial x} + 2k(k-k_0)CC_gA + iA\frac{\partial}{\partial x}(kCC_g) +$$

$$+\frac{\partial}{\partial y}\left(CC_g\frac{\partial A}{\partial y}\right) - k(CC_g)\lambda\,|\,A\,|^2A = 0. \tag{3.97}$$

The initially plane Stokes wave with frequency ω, reference wave number k_0 is approaching from $-\infty$ at a small angle to the x axis; therefore:

$$\varphi(\vec{x}) = A(x,y)\exp(ik_0x), \tag{3.98}$$

and

$$\lambda = k^3 \frac{C}{C_g} \frac{\cosh(4kh) + 8 - 2\tanh^2(kh)}{8\sinh^4(kh)}. \tag{3.99}$$

After linearizing, eq. (3.97) becomes:

$$2ikCC_g \frac{\partial A}{\partial x} + 2k(k - k_0)CC_g A + iA\frac{\partial}{\partial x}(kCC_g) + \frac{\partial}{\partial y}\left(CC_g \frac{\partial A}{\partial y}\right) = 0, \tag{3.100}$$

which is identical to the result obtained by Radder (1979).
On the other hand, the simplification of the nonlinear equation (3.97) to the case of constant depth leads to the nonlinear Schrodinger equation given by Yue and Mei (1980):

$$2ik_0 \frac{\partial A}{\partial x} + \frac{\partial^2 A}{\partial y^2} - k_0\lambda \mid A \mid^2 A = 0, \tag{3.101}$$

which is a nonlinear extention of eq. (3.92).
The limitation to small angles of propagation with respect to prescribed direction is troublesome in the application of the parabolic equation method to waves in the coastal environment. Kirby (1986d) has attempted to remedy this situation by employing the minimax principle, which minimizes the maximum error occuring over a prespecified range of wave directions. Numerical studies shows that the minimax approximations allow for much more successful treatment of large angles of incidence than the previously available methods (Kirby, 1986a). There are several other methods for the development of wide–angle approximation. Through the use of Fourier decomposition, wave model valid to 90° from the assumed wave direction was recently developed by Dalrymple and Kirby (1988).

The results mentioned above are based on the assumption that the re flected wave is absent or negligible. This assumption is certainly valid locally in a slowly varying domain; however, a reflected component may accumulate for waves propagating over long distances. Liu and Tsay (1983) have developed an iterative scheme based on coupled equations for forward and back scattered waves, and have shown that these equations are capable of producing results in agreement with a finite–element solution of Berkhoff equation. A more straightforward approach was developed by Kirby (1986b). He applied the variational principle (1.28) with Lagrangian (1.32). After deriving equations for a general wave motion in two horizontal dimension (x, y) and time, the results were specialized to the case of two waves propagating in an antiparallel direction. The parabolic approximation forms a system of two coupled parabolic equations for the amplitudes envelopes A and B, where:

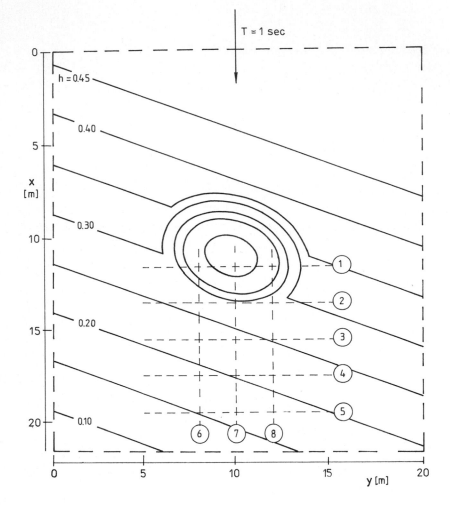

Figure 3.6: Topography and computational domain for experiment of Berkhoff, Booij and Radder (1982)

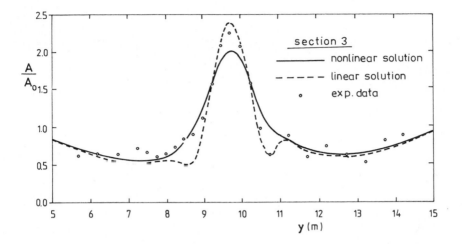

Figure 3.7: Comparison of linear and nonlinear model results to experimental data. Transect 3

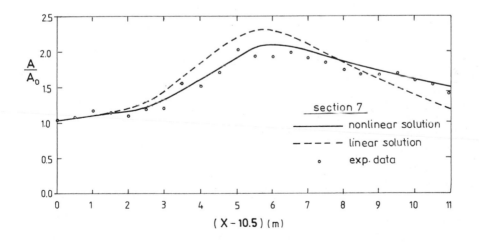

Figure 3.8: Comparison of linear and nonlinear model results to experimental data. Transect 7

$$\tilde{\varphi}^+ = \frac{-ig}{\omega} A \exp(ik_0 x), \qquad \tilde{\varphi}^- = \frac{-ig}{\omega} B \exp(-ik_0 x), \qquad (3.102)$$

$\tilde{\varphi}^+$ - potential of the incident waves, $\tilde{\varphi}^-$ - potential of the reflected waves.

The development of new computational techniques for predicting the characteristics of waves in inhomogenous physical domains has led to the concurrent need for experimental data sets to test the predictions of these models.

Berkhoff et al. (1982) have presented data obtained in the vicinity of a focus and cusped caustic created by an elliptic shoal resting on a plane slope of 1:50. The plane slope rises from a region of constant depth $h = 0.45\ m$ and the entire slope is turned at an angle of 20^0 to a straight wave paddle. The bottom contours are shown in Fig. 3.6. Results of the nonlinear and linear models are presented in comparison to the experimental data in Figs. 3.7 and 3.8 for the labelled transects 3 and 7, respectively. In each figure, nonlinear (3.97) and linear (3.100) model results are indicated by solid and dashed lines, while open circles indicate experiment data points. In transect 3, the cusped caustic is developed. The data falls between predictions of the two models in the region of maximum amplitude. The nonlinear model underpredicts the maximum amplitude by about 8%. Agreement between the nonlinear model and the data is also very good along the longitudinal Section 7, which is just off center of the axis of the focussed region. It can be seen that the nonlinear model predicts both the drop in amplitude in the region of the cusp (when comparing with the linear model predictions) and the slower decay towards the shore. In general, the nonlinear model exhibits closer agreement with the experimental data than do the linear model.

Another examples of the application of the parabolic approximation are now available in literature (Liu, 1986; Liu and Tsay, 1984, 1985; Tsay and Liu, 1982; Kirby, 1986a-d). They indicate that the parabolic approximation should be used for the relatively small areas with typical dimesnions $\sim 10 \div 100$ wave lengths. For the smaller regions (~ 10 wave length) the application of the mild–slope equation is recommended, while for very large areas (> 100 wave lengths) the refraction procedures, supplemented by some averaging techniques can be used (Berkhoff et al., 1982).

Moreover, it should be noted that nonlinear parabolic approximation, based on the Stokes wave model, is valid only when Ursell number $U \leq 75$ ($\Pi \leq 0.30$).

3.5 Propagation of waves over underwater step and channels

The interaction between underwater obstacles and ocean waves is of considerable interest to coastal engineers. Certain geometries are capable of reflecting a significant amount of wave energy and, therefore, protecting the beach. On the other hand, an obstacle influences the propagating waves. In aerial photographs of the sea surface, the formation of the secondary crests in troughs of the main waves was detected in waves propagating from deep water onto the continental shelf (Stoker, 1957). Similar effects have been found in laboratory basin experiments (Bendykowska, 1980; Massel, 1982). Harmonic analysis of simultaneous measurements in an incident region and on a step demonstrates clearly a substantial growth of the higher harmonics in shallow water areas. If the water depths in both areas are small compared to wave length, simple formulas for the transmission and reflection coefficients can be obtained. According to Lamb (1932) we have:

$$|K_R| = \frac{1 - \left(\dfrac{h_t}{h}\right)^{1/2}}{1 + \left(\dfrac{h_t}{h}\right)^{1/2}}, \qquad |K_T| = \frac{2}{1 + \left(\dfrac{h_t}{h}\right)^{1/2}}, \tag{3.103}$$

where: h - water depth in the incident region, h_t - water depth on the stop (Fig. 3.9); $|K_R|, |K_T|$ - modulus of reflection and transmission coefficients, respectively.

In order to investigate the transmission of surface waves over a submerged step, we consider the submerged breakwater which occupies the region $x_p \leq x \leq x_k$, $-h \leq z \leq -h_t$ (Fig. 3.9). It is supposed that a wave of small but finite amplitude is propagating form $x = -\infty$ toward the step. Thus, to second order, the velocity potential of the incident wave should be:

$$\Phi = \Phi^{(1)} + \Phi^{(2)} = \frac{gH_k^{(1)}}{2\omega} \frac{\cosh k(z+h)}{\cosh kh} \sin(kx - \omega t) +$$

$$+ \frac{3}{32}\omega H_k^{(1)2} \frac{\cosh 2k(z+h)}{\cosh 2kh} \sin 2(kx - \omega t), \tag{3.104}$$

where: $H_k^{(1)}$ - height of wave to the first order.
Similarly, in each region should be (Massel, 1983a):

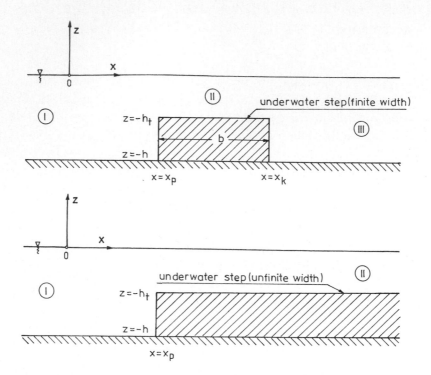

Figure 3.9: Underwater steps considered; definition sketch.

$$
\left.
\begin{aligned}
\Phi &= \Phi^{(1)} + \Phi^{(2)} \\
\zeta &= \zeta^{(1)} + \zeta^{(2)}
\end{aligned}
\right\}.
\tag{3.105}
$$

The indices (1) and (2) denote the linear and second–order solutions, respectively. The first order solution for a submerged step yields the known Mei and Black (1969) solution. Thus, it will be only summarized here for convenience. Let us separate the solution for the first–order velocity potential into propagating and local terms, i.e.:

Region I. $(-\infty < x \leq x_p,\ -h \leq z \leq 0)$.

$$
\Phi_I^{(1)} = Im\left\{ \frac{gH_k^{(1)}}{2\omega} \exp(-i\omega t) \left[\exp(ikx)\frac{\cosh k(z+h)}{\cosh kh} + \right.\right.
$$

$$
\left.\left. + \sum_\alpha R_\alpha \exp[\alpha(x-x_p)]\frac{\cos\alpha(z+h)}{\cos\alpha h} \right] \right\}
\tag{3.106}
$$

The first and second terms on the right–hand side describe the incident and reflected waves, respectively. The infinite sum (over α) is taken for α satisfying the following dispersion relation:

$$\omega^2 = gh \tanh(kh) = -g\alpha \tan(\alpha h),$$
(3.107)

and represents step–induced disturbances in the wave motion.

Region II. $(x_p \leq x \leq x_k, \ -h_t \leq z \leq 0)$.

$$\Phi_{II}^{(1)} \quad = \quad Im \left\{ \frac{gH_k^{(1)}}{2\omega} \exp(-i\omega t) \sum_{\alpha_1} \left[P_{\alpha_1}^{(1)} \exp[-\alpha_1(x - x_p)] + \right. \right.$$

$$\left. \left. + \quad Q_{\alpha_1}^{(1)} \exp[\alpha_1(x - x_k)]] \frac{\cos \alpha_1 (z + h_t)}{\cos(\alpha_1 h_t)} \right\},$$
(3.108)

where:

$$\omega^2 = gk_1 \tanh(k_1 h_t) = -g\alpha_1 \tan(\alpha_1 h_t).$$
(3.109)

The coefficient $P_{k_1}^{(1)}$ is the transmission coefficient from region I to region II and $Q_{k_1}^{(1)}$ is the reflection coefficient from the step junction at $x = x_k$.

Region III. $(x \geq x_k, \ -h \leq z \leq 0)$.

$$\Phi_{III}^{(1)} = Im \left\{ \frac{gH_k^{(1)}}{2\omega} \exp(-i\omega t) \sum_{\alpha} T_\alpha^{(1)} \exp[-\alpha(x - x_k)] \frac{\cos \alpha (z + h)}{\cos(\alpha h)} \right\}$$
(3.110)

The $T_k^{(1)}$ is the total transmission coefficient from the step.
Expressions for the unknown coefficients $R_k^{(1)}, R_\alpha^{(1)}, P_{k_1}^{(1)}, P_{\alpha_1}^{(1)}, Q_{k_1}^{(1)}, Q_{\alpha_1}^{(1)}, T_k^{(1)}$ and $T_\alpha^{(1)}$ are found by involving the following boundary conditions:

$$\Phi_I^{(1)} = \Phi_{II}^{(1)}, \qquad -h_t \leq z \leq 0, \qquad x = x_p,$$
(3.111)

$$\frac{\partial \Phi_I^{(1)}}{\partial x} = \begin{cases} \dfrac{\partial \Phi_{II}^{(1)}}{\partial x}, & -h_t \leq z \leq 0, \qquad x = x_p \\[2mm] 0, & -h \leq z \leq -h_t, \qquad x = x_p \end{cases}$$
(3.112)

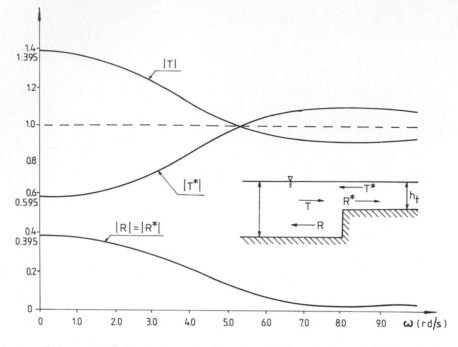

Figure 3.10: Transmission and reflection coefficients for the infinite step.

and

$$\Phi_{II}^{(1)} = \Phi_{III}^{(1)}, \qquad -h_t \leq z \leq 0, \qquad x = x_k \tag{3.113}$$

$$\frac{\partial \Phi_{II}^{(1)}}{\partial x} = \begin{cases} \dfrac{\partial \Phi_{III}^{(1)}}{\partial x}, & -h_t \leq z \leq 0, & x = x_k \\[2mm] 0, & -h \leq z \leq -h_t, & x = x_k \end{cases} \tag{3.114}$$

These equations are solved by expressing $\Phi^{(1)}$ in region I ($x \leq x_p$) and region II ($x_p \leq x \leq x_k$) in terms of $\Phi^{(1)} \big|_{x=x_p}$. The potential $\Phi^{(1)}$ is next determined by the conditions on $\partial \Phi^{(1)}/\partial x$ at $x = x_p$. The same procedure is repeated for the boundary conditions at $x = x_k$. The conditions on $\partial \Phi^{(1)}/\partial x$ at $x = x_p$ and $x = x_k$ yield the infinite set of linear equations for $R_\alpha^{(1)}$ and $T_\alpha^{(1)}$. Fig. 3.10 shows the transmission and reflection coefficients for the infinite step ($x_k \to \infty$; $h = 0.80\ m$, $h_t = 0.15\ m$) versus the frequency ω and for the two directions of incident wave propagation. It should be pointed out that the reflection and transmission coefficients corresponding to these directions of propagation obey the following relations (Massel, 1982):

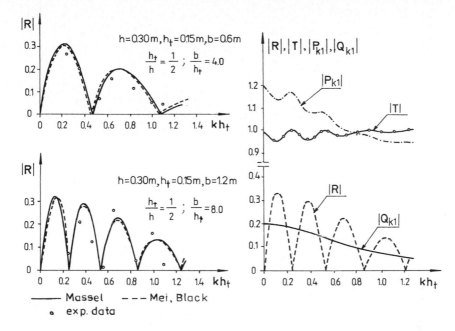

Figure 3.11: Reflection and transmission coefficients versus kh_t for the first harmonic waves. Finite step.

$$
\left.
\begin{array}{rl}
1 - |R_k|^2 &= 1 - |R_k^*|^2 = |\, T_k T_k^* \,| \\[2mm]
\arg T_k &= \arg T_k^*, \qquad \arg R_k + \arg R_k^* = \pi + 2\arg T_k
\end{array}
\right\}.
\tag{3.115}
$$

In Fig. 3.11 the theoretical value of the reflection coefficient $|\, R_k^{(1)} \,|$ for the finite step is compared with experimental data as well as with theory developed by Mei and Black (1969) in which a variational formulation is used. The experiment and both theories confirm the prominent feature of the reflection coefficients, i.e. the oscillating nature resulting from the interaction between two ends of the obstacle. However, the linear approximation of the theory of wave propagation over submerged step is insufficient to predict the higher harmonic generation. Therefore, we formulate the boundary value problem for the velocity potential $\Phi^{(2)}$ as (Massel, 1983a):

$$
\nabla^2 \Phi^{(2)} = 0 \qquad \text{in water layer,}
\tag{3.116}
$$

$$\frac{\partial^2 \Phi^{(2)}}{\partial t^2} + g \frac{\partial \Phi^{(2)}}{\partial z} = - \frac{\partial}{\partial t} \left[\left(\frac{\partial \Phi^{(1)}}{\partial x} \right)^2 + \left(\frac{\partial \Phi^{(1)}}{\partial z} \right)^2 \right] +$$

$$- \zeta^{(1)} \frac{\partial}{\partial z} \left[\frac{\partial^2 \Phi^{(1)}}{\partial t^2} + g \frac{\partial \Phi^{(1)}}{\partial z} \right] \qquad \text{at} \qquad z = 0, \qquad (3.117)$$

$$g\zeta^{(2)} + \frac{\partial \Phi^{(2)}}{\partial t} = - \frac{1}{2} \left[\left(\frac{\partial \Phi^{(1)}}{\partial x} \right)^2 + \left(\frac{\partial \Phi^{(1)}}{\partial z} \right)^2 \right] +$$

$$- \zeta^{(1)} \frac{\partial^2 \Phi^{(1)}}{\partial z \partial t} \qquad \text{at} \qquad z = 0, \qquad (3.118)$$

$$\frac{\partial \Phi^{(2)}}{\partial z} = 0 \qquad \text{at} \qquad \text{bottom} \qquad (3.119)$$

Moreover, the conditions expressing the continuity of the potentials $\Phi^{(2)}$ and horizontal velocities $\partial \Phi^{(2)} / \partial x$ at the transitions are to be satisfied. Implying the fact that the left-hand sides of eqs. (3.116) - (3.119) are linear in $\Phi^{(2)}$ and $\zeta^{(2)}$, we assume:

$$\left. \begin{array}{rcl} \Phi^{(2)} & = & \Phi^{(2)S} + \Phi^{(2)F} \\ \zeta^{(2)} & = & \zeta^{(2)S} + \zeta^{(2)F} \end{array} \right\}. \qquad (3.120)$$

The first terms on the right–hand side of eq. (3.120) satisfy eqs. (3.116) - (3.120) disregarding the junction conditions at $x = x_p$ and $x = x_k$. The remaining parts of the second–order solutions $\Phi^{(2)F}$ and $\zeta^{(2)F}$ must satisfy the homogoneous form of eqs. (3.116) - (3.119) and the boundary conditions at step junction. Therefore, $\Phi^{(2)F}$ represents the free waves.

In Region I $(-\infty < x \le x_p, \ -h \le z \le 0)$, the following forms for the potential $\Phi^{(2)S}$ and $\Phi^{(2)F}$ are assumed (Massel, 1982):

$$\Phi_I^{(2)S} = \frac{3}{32} \omega H_k^{(1)2} \frac{\cosh 2k(z+h)}{\sinh^4(kh)} \sin 2(kx - \omega t) +$$

$$- \frac{3}{32} \omega H_{kr}^{(1)2} \frac{\cosh 2k(z+h)}{\sinh^4(kh)} \sin 2(kx + \omega t - \psi_I^{(1)}), \qquad (3.121)$$

and

$$\Phi_I^{(2)F} = Im\left\{\sum_\beta R_\beta^{(2)} \frac{\cos\beta(z+h)}{N_\beta} \exp[\beta(x-x_p) - 2i\omega t]\right\}. \tag{3.122}$$

The first term in eq. (3.121) represents the second Stokes harmonic of the incident wave. The second term describes the second Stokes harmonic of the reflected waves. Eq. (3.122) represents the free waves with the frequency 2ω which propagate out of the step in the direction $x \to -\infty$. The wave number β satisfies the following dispersion relation:

$$4\omega^2 = gl\tanh(lh) = -g\beta\tan(\beta h), \tag{3.123}$$

while N_β is the factor of normalization:

$$N_\beta^2 = \frac{h}{2}\left(1 + \frac{\sin 2\beta h}{2\beta h}\right). \tag{3.124}$$

In Regions II and III, the potential functions can be formulated in a similar way (Massel, 1982, 1983a). Therefore, they will be omitted here. The local terms in (3.122) vanish within a short distance from the depth discontinuities ($x = x_p$ and $x = x_k$). Therefore, in Region I the following free waves are presented:

- basic first harmonic with ampl. $0.5H_k^{(1)}$,
- reflected first harmonic with ampl. $0.5H_k^{(1)}\mid R_k^{(1)}\mid$,
- free second harmonic with ampl. $0.5H_k^{(1)}\mid R_l^{(2)}\mid$,

and the following forced waves are propagating:
- Stokes harmonic related to the basic first harmonic,
- Stokes harmonic related to the reflected first harmonic,

The fact that free waves of frequency 2ω, have a wave number $l \geq 2k$ ($2k$ - Stokes harmonic wave number) accounts for the oscillation of the second harmonic amplitude. The free second harmonic and the Stokes harmonic associated with the first harmonic, reflected from the step, cause the amplitude to be spatially periodic in $L_t^{(-)}$, where:

$$L_t^{(-)} = \frac{2\pi}{l - 2k}. \tag{3.125}$$

However, the free second harmonic and the Stokes harmonic associated with the incident first harmonic form the spatial variation in the scale:

$$L_t^{(+)} = \frac{2\pi}{l + 2k}.$$

(3.126)

In Region III, the two free harmonics are:

    - first harmonic transmitted from area I with ampl. $0.5H_k^{(1)} \mid T_k^{(1)} \mid$
    - free second harmonic with ampl. $0.5H_k^{(1)} \mid T_l^{(2)} \mid$
    and one forced wave:
    - Stokes harmonic related to the first harmonic.

These produce a spatial variation of the first and second harmonics in the same scale:

$$L_t^{(-)} = \frac{2\pi}{l - 2k}.$$

(3.127)

In the theory presented, the potentials $\Phi^{(2)F}$ are introduced in order to satisfy the boundary conditions at the step junctions. Physically they represent the free harmonics which are propagating up and down stream of the step. It appears, that the incident waves once transmitted past the upstream junction generate second and higher harmonics. These harmonics are then propagated into deeper water as free wave trains. Owing to spatial variation of harmonic amplitudes, the amount of higher harmonics in these regions depends on the step parameters.

It should be noted that theory given above is valid for Ursell number (Svendsen and Jonsson, 1982):

$$U = \frac{H}{h}\left(\frac{L}{h}\right)^2 \le \frac{16}{3}\pi^2 \qquad \text{or} \qquad \Pi \le \frac{2}{3\pi} = 0.212.$$

(3.128)

When comparing the theory with experiments performed in wave channel, the contamination of observed waves by the higher harmonics produced by wave–maker and the interaction of the basic harmonic with the second free harmonic should be taken into account (Massel, 1981, 1983b). Particularly, the variation of the 1st harmonic in Region III, observed in experiments, can be considered as a result of the nonresonant interaction between free harmonics, i.e.: 1st harmonic with frequency ω and wave number k and 2nd free harmonic with frequency 2ω and wave number l. Then, the nonlinear theory of the second order suggest that this interaction generates the forced waves with a constant amplitude (Massel, 1981). Combining the ordinates of 1st harmonic and that resulting from the nonresonant interaction gives:

$$\zeta = a_k^{(1)} \cos(kx - \omega t) + a_{kl}^{(1,2)} \cos[(l - k)x - \omega t],$$

(3.129)

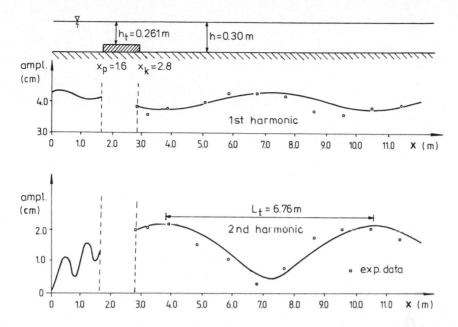

Figure 3.12: Second harmonic amplitude oscillation for paddle generated waves. Theory and experiment.

in which: $a_{kl}^{(1,2)}$ - amplitude of the interaction wave with frequency ω and wave number $(l - k)$.

If we keep the phase constant and equal to $\chi - kr - \omega t = 0$, from (3.129) we obtain:

$$\zeta = a_k^{(1)} + a_{kl}^{(1,2)} \cos[(l - 2k)x], \qquad (3.130)$$

Thus again:

$$L_t^{(-)} = \frac{2\pi}{l - 2k}. \qquad (3.131)$$

Fig. 3.12 presents a comparison of experimental data provided by Bendykowska (1980) with the theory in which the secondary wave generation by paddle motion and interaction between basic and secondary free harmonics are included. A complete, second–order theory, including these mechanisms as well as the higher harmonics generated by underwater steps has been developed elsewhere (Massel, 1981, 1983b).

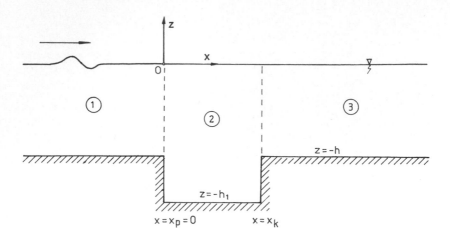

Figure 3.13: Rectangular trench. Definition sketch.

Let us now consider wave propagation over underwater trenches. Interest in this problem is largely due to phenomena associated with passage of waves over submarine trenches in the ocean and wave propagation across navigational channels, where changes in water depth are commonly the case. The underwater trench induces substantial changes in the waves propagating over it. The incident wave energy is partly reflected and partly transmitted. Moreover, close to the trench, local disturbances are observed. Several methods based on matched eigenfunction expansions are developed to study the diffraction of a linear, steady train of waves on a submerged channel. As in the case of the submerged step, discussed above, the solution for each constant–depth subdomain is presented in the terms of eigenfunction expansions of the velocity potential; the solutions are then matched at the vertical boundaries, resulting in sets of linear equations. Lee and Ayer (1981) applied this method for the transmission of waves over a channel of rectangular cross–section. Kirby and Dalrymple (1983a) have extended Lee and Ayer's work to the case of obliquely incident waves. For the normal approach to the rectangular trench (see Fig. 3.13) the solution can easily be obtained by the expansions method given above (Massel, 1985). In Fig. 3.14 the resulting reflection and transmission coefficients are presented versus (h/L). For comparison, Lee and Ayer's theoretical and experimental data are given. It should be noted that in their theoretical solution, the area of motion was divided, by drawing a horizontal line, into two subdomains, namely an infinite rectangular region of constant depth and a finite rectangular region representing the trench itself. In the more general case of obliquely incident surface waves (see Fig. 3.15), the velocity potentials $\Phi_i(x, y, z, t)$ in each region can be assumed

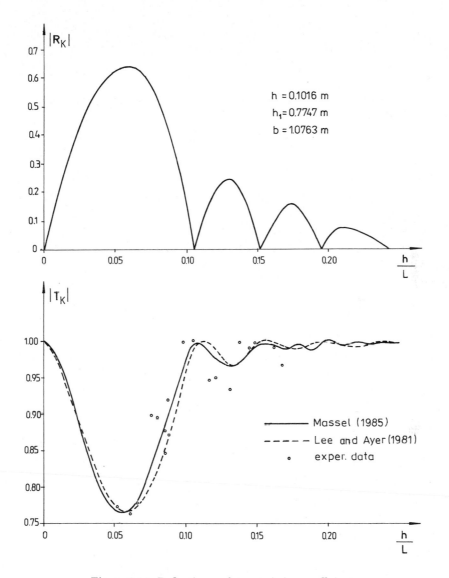

Figure 3.14: Reflection and transmission coefficients

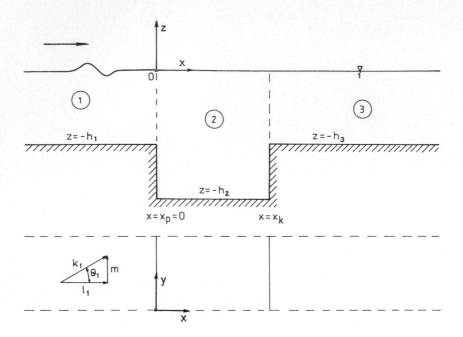

Figure 3.15: Rectangular trench. Oblique incidence

as:

$$\Phi_i(x, y, z, t) = \varphi_i(x, z) \exp[i(my - \omega t)], \qquad i = 1, 2, 3 \tag{3.132}$$

where:

$$m = k_1 \sin \theta_1 = const, \tag{3.133}$$

results from the Snell's law (3.7) and:

$$\varphi_i(x, z) = \frac{g A_i^{\pm}}{2\omega} \cosh k_i(z + h_i) \exp(\pm l_i x) + \text{non} - \text{propagating modes.} \tag{3.134}$$

Here:
A_1^+ - amplitude of the incident wave,
A_1^- - amplitude of the reflected wave in the region 1,
A_2^+ - amplitude of wave propagating in region 2 in the $+x$ direction,
A_2^- - amplitude of wave reflected from junction $x = x_k$ and propagating in the $-x$ direction,
A_3^+ - amplitude of wave transmitted in region 3, and

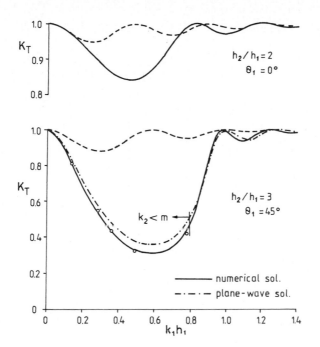

Figure 3.16: Transmission coefficient versus $k_1 h_1$. (From Kirby and Dalrymple, 1983a)

$$l_i = (k_i^2 - m^2)^{1/2}, \qquad \omega^2 = gk_i \tanh(k_i h_i). \tag{3.135}$$

The intensity of reflection or transmission depends on the values of θ_1, h_2/h_1 and h_2/h_3. If for a range of values of θ_1 and h_2/h_1, we get $m > k_2$ (l_2 is purely imaginary), the boundary between regions 1 and 2 becomes reflective to the incident plane wave, and the wave forms corresponding to $A_2^\pm$ change from sinusoidal to exponential dependency in the cross–trench direction. Wave energy can be transmitted from region 2 to 3 only if $m < k_3$. In the event that $m > k_3$, total reflection occurs.

Solutions to the full problem must satisfy matching conditions at the junctions $x = x_p$ and $x = x_k$; the resulting matrix equation is solved numerically. In Fig. 3.16 the results for symmetric trench ($h_1 = h_3, h_2/h_1 = 3, (x_k - x_p)/h_1 = 10$) and the incidence wave angle $\theta_1 = 45^o$ is shown. The wave number $l_2 = 0$ at a value of $k_1 h_1 = 0.792$ (Kirby, Dalrymple, 1983a). Hence, the reduction in transmission across the trench caused by the totally reflective local barrier at $x = x_p$ can be easily observed (min $K_T \approx 0.31$). The recovery of the transmission coefficient to 1.0 in the limit $k_1 h_1 \to 0$ is due to vanishing of the trench width, relative to the incident wave length. In the same Figure, the results based on the plane–wave approximation (the non–propagating modes are neglected) are given. The comparison indicates that this approximation yields reasonably accurate results for relatively small depth differences.

3.6   Scattering of surface waves by periodic sandbars

On natural sandy coasts, with mild slopes, nearly periodic longshore under-water bars are often observed. The number of bars varies widely from one to a dozen or so, depending on the hydro and litho–dynamical characteristics of the coastal zone. When surface waves are incident in a region of undulating seabed topography, it is well known that wave energy may be scattered by the bedforms. In particular, when the long–crested waves are incident upon purely transverse forms, there are only two types of interaction, namely back–scatter (wave reflection) and forward–scatter (wave transmission). Let $h'(\vec{x})$ denote the total still–water depth, and let (Fig. 3.17):

$$h'(\vec{x}) = h(\vec{x}) - \eta(\vec{x}),\tag{3.136}$$

in which $h(\vec{x})$ is a slowly varying depth, satisfying the mild–slope assumption:

$$\mu = \frac{\nabla_h h}{kh} \ll 1.0,\tag{3.137}$$

and $\eta(\vec{x})$ represents rapid undulations about the mean level with the amplitude scale, i.e.:

$$O(k\eta) \approx O(\mu).\tag{3.138}$$

Following mainly to Kirby (1986b) we consider the Lagrangian (1.32) in the form:

$$\mathcal{L} = \int_{-h'}^{\zeta} \rho \left\{ \frac{1}{2}(\nabla_h \Phi)^2 + \frac{1}{2}\left(\frac{\partial \Phi}{\partial z}\right)^2 + gz \right\} dz.\tag{3.139}$$

Expanding the Lagrangian about $z = -h$ and retaining the $O(\epsilon^2)$ contributions to the linear motion we get (ϵ - wave steepness scale):

$$\mathcal{L} = \frac{1}{2}g\zeta^2 + \zeta \frac{\partial \Phi}{\partial t}\big|_{z=0} + \int_{-h}^{0} \frac{1}{2}(\nabla_h \Phi)^2 dz +$$

$$- \frac{1}{2}(\nabla_h \Phi)^2 \eta\big|_{z=-h} + \int_{-h}^{0} \frac{1}{2}\left(\frac{\partial \Phi}{\partial z}\right)^2 dz - \frac{1}{2}\left(\frac{\partial \Phi}{\partial z}\right)^2 \eta\big|_{z=-h}\tag{3.140}$$

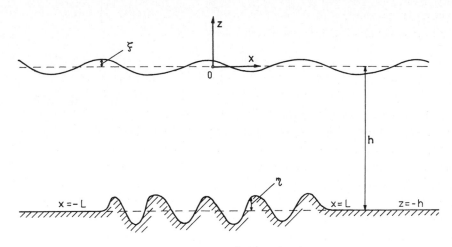

Figure 3.17: Definition sketch

Similarly to eq. (3.56) we represent the potential Φ in the form:

$$\Phi = \frac{\cosh k(z+h)}{\cosh kh}\, \tilde{\varphi}(\vec{x}, t). \tag{3.141}$$

Substituting (3.141) into (3.140) and integrating we obtain:

$$\mathcal{L} = \frac{1}{2}g\zeta^2 + \zeta\frac{\partial\tilde{\varphi}}{\partial t} + \frac{1}{2}\left(\frac{CC_g}{g}\right)(\nabla_h\tilde{\varphi})^2 +$$

$$+ \frac{1}{2}\left(\frac{\omega^2 - k^2 CC_g}{g}\right)\tilde{\varphi}^2 - \frac{\eta}{2\cosh^2(kh)}(\nabla_h\tilde{\varphi})^2 \tag{3.142}$$

The variation of $\mathcal{L}$ with respect to $\tilde{\varphi}$ and ζ and subsequent elimination of ζ yields the following modified time–dependent mild–slope equation:

$$\frac{\partial^2\tilde{\varphi}}{\partial t^2} - \nabla_h \cdot (CC_g\nabla_h\tilde{\varphi}) + (\omega^2 - k^2 CC_g)\tilde{\varphi} +$$

$$+ \frac{g}{\cosh^2(kh)}\nabla_h \cdot (\eta\nabla_h\tilde{\varphi}) = O(\epsilon, \mu^2). \tag{3.143}$$

Neglecting the term in μ yields a time–dependent form of the mild–slope equation (3.65) for the slowly varying bottom alone. We apply now eq. (3.143) to the case of waves normally incident on a finite ripple path. The bed topography is given by $h'(x) = h - \eta(x)$ where:

$$\eta(x) = \begin{cases} 0 & x < 0 \\ b\sin(\lambda x) & 0 \le x \le nl \\ 0 & x > nl \end{cases} \tag{3.144}$$

in which l is the ripple length, b is the amplitude of the bottom undulation and n is the number of ripples.

Taking this into account, eq. (3.143) can be simplified to:

$$\frac{d^2\varphi}{dx^2} + k^2\varphi - \frac{g}{CC_g\cosh^2 kh}\frac{d}{dx}\left(\eta\frac{d\varphi}{dx}\right) = 0, \tag{3.145}$$

where:

$$\tilde{\varphi}(x,t) = \varphi(x)\exp(-i\omega t). \tag{3.146}$$

Computed results for reflection coefficient $|K_R|$ are given by the solid curve in Fig. 3.18 in comparison with the laboratory data by Davis and Heathershaw (1984) for $n = 4$ and $b/h = 0.32$. They clearly demonstrate a strong resonance in the neighbourhood of:

$$\frac{2k}{\lambda} = 1, \tag{3.147}$$

i.e. bottom undulation has a wave length of one half the surface wave length leading to greatly enhanced reflection. This kind of resonant reflection is known as Bragg reflection in crystallography. With such a matching of phases, waves are reflected from successive bars in phase and therefore reinforce one another, resulting in strong total reflection (Mei, 1985). The existence of the Bragg scattering mechanism provides a possible means for constructing coastal protection measures which are relatively low in profile in comparison with local water depth. Mei (1985) and Hara and Mei (1987) have examined that the detuned interaction, where wave frequency ω is allowed to deviate slightly from resonant frequency ω_r by an amount $\epsilon\Omega$, where:

$$\Omega = C_g K. \tag{3.148}$$

The corresponding detuned wave number is then $k + \epsilon K$, where K is of order unity. The scattering process is found to depend critically on whether the modulational frequency lies above or below a cut off frequency Ω_0 which couples amplitudes of the incident and reflected waves:

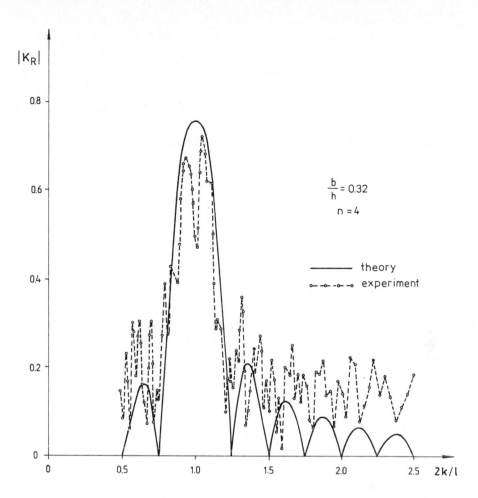

Figure 3.18: Results for the reflection coefficients of the ripple patch. (From Davis and Heathershaw, 1984 for $n = 4$, $b/h = 0.32$)

$$\Omega_0 = \frac{\omega k b}{2 \sinh 2kh} = \frac{g k^2 b}{4 \omega \cosh^2 kh}, \qquad (3.149)$$

The resonant-interaction theory extended by Hara and Mei (1987) to second order in η and their experiments illustrate the existence of a cutoff condition for frequency Ω. If the detuning frequency Ω is above cutoff frequency $\Omega_0, \Omega^2 > \Omega_0^2$, the wave envelope is oscillating in x, while for $\Omega^2 < \Omega_0^2$, the amplitude is decaying monotonically in x away from the incident edge of the ripple path.

The application of the Bragg scattering mechanism to the oblique incidence on a strip of infinitely long bars or on seabed with a mean slope was discussed recently by Mei et al. (1988) and the influence of an induced circulation on the scattering of surface waves by sandbars was examined by Kirby (1988). The presence of a current is found to shift resonant frequencies by possible significant amounts and is also found to enhance reflection of waves by bar fields.

The theories reported above indicate that the Bragg scattering mechanism provides a sufficient reflection which can initiate new offshore sandbars through mass transport, once there are already enough bars on the beach. Moreover, according to Mei (1985), a breakpoint bar generates a sufficient amount of reflection to start the first few bars, thereby setting the stage for resonant reflection for more bars. On the natural beach, the mechanism proposed here may be one of the most important in the underwater bars formation and sediment transport. However, these problems are out of the scope of this book; therefore they will not be discussed here.

It will be well to add that on natural beaches, the longshore sandbars can also be found on rather small water depth. In this depth range, water waves tend to become nonlinear and the nonlinear shallow–water equations are more suitable for describing the resonant interaction between waves and rippled beds. Such equations will be developed in Chapter 4.

3.7   References

Behrendt, L. and Jonsson, I.G., 1984. The physical basis of the mild–slope wave equation. *Proc. 19th Coastal Eng. Conf.,* 1: 941–954.

Bendykowska, G., 1980. Transformacja fali spowodowana przejsciem nad prostokatnym, zatopionym progiem. *Arch.Hydrot.,* XXVII: 171–187 (in Polish).

Berkhoff, J.C.W., 1972. Computation of combined refraction–diffraction. *Proc. 13th Coastal Eng. Conf.,* 1: 471–490.

Berkhoff, J.C.W., 1976. Mathematical models for simple harmonic linear water waves. Wave diffraction and refraction. Delft Hydr. Lab., Publ., 163, 103 pp.

Berkhoff, L.C.W., Booij, N. and Radder, A.C., 1982. Verification of numerical wave propagation models for simple harmonic linear water waves. *Coastal Eng.,* 6: 255–279.

Boczar–Karakiewicz, B., Paplinska, B. and Winiecki, J., 1981. Atlas form rewowych. Internal Rep., IBW–PAN, Gdansk, 52 pp (in Polish).

Booij , N., 1981. Gravity waves on water with non–uniform depth and current. Dept. Civil Eng., Delft Univ. of Tech., Rep. 81-1, 130 pp.

Booij, N., 1983. A note on the accuracy of the mild–slope equation. *Coastal Eng.,* 7: 191–203.

Chao, Y.Y., 1971. An asymptotic evaluation of the wave field near a smooth caustic. *Jour. Geoph. Res.,* 76: 7401–7408.

Cokelet, E.D., 1977. Steep gravity waves in water of arbitrary uniform depth. *Phil. Trans. Roy. Soc. London,* A286: 183–230.

Copeland, G.J.M., 1985. A practical alternative to the "mild-slope" wave equation. *Coastal Eng.,* 9: 125–149.

Dalrymple, R.A. and Kirby, J.T., 1988. Models for very wide–angle water waves and wave diffraction. *Jour. Fluid Mech.,* 192: 33–50.

Davis, A.G. and Heathershaw, A.D., 1984. Surface–wave propagation over sinusoidally varying topography. *Jour. Fluid Mech.,* 144: 419–443.

Dean, R.G. and Dalrymple, R.A., 1984. *Water wave mechanics for engineers and scientists.* Prentice–Hall Inc., Englewood Cliffs, 353 pp.

Dingemans, M.W., 1985. Evaluation of two–dimensional horizontal wave propagation models. Delft Hydr. Lab., Rep. W301, part 5, 117 pp.

Hara, T. and Mei, C.C., 1987. Bragg scattering of surface waves by periodic bars. *Jour. Fluid Mech.,* 178: 221–241.

Kamphius, J.W., 1975. Friction factor under oscillatory waves. *Proc. ASCE, Jour. Waterw. and Harb. Div.,* 101: 135–144.

Kirby, J.T., 1984. A note on linear surface wave current interaction over slowly varying topography. *Jour. Geoph. Res.,* 89: 745–747.

Kirby, J.T., 1986a. Higher–order approximations in the parabolic equation method for water waves. *Jour. Geoph. Res.,* 91: 933–952.

Kirby, J.T., 1986b. On the gradual reflection of weakly nonlinear Stokes waves in regions with varying topography. *Jour. Fluid Mech.,* 162: 187–209.

Kirby, J.T., 1986c. Open boundary condition in parabolic equation method. *Proc. ASCE, Jour. Waterway, Port, Coastal and Ocean Eng.,* 112: 460–465.

Kirby, J.T., 1986d. Rational approximations in the parabolic equation method for water waves. *Coastal Eng.,* 10: 355–378.

Kirby, J.T., 1988. Current effects on resonant reflection of surface water waves by sand bars. *Jour. Fluid Mech.,* 186: 501–520.

Kirby, J.T. and Dalrymple, R.A., 1983a. Propagation of obliquely incident water waves over a trench. *Jour. Fluid Mech.,* 133: 47–63.

Kirby, J.T. and Dalrymple, R.A., 1983b. A parabolic equation for the combined refraction–diffraction of Stokes waves by mildly varying topography. *Jour. Fluid Mech.,* 136: 453–466.

Lamb, H., 1932. *Hydrodynamics.* Dover Publ., New York, 738 pp.

Lee, J.J. and Ayer, R.M., 1981. Wave propagation over a rectangular trench. *Jour. Fluid Mech.,* 110: 335–347.

Le Mehaute, B. and Webb, L.M., 1964. Periodic gravity waves over a gentle slope at a third order of approximation. *Proc. 10th Coastal Eng. Conf.,* 1: 23–40.

Le Mehaute, B. and Wang, J.D., 1980. Transformation of monochromatic waves from deep to shallow water. Coastal Eng. Res. Center, Tech. Rep., 80-2, 43 pp.

Lighthill, J., 1978. *Waves in fluids.* Camb. Univ. Press, 598 pp.

Liu, P.L.F., 1986. Viscous effects on evolution of Stokes waves. *Proc. ASCE, Jour. Waterway, Port, Coastal and Ocean Eng.,* 112: 55–63.

Liu, P.L.F. and Tsay, T.K., 1983. On weak reflection of water waves. *Jour. Fluid Mech.,* 131: 59–71.

Liu, P.L.F. and Tsay, T.K., 1984. Refraction–diffraction model for weakly nonlinear water waves. *Jour. Fluid Mech.,* 141: 265–274.

Liu, P.L.F. and Tsay, T.K., 1985. Numerical prediction of wave transformation. *Proc. ASCE, Jour. Waterway, Port, Coastal and Ocean Eng.,* 111: 843–855.

Longuet–Higgins, M.S., 1983. Wave set–up, percolation and undertow in the surf zone. *Proc. Roy. Soc. London,* A390: 283–291.

Madsen, P.A. and Larsen, J., 1987. An efficient finite–difference approach to the mild–slope equation. *Coastal Eng.*, 11: 329–351.

Massel, S., 1981. On the nonlinear theory of mechanically generated waves in laboratory channels. *Mitteilungen des Leichtweiss- Inst. fur Wasserbau*, TU Braunschweig, Heft 70: 331–376.

Massel, S., 1982. Zmiany w strukturze czestotliwosciowej falowania rozprzestrzeniajacego sie nad przeszkodami podwodnymi. *Arch. Hydrot.*, XXIX: 357–388 (in Polish).

Massel, S., 1983a. Harmonic generation by waves propagating over a submerged step. *Coastal Eng.*, 7: 357–380.

Massel, S., 1983b. Transformacja fali wodnej (generowanej mechanicznie) nad progiem podwodnym, w ujeciu nieliniowym. *Arch. Hydrot.*, XXX: 3–20 (in Polish).

Massel, S., 1985. Propagacja fal powierzchniowych nad kanalem podwodnym o przekroju prostokatnym. *Arch. Hydrot.*, XXXI: 3–29 (in Polish).

Mei, C.C., 1983. *The applied dynamics of ocean surface waves*. A Wiley-Inter-Science Publication, New York, 734 pp.

Mei, C.C., 1985. Resonant reflection of surface water waves by periodic sandbars. *Jour. Fluid Mech.*, 152: 315–335.

Mei, C.C. and Black, J.L., 1969. Scattering of surface waves by rectangular obstacles in waters of finite depth. *Jour. Fluid Mech.*, 38: 499–511.

Mei, C.C., Hara, T. and Naciri, M., 1988. Note on Bragg scattering of water waves by parallel bars on the seabed. *Jour. Fluid Mech.*, 186: 147–162.

Meyer, R.E., 1979. Theory of water–wave refraction. *Adv. Appl. Mech.*, 19: 53–141.

Radder, A.C., 1979. On the parabolic equation method for water wave propagation. *Jour. Fluid Mech.*, 95: 159–176.

Sakai, T. and Battjes, J.A., 1980. Wave shoaling calculated from Cokelet' theory. *Coastal Eng.*, 4: 65–84.

Selezov, I.T., Sidorchuk, W.N. and Yakovlev, V.V., 1983. *Transformatsiya voln v pribrezhnoy zone shelfa*. Izd. Naukowa Dumka, Kiyev, 205 pp (in Russian).

Shore Protection Manual, 1977. U.S. Army, Coastal Eng. Res. Center, Washington, D.C., I–III.

Stoker, J.J., 1957. *Water waves.* Interscience Publ., New York, 567 pp.

Svendsen, I.A. and Jonsson, I.G., 1982. *Hydrodynamics of coastal waters.* Technical Univ. of Denmark, 285 pp.

Tsay, T.K. and Liu, P.L.F., 1982. Numerical solution of water–wave refraction and diffraction problems in the parabolic approximation. *Jour. Geoph. Res.,* 87: 7932–7940.

Yue, D.K.P. and Mei, C.C., 1980. Forward diffraction of Stokes waves by a thin wedge. *Jour. Fluid Mech.,* 99: 33–52.

Chapter 4

LONG SURFACE WAVES

4.1 General remarks

In many applications, in coastal regions, the wave length L is much larger than the water depth h. Moreover, the wave height H is an appreciable fraction of h, so that Ursell parameter U (or parameter Π) becomes too large to allow Stokes theory to be used. In Section 2.3.1 it was shown that for $\frac{h}{L} < 10$, the Stokes solution (even of the 5 order of approximation) yields incorrect results and it should be replaced by the theory of long waves. To formulate the problem for the long waves uniquely we define two small parameters δ and ν by:

- relative wave length: $\delta = kh$,
- relative wave amplitude: $\nu = \frac{A}{h}$.

In fact, the Ursell and Goda parameters are the combination of δ and ν, as:

$$U = \frac{H}{h}\left(\frac{L}{h}\right)^2 = 2(2\pi)^2 \frac{\frac{A}{h}}{(kh)^2} \approx 75 \frac{\nu}{\delta^2},\tag{4.1}$$

and

$$\Pi = \left(\frac{1}{2\pi}\right)^3 U = \frac{1}{\pi} \frac{\frac{A}{h}}{(kh)^2} \approx 0.30 \frac{\nu}{\delta^2}.\tag{4.2}$$

The ν is a typical scale of the motion nonlinearity, while the denominator of eq. (4.1) expresses the rate of the dispersion. Therefore, the value of ratio $A/[h(kh)^2] \sim O(1)$ or $U \sim O(75)$ represents a balance between nonlinearity and dispersion. In general, the parameters δ and ν allow to distinguish between three different important models of the long waves:

a) non–linear shallow water equation for which:
$\nu \gg \delta^2$, $\nu = O(1)$ or $U \gg 75$;

b) Boussinesq and Korteweg–de Vries equations for which:
$$O(\nu) \approx O(\delta^2) < 1 \quad \text{or} \quad U \sim O(75);$$
c) linearized long wave equation when:
$$\nu \ll \delta^2 \quad \text{or} \quad U \ll 75$$

For the later convenience we discuss first the case b).

4.2 Boussinesq solution $(O(\nu) = O(\delta^2))$. Solitary waves

Consider the functional (1.28) for two–dimensional motion in the form:

$$\delta J = \delta \int_t \int_x \int_{-h}^{\zeta(x,t)} \left\{ \frac{\partial \Phi}{\partial t} + \frac{1}{2} \left[\left(\frac{\partial \Phi}{\partial x} \right)^2 + \left(\frac{\partial \Phi}{\partial z} \right)^2 \right] + gz \right\} dxdzdt = 0. \qquad (4.3)$$

The shallow water theory, with the velocity $\left(\frac{\partial \Phi}{\partial x} \right)$, which is approximately independent of z, suggests the following expansion:

$$\Phi = \sum_0^\infty (-1)^m \frac{(z+h)^{2m}}{(2m)!} \frac{\partial^{2m} \Phi_b}{\partial x^{2m}}, \qquad (4.4)$$

where Φ_b is the velocity potential at $z = -h$.

Sometimes it is more convenient to consider the velocity $\bar{u} = \frac{\partial \Phi}{\partial x}$ which is average velocity over the water depth. Therefore eq. (4.4) takes the form:

$$\bar{\Phi} = \Phi_0 - \frac{1}{6}(h + \zeta)^2 \nabla^2 \Phi_0 + O(\delta^4). \qquad (4.5)$$

Substituting the representation (4.5) into functional (4.3) and taking the variations in $\delta \bar{\Phi}$ and $\delta \zeta$ yields the system of the two equations (Whitham, 1967):

$$\left. \begin{array}{l} \dfrac{\partial \zeta}{\partial t} + \dfrac{\partial}{\partial x} \left[(h + \zeta) \dfrac{\partial \bar{\Phi}}{\partial x} \right] = 0 \\[4mm] \dfrac{\partial \bar{\Phi}}{\partial t} + \dfrac{1}{2} \left(\dfrac{\partial \bar{\Phi}}{\partial x} \right)^2 + g(\zeta + h) - \dfrac{1}{3} h^2 \dfrac{\partial^3 \bar{\Phi}}{\partial x^2 \partial t} = 0 \end{array} \right\}, \qquad (4.6)$$

or:

$$\left. \begin{array}{l} \dfrac{\partial \zeta}{\partial t} + \dfrac{\partial}{\partial x} [(h + \zeta) \bar{u}] = 0 \\[4mm] \dfrac{\partial \bar{u}}{\partial t} + \bar{u} \dfrac{\partial \bar{u}}{\partial x} + g \dfrac{\partial \zeta}{\partial x} = \dfrac{1}{3} h^2 \dfrac{\partial^3 \bar{u}}{\partial x^2 \partial t} \end{array} \right\}. \qquad (4.7)$$

Equations (4.6) are called the Boussinesq equations (Peregrine, 1972; Mei, 1983). A number of other sets of Boussinesq–like equations (under the condition that $O(\nu) \approx O(\delta^2)$) can be derived. However, the essence of the method lies in eliminating the z - dependence and adopting the characteristic horizontal velocity either as the velocity at $z = 0$ or at the bottom $z = -h$ or the velocity averaged over the depth. In the original Boussinesq work, the horizontal velocity u_b at the bottom was used (Miles, 1981). Eliminating $\bar{u}$ in (4.7) yields (Vis and Dingemans, 1978):

$$\frac{\partial^2 \zeta}{\partial t^2} - gh \frac{\partial^2 \zeta}{\partial x^2} = gh \frac{\partial^2}{\partial x^2} \left\{ \frac{3}{2} \frac{\zeta^2}{h} + \frac{1}{3} h^2 \frac{\partial^2 \zeta}{\partial x^2} \right\}. \tag{4.8}$$

Consider the periodic solution of eq. (4.7):

$$\zeta = A \exp[i(kx - \omega t)], \qquad \bar{u} = U \exp[i(kx - \omega t)]. \tag{4.9}$$

After neglecting the nonlinear terms in eq. (4.7), the dispersion relation becomes:

$$C = \frac{\sqrt{gh}}{\left[1 + \frac{1}{3}(kh)^2\right]^{1/2}} \approx \sqrt{gh} \left[1 - \frac{1}{3}(kh)^2 + \ldots\right]^{1/2}. \tag{4.10}$$

The term $(1/3)(kh)^2$ represents the dispersion of wave motion.

The elementary solution of the Boussinesq equation is the solitary wave (Miles, 1980). For small value of ν, the profile becomes simply (Fig. 4.1):

$$\zeta(x, t) = H \cosh^{-2} \left\{ \sqrt{\frac{3H}{4h^3}} (x - Ct) \right\}, \tag{4.11}$$

in which:

$$C = \sqrt{gh} \left(1 + \frac{H}{h}\right)^{1/2} \approx \sqrt{gh} \left[1 + \frac{1}{2}\left(\frac{H}{h}\right)\right]. \tag{4.12}$$

Thus, the higher the peak, the narrower the profile (Fig. 4.1) and the wave speed increases with amplitude, The higher order solution for solitary waves was obtained by Fenton (1972). Following an expansion procedure due to Benjamin and Lighthill he has derived an exact operator equation for the free surface of the permanent long waves. In particular he worked out the case of solitary waves up to ninth order. Here, we limit our attention up to the third

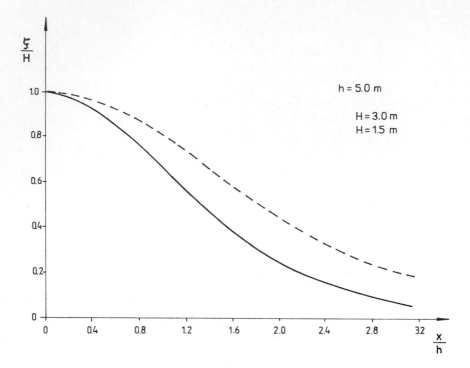

Figure 4.1: Solitary wave profiles with various amplitudes.

order in ν. After the introducing the notation:

$$S = \cosh^{-1}[\alpha(x - Ct)], \qquad T = \tanh[\alpha(x - Ct)], \tag{4.13}$$

the third order solitary wave solution takes the form (Fenton, 1972):

$$\frac{\zeta(x,t)}{h} = \nu S^2 - \frac{3}{4}\nu^2 S^2 T^2 + \nu^3 \left(\frac{5}{8} S^2 T^2 - \frac{101}{80} S^4 T^2\right) + O(\nu^4), \tag{4.14}$$

$$\frac{C^2}{gh} = 1 + \nu - \frac{1}{20}\nu^2 - \frac{3}{70}\nu^3 + O(\nu^4), \tag{4.15}$$

where:

$$\alpha = \left(\frac{3}{4}\frac{\nu}{h^2}\right)^{1/2} \left[1 - \frac{5}{8}\nu + \frac{71}{128}\nu^2\right]. \tag{4.16}$$

The velocity components and pressure at any point in the fluid are:

$$\frac{u}{\sqrt{gh}} = 1 + \frac{1}{2}\nu - \frac{3}{20}\nu^2 + \frac{3}{56}\nu^3 - \nu S^2 +$$

$$+ \nu^2 \left[-\frac{1}{4}S^2 + S^4 + (z+h)^2 \left(\frac{3}{2}S^2 - \frac{9}{4}S^4 \right) \right] +$$

$$+ \nu^3 \left[\frac{19}{40}S^2 + \frac{1}{5}S^4 - \frac{6}{5}S^6 + (z+h)^2 \left(-\frac{3}{2}S^2 - \frac{15}{4}S^4 + \frac{15}{2}S^6 \right) + \right.$$

$$+ \left. (z+h)^4 \left(-\frac{3}{8}S^2 + \frac{45}{16}S^4 - \frac{45}{16}S^6 \right) \right] + O(\nu^4), \tag{4.17}$$

$$\frac{w}{\sqrt{gh}} = (3\nu)^{1/2}(z+h)T \left\{ -\nu S^2 + \nu^2 \left[\frac{3}{8}S^2 + 2S^4 + \right. \right.$$

$$+ (z+h)^2 \left(\frac{1}{2}S^2 - \frac{3}{2}S^4 \right) \right] + \nu^3 \left[\frac{49}{640}S^2 - \frac{17}{20}S^4 + \right.$$

$$- \frac{18}{5}S^6 + (z+h)^2 \left(-\frac{13}{16}S^2 - \frac{25}{16}S^4 + \frac{15}{2}S^6 \right) +$$

$$+ \left. \left. 3(z+h)^4 \left(-\frac{3}{40}S^2 + \frac{9}{8}S^4 - \frac{9}{16}S^6 \right) \right] \right\} + O(\nu^{9/2}), \tag{4.18}$$

and

$$\frac{p}{\rho gh} = 1 - (z+h) + \nu S^2 + \nu^2 \left[\frac{3}{4}S^2 - \frac{3}{2}S^4 + \right.$$

$$+ (z+h)^2 \left(-\frac{3}{2}S^2 + \frac{9}{4}S^4 \right) \right] + \nu^3 \left[-\frac{1}{2}S^2 - \frac{19}{20}S^4 + \frac{11}{5}S^6 + \right.$$

$$+ (z+h)^2 \left(\frac{9}{4}S^2 + \frac{39}{8}S^4 - \frac{33}{4}S^6 \right) +$$

$$+ \left. (z+h)^4 \left(\frac{3}{8}S^2 - \frac{45}{16}S^4 + \frac{45}{16}S^6 \right) \right] + O(\nu^4). \tag{4.19}$$

The Fenton solution yields that the maximum of the solitary wave height $H_{max} = 0.85h$ and its propagation velocity $C^2_{max} = 1.7gh$, while the classical McCowan results yield $H_{max} = 0.78, and \quad C^2_{max} = 1.56gh$.

For many purposes, particularly for calculations of the dynamics of wave breaking in shallow water, certain integral properties of the motion such as the total energy, mass and momentum are particularly interesting. Let M, E_p and E_k denote the excess mass of the wave, its potential and kinetic energy, respectively. Between the above quantities there are the following known relations (Longuet–Higgins, 1974; Longuet–Higgins and Fenton, 1974):

$$
\left.
\begin{aligned}
3E_p &= (C^2 - gh)M \\[2mm]
dE &= C\,d\left(\frac{2E_k}{C}\right) \\[2mm]
E &= E_p + E_k
\end{aligned}
\right\}.
\tag{4.20}
$$

For waves of small amplitude ($\nu \ll 1.0$) we obtain:

$$
\left.
\begin{aligned}
E_k &\approx E_p \approx \rho g\,\frac{8}{3\sqrt{3}}(H^3 h)^{1/2} \\[2mm]
M &\quad \left(\frac{16}{3}h^3 H\right)^{1/2}
\end{aligned}
\right\}
\tag{4.21}
$$

For the solitary waves of maximum amplitude, the expansion in power series in ν is not the suitable technique for obtaining the accurate results. At this value of ν, the surface develops a sharp crest with an angle of $120°$ which corresponds to a singularity in the vicinity of the radius of convergence. Longuet–Higgins and Fenton (1974) have discovered that a change of independent variable from ϵ to $\omega = 1 - \left(u_c^2/gh\right)$ (u_c - particle speed at the wave crest), and the summation by Pade approximations leads to rapidly converging results. It should be noted that the range of variation of ω is precisely known, i.e. $0 \le \omega \le 1.0$.

The most striking feature of the behaviour of M, E_k, E_p and C is that while ν is increasing throughout the entire range, all above parameters have maxima between $\omega = 0.8$ and $\omega = 1.0$. This has been shown very clearly in Fig. 4.2 for mass M and total energy E as a function of the wave height ν. Numerical calculations due to Longuet–Higgins and Fenton (1974) indicate that the maximum values are:

$M = 2.036h^2$ at $\nu = 0.734,$

$E = E_p + E_k = 1.0040\rho gh^3$ at $\nu = 0.778,$

$C = 1.294\sqrt{gh}$ at $\nu = 0.790,$

while at the limiting value of $\omega_{max} = 1.0$ or $\nu_{max} = 0.827$ (the stagnation point) we have:

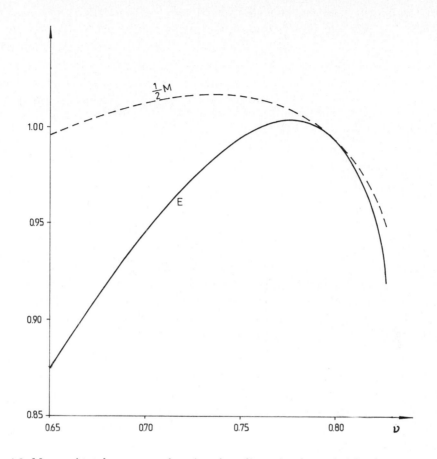

Figure 4.2: Mass and total energy as a function of nondimensional wave height. (From Longuet–Higgins and Fenton, 1974)

$$M = 1.897h^2, \qquad E = 0.918\rho g h^3 \qquad \text{and} \qquad C = 1.285\sqrt{gh}.$$

The fact that the highest solitary wave is not the most energetic helps to explain the qualitative difference between plunging and spilling breakers. This problem is discussed in detail in Section 5.2. Solitary waves represent the limit form of the solution of eqs. (4.7). Usually the general solution of the Bouissinesq equations is obtained by numerical methods (Abbott et al., 1978; Schaper and Zielke, 1984).

Omitting the dispersion term in (4.7) we obtain the nonlinear shallow water equations of finite amplitude as:

$$
\left.\begin{aligned}
\frac{\partial \zeta}{\partial t} + \frac{\partial}{\partial x}[(h + \zeta)\bar{u}] = 0 \\
\frac{\partial \bar{u}}{\partial t} + \bar{u}\frac{\partial \bar{u}}{\partial x} + g\frac{\partial \zeta}{\partial x} = 0
\end{aligned}\right\}.
\tag{4.22}
$$

We now write the equations in terms of the wave speed $C = \sqrt{g(\zeta + h)}$. After some manipulations we obtain:

$$
\left.\begin{aligned}
\left[\frac{\partial}{\partial t} + (\bar{u} + C)\frac{\partial}{\partial x}\right](\bar{u} + 2C) &= 0 \\
\left[\frac{\partial}{\partial t} + (\bar{u} - C)\frac{\partial}{\partial x}\right](\bar{u} - 2C) &= 0
\end{aligned}\right\}.
\tag{4.23}
$$

Therefore, $\bar{u} + 2C = const$ on the curve, on which $\frac{dx}{dt} = \bar{u} + C$ and $\bar{u} - 2C = const$ on the curve, on which $\frac{dx}{dt} = \bar{u} - C$. Both curves are the characteristics of the differential equations system (4.22). Considering the propagation of disturbances along the characteristics we can find in some circumstances that the characteristics cross, giving two depths and speeds at the same point. The equations (4.23) are too simple to describe the complicated flow in such region. However, the mathematical approximation is possible by introducing into the flow a sudden increase in depth as in Fig. 4.3. This we call a *bore* or, if stationary, a *hydraulic jump* (Wehausen and Laitone, 1960). It is equivalent to a shock wave in a supersonic gas flow. The conservation conditions for mass and momentum across the bore provide the rate of energy loss in the form:

$$
\Delta E = \frac{1}{4} g \frac{(h_1 - h_0)^3}{h_1 h_0}.
\tag{4.24}
$$

Since energy can not be added, the flow through the bore must always be from the low side to the high side (Fig. 4.3).

4.3   Korteweg–de Vries equation. Cnoidal waves

The Boussinesq equations (4.7) describe both right and left–running waves. If we specify the direction of propagation, eq. (4.7) may be reduced to the equation which is commonly called the *Korteweg–de Vries equation* (or KdV). Let $\zeta = \zeta(x - Ct)$ and $\bar{u} = \bar{u}(x - Ct)$. Additionally we introduce the following nondimensional parameters:

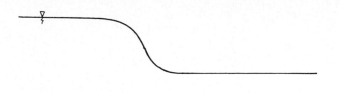

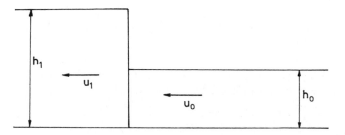

Figure 4.3: The bore. Surface shapes and mathematical approximation.

$$\tilde{x} = kx, \qquad \tilde{z} = h^{-1}z, \qquad \tilde{t} = kC_0 t, \qquad \tilde{\zeta} = A^{-1}\zeta, \qquad \bar{u} = \frac{AC_0}{h}\tilde{\bar{u}}, \qquad (4.25)$$

in which: $C_0 = \sqrt{gh}$.

Substituting (4.25) into (4.7) we obtain (the " $\sim$ " subscript were dropped):

$$\left.\begin{array}{l} \dfrac{\partial \bar{u}}{\partial t} + \nu\bar{u}\dfrac{\partial \bar{u}}{\partial x} + \dfrac{\partial \zeta}{\partial x} = \dfrac{1}{3}\delta^2 \dfrac{\partial^3 \bar{u}}{\partial x^2 \partial t} \\[3mm] \dfrac{\partial \zeta}{\partial t} + \dfrac{\partial}{\partial x}[(1+\nu\zeta)\bar{u}] = 0 \end{array}\right\}. \qquad (4.26)$$

Following the Crapper (1984) approach, we look for a solution of the form:

$$\bar{u} = \zeta + \nu B_1 + \delta^2 B_2, \qquad (4.27)$$

with unknown functions B_1 and B_2.

Equations (4.26) become:

$$\left.\begin{array}{l} \dfrac{\partial \zeta}{\partial t} \, + \, \nu \dfrac{\partial B_1}{\partial t} + \delta^2 \dfrac{\partial B_2}{\partial t} + \dfrac{1}{2}\,\nu\,\dfrac{\partial \zeta^2}{\partial x} + \dfrac{\partial \zeta}{\partial x} - \dfrac{\delta^2}{3}\dfrac{\partial^3 \zeta}{\partial x^2 \partial t} = 0 \\[4mm] \dfrac{\partial \zeta}{\partial t} \, + \, \dfrac{\partial \zeta}{\partial x} + \nu \dfrac{\partial B_1}{\partial x} + \delta^2 \dfrac{\partial B_2}{\partial x} + \nu \dfrac{\partial \zeta^2}{\partial x} = 0 \end{array}\right\}. \tag{4.28}$$

For these equations to be consistent we require to zero order:

$$\frac{\partial \zeta}{\partial t} = -\frac{\partial \zeta}{\partial x}, \qquad B_1 = -\frac{1}{4}\zeta^2, \qquad B_2 = \frac{1}{6}\frac{\partial^2 \zeta}{\partial x^2}. \tag{4.29}$$

With (4.29), eqs. (4.28) are:

$$\frac{\partial \zeta}{\partial t} + \frac{\partial \zeta}{\partial x} + \frac{3}{2}\nu\zeta\frac{\partial \zeta}{\partial x} + \frac{\delta^2}{6}\frac{\partial^3 \zeta}{\partial x^3} = 0. \tag{4.30}$$

In physical variables, eq. (4.30) takes the form:

$$\frac{\partial \zeta}{\partial t} + \sqrt{gh}\left(1 + \frac{3}{2}\frac{\zeta}{h}\right)\frac{\partial \zeta}{\partial x} + \frac{h^2}{6}\sqrt{gh}\,\frac{\partial^3 \zeta}{\partial x^3} = 0, \tag{4.31}$$

which is the Korteweg–de Vries equation. It should be noted that the products and squares of ν and δ^2 are assumed to be small enough to be neglected. Moreover, the ratio ν/δ^2 has to be of order one. It can be shown that the Korteweg–de Vries equation in the terms of the velocity $\bar{u}$ has the form (Selezov et al., 1983):

$$\frac{\partial \bar{u}}{\partial t} + \left(\sqrt{gh} + \frac{3}{2}\bar{u}\right)\frac{\partial \bar{u}}{\partial x} + \frac{h^2}{6}\sqrt{gh}\,\frac{\partial^3 \bar{u}}{\partial x^3} = 0. \tag{4.32}$$

The Korteweg–de Vries (KdV) equation serves as a model equation for any physical system for which the dispersion relation is approximated by (Miles, 1981):

$$\omega = k\sqrt{gh}(1 - \beta k^2), \qquad \beta \ll L^2, \tag{4.33}$$

($\beta = (1/6)h^2$ for gravity waves).
Prior to compare the other equations, related to equations (4.31) and (4.32), we introduce the normal form of KdV equation:

$$\frac{\partial \eta}{\partial \tau} + \eta\frac{\partial \eta}{\partial \xi} + \frac{\partial^3 \eta}{\partial \xi^3} = 0 \qquad \eta = \eta(\xi, \tau), \tag{4.34}$$

through the transformation:

$$
\left.
\begin{array}{ll}
\xi = U_1^{1/2}\left\{\dfrac{x-(1+\alpha)\sqrt{gh}\,t}{l}\right\}, & \tau = \dfrac{1}{6}U_1^{3/2}\left(\dfrac{h}{l}\right)^2\dfrac{\sqrt{gh}\,t}{l} \\[4mm]
\eta = \dfrac{3\zeta-2\alpha h}{A}, & U_1 = \dfrac{3Al^2}{h^3}
\end{array}
\right\},
\tag{4.35}
$$

in which $A \ll h$ is an amplitude scale, $l \gg h$ is a horizontal scale, and $\alpha = O(A/h)$ is an arbitrary constant. It should be noted that U_1 expresses the ratio of nonlinearity and dispersion and it is proportional to the Ursell number (2.6). If we replace the term $\eta(\partial\eta/\partial\xi)$ in (4.34) by the term of higher nonlinearity, we get the modified KdV equation in the form:

$$
\frac{\partial\eta}{\partial\tau} + \eta^2\frac{\partial\eta}{\partial\xi} + \frac{\partial^3\eta}{\partial\xi^3} = 0.
\tag{4.36}
$$

Replacing the dispersion term $(h^2/6)\sqrt{gh}\,(\partial^3\zeta/\partial x^3)$ by $(-h^2/6)(\partial^3\zeta/\partial x^2\partial t)$ in eq. (4.31), yields so called BBM equation (or regularized long–wave equation):

$$
\frac{\partial\zeta}{\partial t} + \sqrt{gh}\left(1+\frac{3}{2}\frac{\zeta}{h}\right)\frac{\partial\zeta}{\partial x} - \frac{h^2}{6}\frac{\partial^3\zeta}{\partial x^2\partial t} = 0,
\tag{4.37}
$$

associated with the dispersion relation:

$$
\omega = k\sqrt{gh}\left[1+\frac{1}{6}(kh)^2\right]^{-1},
\tag{4.38}
$$

which is more advantageous for numerical work.

The Korteweg–de Vries equation is closely related to the evolution of a weakly dispersive and weakly nonlinear waves. The corresponding modifications of the KdV equation are given by Miles (1981); they are also discussed in Section 4.5. The Korteweg–de Vries equation has the periodic and stable solutions. Besides the solitary wave discussed in Section 4.2, another permanent form of the surface profile of progressive wave being the solution of KdV equation, is that described by the elliptic function $cn(\Theta; m)$; thus this wave is called the *cnoidal* wave. It is convenient to adopt a steady reference frame, located at the sea bed (Fig. 4.4), which moves at speed C with the wave crest. Then, the major parameters for the first order cnoidal waves can be summarized as follows (Sobey et al., 1987):

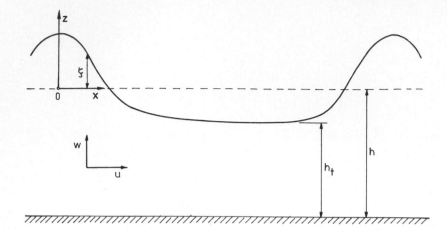

Figure 4.4: Cnoidal wave. Definition sketch.

$$\frac{\zeta(x,t)}{h} = \left(\frac{h_t}{h} - 1\right) + \frac{H}{h} cn^2(\theta; m),$$ (4.39)

$$\frac{u(x,z,t)}{\sqrt{gh_t}} = -1 - \left(\frac{H}{mh_t}\right)\left[-\frac{1}{2} + m - m\, cn^2(\theta; m)\right],$$ (4.40)

$$\frac{w(x,z,t)}{\sqrt{gh_t}} = 2\alpha\left(\frac{H}{h_t}\right) cn(\theta; m)\, sn(\theta; m)\, dn(\theta; m)\left(\frac{z+h}{h_t}\right),$$ (4.41)

$$p(x,z,t) = \rho g(\zeta - z),$$ (4.42)

in which:

$$\frac{h_t}{h} = 1 + \frac{H}{mh}\left(1 - m - \frac{\mathbf{E}}{\mathbf{K}}\right),$$ (4.43)

$$\alpha = \left(\frac{3}{4m}\frac{H}{h_t}\right)^{1/2}; \qquad \theta = \frac{\alpha L}{h_t}\left(\frac{x}{L} - \frac{t}{T}\right) = 2\mathbf{K}(m)\left(\frac{x}{L} - \frac{t}{T}\right),$$ (4.44)

where (Abramowitz and Stegun, 1975):
K - complete elliptic integral of the first kind,

$$\mathbf{K}(m) = \int_0^{\pi/2} \frac{du}{\sqrt{1 - m\sin^2 u}},$$ (4.45)

E - complete elliptic integral of the second kind,

$$\mathbf{E}(m) = \int_0^{\pi/2} \sqrt{1 - m \sin^2 u} \; du, \tag{4.46}$$

m - parameter of the elliptic functions and integrals, $0 \le m \le 1$.
In practice, the parameter m is not known "*a priori*" but must be obtained
as a function of wave height H, wave length L (or wave period T) and water
depth h. Therefore, two different cases can be discussed:

<u>H, h, L are known,</u>　　　　　　　　<u>H, h, T are known</u>

$$U = \left(\frac{H}{h}\right)\left(\frac{L}{h}\right)^2 = \frac{16}{3} m \mathbf{K}^2; \qquad \tilde{U} = \left(\frac{H}{h}\right)\left(\frac{gT^2}{h}\right) = \frac{16}{3} m \mathbf{K}^2, \tag{4.47}$$

which are transcendental in m. If $m \to 1.0$, $U(\tilde{U}) \to \infty$, and we get the
solitary wave described in Section 4.2.

The cnoidal wave theory has never become popular among engineers, and
oceanographers who seem to be repelled by the unfamilar elliptic functions
involved which are quite tedious to use for practical purposes. In order to simplify the calculations, the following approximations are introduced for small
$m(m \le 0.5)$ and for m close to unity $(m > 0.5)$:

<u>$m \le 0.5$ (according to Abramowitz and Stegun, 1975)</u>

$$q = \exp\left(-\frac{\pi \mathbf{K}'}{\mathbf{K}}\right) = \frac{m}{16} + 8\left(\frac{m}{16}\right)^2 + 84\left(\frac{m}{16}\right)^3 + 992\left(\frac{m}{16}\right)^4 + \dots, \tag{4.48}$$

$$\mathbf{K} = \frac{\pi}{2} + 2\pi \sum_{s=1}^{\infty} \frac{q^s}{1 + q^{2s}}, \tag{4.49}$$

$$\frac{\mathbf{E}}{\mathbf{K}} = \frac{1}{3}(2 - m) + \left(\frac{\pi}{\mathbf{K}}\right)^2 \left[\frac{1}{12} - 2\sum_{s=1}^{\infty} q^{2s}\left(1 - q^{2s}\right)^{-2}\right], \tag{4.50}$$

$$sn(\theta; m) = \frac{2\pi}{\sqrt{m} \; \mathbf{K}} \sum_{n=0}^{\infty} \frac{q^{n+1/2}}{1 - q^{2n+1}} \sin(2n + 1)\eta, \tag{4.51}$$

$$cn(\theta; m) = \frac{2\pi}{\sqrt{m} \; \mathbf{K}} \sum_{n=0}^{\infty} \frac{q^{n+1/2}}{1 + q^{2n+1}} \cos(2n + 1)\eta, \tag{4.52}$$

$$dn(\theta; m) = \frac{\pi}{2\mathbf{K}} + \frac{2\pi}{\mathbf{K}} \sum_{n=1}^{\infty} \frac{q^n}{1 + q^{2n}} \cos 2n\eta, \tag{4.53}$$

$$m_1 = 1 - m; \qquad \eta = \frac{\pi\theta}{2\mathbf{K}}. \tag{4.54}$$

$$m > 0.5 \text{ (according to Fenton, Gardiner–Garden, 1982)}$$

$$w = \frac{\pi\theta}{2\mathbf{K}'}, \tag{4.55}$$

$$\epsilon_1 = \frac{1 - m^{1/4}}{2(1 + m^{1/4})}, \tag{4.56}$$

$$q_1 = \epsilon_1 + 2\epsilon_1^5 + 15\epsilon_1^9 + \dots, \tag{4.57}$$

$$\mathbf{K}' = \frac{\pi}{2}\left(1 + 2q_1 + 2q_1^4 + \dots\right)^2, \tag{4.58}$$

$$\mathbf{K} = \frac{\mathbf{K}'}{\pi}\ln\frac{1}{q_1}, \tag{4.59}$$

$$\frac{\mathbf{E}'}{\mathbf{K}'} = \frac{1}{3}(1 + m) + \left(\frac{\pi}{\mathbf{K}'}\right)^2\left[\frac{1}{12} - 2\sum_{s=1}^{\infty}q_1^{2s}(1 - q_1^{2s})^{-2}\right], \tag{4.60}$$

$$\frac{\mathbf{E}}{\mathbf{K}} = 1 + \frac{\pi}{2\mathbf{K}\mathbf{K}'} - \frac{\mathbf{E}'}{\mathbf{K}'}, \tag{4.61}$$

$$cn(\theta; m) = \frac{1}{2}\left(\frac{m_1}{mq_1}\right)^{1/4}\frac{1 - 2q_1\cosh(2w) + 2q_1^4\cosh(4w) + \dots}{\cosh(w) + q_1^2\cosh(3w) + q_1^6\cosh(5w) + \dots}, \tag{4.62}$$

$$sn^2(\theta; m) = 1 - cn^2(\theta; m), \tag{4.63}$$

$$dn^2(\theta; m) = 1 - m\,sn^2(\theta; m). \tag{4.64}$$

It should be pointed out that for the long waves ($U > 75$), the parameter m is usually greater than 0.5; the value $m = 0.5$ is associated with $U \approx 10.0$. Osborne et al. (1985), using the theory of the infinite–interval spectral transform, obtained a simple approximation to the small–amplitude periodic solution of the KdV equation in the form:

$$\zeta(x, t) = \zeta_M\left[\frac{(1 + a_0^2)\cos\theta - 2a_0}{(1 + a_0^2 - 2a_0\cos\theta)^2}\right], \tag{4.65}$$

in which $a_0 = 3\zeta_M/8(kh)^2$ must be small and $0 \le \theta = kx - \omega t \le 2\pi$; wave number k and frequency ω are given by dispersion relation (4.33).

Comparison of a cnoidal wave with eq. (4.65) indicates that this approximation is remarkably close to the cnoidal wave even for large amplitude ($U \approx 150$).

In Section 2.3.1 it was found that the fifth–order Stokes wave solution has a limited range of application and should be used for $L/h \le 1/10$. Therefore, an accurate theory for shallow water waves, which should be in preference to Stokes wave theory for greater wave length is of great interest. At present, the higher–order solutions for cnoidal waves are available due to Laitone (1962), Fenton (1979) and Isobe and Kraus (1983). In particular, Fenton (1979), using a Rayleigh–Boussinesq series, discovered a recursion relationship for a solution of any order, however, the effective formulas to fifth order only, are given when water depth, wave height and wave length (or wave period) are known. In the following we list them for some wave parameters in the form suitable for numerical manipulations:

wave profile

$$\frac{\zeta}{h_t} = \sum_{i=1}^{5} \sum_{j=0}^{i} \sum_{k=1}^{j} \eta_{ijk} \left(\frac{H}{mh_t}\right)^i m^j cn^{2k}(\theta; m),$$
(4.66)

phase velocity

$$\frac{C}{\sqrt{gh_t}} = 1 + \sum_{i=1}^{5} \sum_{j=0}^{i} \sum_{k=0}^{1} c_{ijk} \left(\frac{H}{mh_t}\right)^i m^j \left(\frac{\mathbf{E}}{\mathbf{K}}\right)^k,$$
(4.67)

water depth h_t

$$\frac{h_t}{h} = 1 + \sum_{i=1}^{5} \sum_{j=0}^{i} \sum_{k=0}^{i} h_{ijk} \left(\frac{H}{mh}\right)^i m^j \left(\frac{\mathbf{E}}{\mathbf{K}}\right)^k,$$
(4.68)

parameter α

$$\alpha = \left(\frac{3}{4}\frac{H}{mh_t}\right)^{1/2} \left\{1 + \sum_{i=1}^{4} \sum_{j=0}^{i} \alpha_{ij} \left(\frac{H}{mh_t}\right)^i m^j\right\}$$
(4.69)

The coefficients $\eta_{ijk}, c_{ijk}, h_{ijk}$ and α_{ij} are listed in Fenton (1979) paper. The higher–order solution of the cnoidal wave enables us to solve for m, provided the three quantities H, h and L are known. Thus, instead of eq. (4.47) we have (Fenton, 1979):

H, h and L are known

$$\left(\frac{H}{h}\right)\left(\frac{L}{h}\right)^2 = \frac{16}{3}m\mathbf{K}^2\left\{1 + \sum_{i=1}^{4}\sum_{j=0}^{i}\sum_{k=0}^{i}\lambda_{ijk}\left(\frac{H}{mh}\right)^i m^j\left(\frac{\mathbf{E}}{\mathbf{K}}\right)^k\right\}^2, \tag{4.70}$$

H, h and T are known

$$\left(\frac{H}{h}\right)\left(\frac{gT^2}{h}\right) = \frac{16}{3}m\mathbf{K}^2\left\{1 + \sum_{i=1}^{4}\sum_{j=0}^{i}\sum_{k=0}^{i}\tau_{ijk}\left(\frac{H}{mh}\right)^i m^j\left(\frac{\mathbf{E}}{\mathbf{K}}\right)^k\right\}^2. \tag{4.71}$$

The coefficients λ_{ijk} and τ_{ijk} and the calculation of other wave parameters, i.e.: velocities, accelerations, dynamical pressure, and integral properties of cnoidal waves (wave impulse, kinetic and potential energy, radiation stress and momentum flux) are available in the Fenton (1979) paper.

Comparison of higher–order solution of the cnoidal waves with other high-order theories and experimental data indicates that the fifth–order cnoidal wave solution is applicable for L/h or $(g/h)^{1/2}T$ greater than 8. For shorter waves, fifth–order Stokes wave theory is preferable. Whithin this range, the integral quantities are given with high accuracy up to H/h of about 0.65 ($U \geq 42$) while the fluid flow beneath waves is described accurately for $H/h \geq 0.5(U \geq 32)$. Thus, in the Ursell number range $32 \leq U \leq 75$, we have two theories at our disposal. In Fig. 4.5, the comparison of the wave profiles predicted by the linear wave theory and by two approximations (first and fifth order) of the cnoidal wave is presented.

It should be noted that the cnoidal theory in two limit cases $m = 0$ and $m = 1$ yields the known solutions. At $m \to 1.0$ is (Abramowitz and Stegun, 1975):

$\mathbf{K}(1.0) \to \infty, \quad \mathbf{E} \to 1.0 \quad$ and $\quad cn^2(\theta; 1.0) \to \cosh^{-2}(\theta).$

Thus, we have:

$$\zeta(x, t) = H\cosh^{-2}\left\{\left(\frac{3H}{4h^3}\right)^{1/2}(x - Ct)\right\}, \tag{4.72}$$

which is the solitary wave profile.
At $m \to 0$ we have:

$\mathbf{K}(0) \to \pi/2, \quad \mathbf{E}(0) \to \pi/2,$

$sn(\theta) \to \sin\theta, \quad cn(\theta) \to \cos\theta, \quad dn(\theta) \to 1.0.$

Therefore, the profile of the cnoidal wave is approaching to the sinusoidal

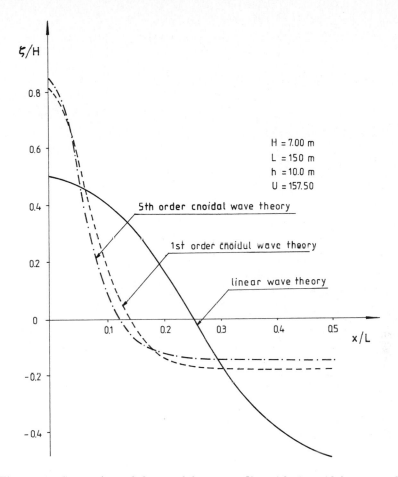

Figure 4.5: Comparison of the cnoidal wave profiles with sinusoidal wave profile.

profile.

It should be added that the cnoidal waves, propagating in a constant water depth without changing its shape, can be expressed in the Fourier series (Sarpkaya and Isaacson, 1981):

$$\zeta(x,t) = \frac{1}{2} \sum_n \zeta_n(x) \exp(-int), \qquad (4.73)$$

in which the amplitude of each harmonic is found to be:

$$A_n = \frac{8}{3} k_1^2 h^3 n r^n (1 - r^{2n})^{-1}, \qquad r = \exp\left[-\frac{\pi \mathbf{K}'(m)}{\mathbf{K}(m)}\right], \qquad (4.74)$$

where: k_1 - wave number of the first harmonic.

The expansion (4.73) is useful in some practical applications.

4.4   Linearized long wave equation

Under the assumption that $\nu \ll \delta^2$ and $\nu \ll 1$ (or $U \ll 75$), the non–linear terms in Boussinesq equation (4.7) are negligible, and we get:

$$\left.\begin{aligned}
\frac{\partial \zeta}{\partial t} + h\frac{\partial \bar{u}}{\partial x} &= 0 \\
\frac{\partial \bar{u}}{\partial t} + g\frac{\partial \zeta}{\partial x} &= 0
\end{aligned}\right\}. \tag{4.75}$$

The solution of eq. (4.75) can be taken in the form:

$$\left.\begin{aligned}
\zeta &= f_1(x - Ct) + f_2(x - Ct) \\
\bar{u} &= \sqrt{gh}\,[f_1(x - Ct) + f_2(x - Ct)]
\end{aligned}\right\}, \tag{4.76}$$

in which the functions f_1 and f_2 depend on the initial and boundary conditions.

Eliminating velocity $\bar{u}$ from eqs. (4.75), we obtain:

$$\frac{\partial^2 \zeta}{\partial t^2} = g\frac{\partial}{\partial x}\left(h\frac{\partial \zeta}{\partial x}\right). \tag{4.77}$$

When:

$$\zeta = \zeta_0(x)\exp(-i\omega t), \tag{4.78}$$

eq. (4.77) takes the form:

$$\frac{\partial}{\partial x}\left(gh\frac{\partial \zeta_0}{\partial x}\right) + \omega^2 \zeta_0 = 0. \tag{4.79}$$

4.5   Propagation of long waves over shoaling bottom

The discussion so far has been based on solutions for uniform depth and two dimensions. The problem of propagation of long waves on non–uniform depth can be solved either by means of a direct numerical solution of the Boussinesq equations (4.7) or by derivation of the evolution equations for the wave parameters so that eqs. (4.7) are satisfied in an approximate way.

Peregrine (1967) treated the problem using an approach, which essentially consists of solving the momentum and continuity equations by a perturbation scheme. He used the average velocity $\bar{u}$, as the velocity variable and obtained the following equations:

$$
\begin{aligned}
\frac{\partial \bar{u}}{\partial t} + (\bar{u} \cdot \nabla_h)\bar{u} + g\nabla_h \zeta &= \frac{1}{2}h\nabla_h\left[\nabla_h \cdot \left(h\frac{\partial u}{\partial t}\right)\right] - \frac{1}{6}h^2\nabla_h\left(\nabla_h \cdot \frac{\partial \bar{u}}{\partial t}\right) \\
\frac{\partial \zeta}{\partial t} + \nabla_h[\bar{u}(h + \zeta)] &= 0
\end{aligned}
\right\} \quad .(4.80)
$$

The approximations required to derive these equations are the same as for uniform depth, except that new length scales associated with variations in depth must be considered, eq. $\alpha = max(\nabla_h h)$. In derivation of (4.80), it is necessary to assume $\alpha \ll \delta$. If the slope is small enough for reflection to be negligible, the variable–coefficient Korteweg–de Vries equation may be used.

Ostrovskiy and Pelinovskiy (1970, 1975) (see also Pelinovskiy, 1982) considered quasi–plane waves propagating over on slowly changing water depth. Thus, the velocity $\bar{u}$ and surface ordinate ζ are supposed to be functions of:

$$
s = \tau(\vec{x}) - t \quad \text{and} \quad \vec{x}, \tag{4.81}
$$

where the variation with respect to s is faster than that with respect to $\vec{x}$. A multiple scale technique can now be applied and the equation which describes the evolution of wave takes the form (Pelinovskiy, 1982):

$$
\sqrt{gh}\,\frac{\partial \zeta}{\partial l} + \frac{3\zeta}{2h}\frac{\partial \zeta}{\partial s} + \frac{h}{6g}\frac{\partial^3 \zeta}{\partial s^3} + \frac{\sqrt{gh}}{4hb^2}\zeta\frac{d(hb^2)}{dl} = 0, \tag{4.82}
$$

in which: l - distance along the ray, b - distance between rays; or:

$$
\frac{\partial \zeta}{\partial t} + \sqrt{gh}\left(1 + \frac{3}{2}\frac{\zeta}{h}\right)\frac{\partial \zeta}{\partial l} + \frac{h^2\sqrt{gh}}{6}\frac{\partial^3 \zeta}{\partial l^3} + \frac{\sqrt{gh}}{4hb^2}\frac{d(hb^2)}{dl}\zeta = 0. \tag{4.83}
$$

When $h = $ const, eq. (4.83) is reducing to the KdV equation (4.31). Consider the case of straight isobaths, parallel to the coastline ($h = h(x)$) and denote the angle between wave ray and isobath at some depth h_0 by ($\frac{\pi}{2} - \alpha$). Then, eq. (4.82) yields:

$$
\frac{\partial \zeta}{\partial x} + \frac{1 - 2\left(\frac{h}{h_0}\right)\sin^2\alpha}{4hb^2}\zeta\frac{dh}{dx} + \frac{\zeta}{h\sqrt{ghb}}\frac{d\zeta}{dx} + \frac{h^2}{6b^2(gh)^{3/2}}\frac{\partial^3 \zeta}{\partial s^3} = 0, \tag{4.84}
$$

where:

$$\tau = \int_0^x \sqrt{\frac{1 - h\sin^2/h_0}{gh}}\, dx + \frac{\sin \alpha}{\sqrt{gh_0}} y + const, \tag{4.85}$$

and

$$b = \left(1 - \frac{h}{h_0}\sin^2\alpha_0\right)^{1/2}. \tag{4.86}$$

The solution for the non–horizontal bottom equation (4.83) is taken to be a cnoidal wave with slowly varying parameters. This is similar to the linear refraction theory where a sine wave is considered. Due to the slowness of the change of the wave parameters, the following conservation equation for the energy flux is obtained:

$$b(gh)^{1/2}\, \overline{\zeta^2} = const. \tag{4.87}$$

Therefore, the variation of the wave height satisfies the following equation (Ostrovskiĵ and Pelinovskiy, 1975):

$$\left.\begin{aligned}
h^{9/2}Y(m)b &= const \\[2mm]
Y(m) &= \mathbf{K}^4(m)\left[\frac{4 - 2m}{3}\frac{\mathbf{E}}{\mathbf{K}} - \frac{1 - m}{3} - \left(\frac{\mathbf{E}}{\mathbf{K}}\right)^2\right] \\[2mm]
H &= 4\omega^2 h^2 m\frac{\mathbf{K}^2}{3\pi^2 g}
\end{aligned}\right\}. \tag{4.88}$$

The limiting values of the cnoidal wave are:
- linear (sine) waves ($m \to 0$, $\mathbf{K} \sim \pi/2$, $Y \sim m^2$):

$$H\, h^{1/4}b^{1/2} = const, \tag{4.89}$$

which is the Green' law (see also (3.17);
- solitary wave ($m \to 1$, $\mathbf{K} \to \infty$, $Y \sim \mathbf{K}^3$):

$$Hhb^{2/3} = const. \tag{4.90}$$

In the vicinity of the breaking point, when the wave steepness is increasing considerably, eq. (4.84) is not applicable and higher–order theories should be used (Svendsen and Hansen, 1978; Miles, 1979; Stiassnie and Peregrine, 1980; Badawi, 1981).

Consider now the variations of water depth which are significant within the wave length (e.g. underwater step or underwater channel). If $\nu \ll \delta^2$ and $\nu \ll 1.0$, the long wave propagation can be described by eqs. (4.75). In the two–dimensional case, the reflection and transmission coefficients are given by (3.103), i.e.:

$$| K_R | = \frac{1 - \left(\dfrac{h_t}{h}\right)}{1 + \left(\dfrac{h_t}{h}\right)^2}, \qquad |K_T| = \frac{2}{1 + \left(\dfrac{h_t}{h}\right)^2}. \tag{4.91}$$

When $h_t/h \ll 1.0$, eq. (4.91) yields:

$$| K_R | \approx 1.0 - 2.0\left(\frac{h_t}{h}\right)^{1/2}, \qquad | K_T | \approx 2\left[1 - \left(\frac{h_t}{h}\right)^{1/2}\right]. \tag{4.92}$$

It is worth to note that eq. (4.91) stems from general method developed in Section 3.5 when $kh \to 0$ and $k_1 h_t \to 0$. The coefficient $| K_R |$ is very close to unity while the amplitude of the transmitted wave is increasing to double value of the incident wave amplitude. However, very small energy is transmitted over the step (energy flux $\bar{I} \sim h_t^{1/2}$). The coefficients of reflection and transmission over the step of the finite width can be found in a similar way (Miles, 1982; Massel, 1983, 1985).

Consider now a plane wave incident at an angle θ_1 with respect to the depth discontinuity (Fig. 4.6); the axis is normal to it. Therefore we have:

$$\left.\begin{aligned} \zeta_I &= A[\exp(i\alpha_1 x) + K_R \exp(-i\alpha_1 x)]\exp[i(\beta y - \omega t)] \\ k_1^2 &= \alpha_1^2 + \beta^2 \end{aligned}\right\}, \qquad x < 0 \tag{4.93}$$

and:

$$\left.\begin{aligned} \zeta_{II} &= A K_T \exp i(\alpha_2 x)\exp[i(\beta y - \omega t)] \\ k_2^2 &= \alpha_2^2 + \beta^2 \end{aligned}\right\}, \qquad x > 0, \tag{4.94}$$

in which: $\theta_1 = \arctan\left(\frac{\beta}{\alpha_1}\right)$ and α_1 is x component of the wave number of the incident wave while α_2 is x component of the wave number of the transmitted wave and β is y component of the wave number.

The reflection and transmission coefficients are determined by matching the

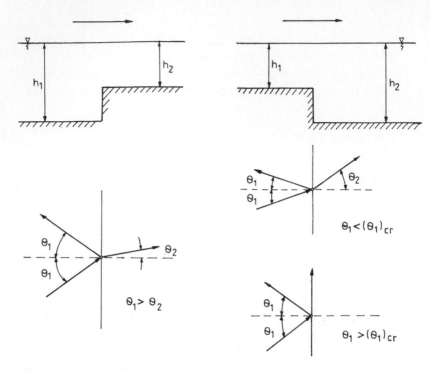

Figure 4.6: Directions of wave vectors at a step.

surface height and the volume flux at $x = 0$. Thus (Mei, 1983):

$$K_T = \frac{2\alpha_1 h_1}{\alpha_1 h_1 + \alpha_2 h_2}, \qquad K_R = \frac{\alpha_1 h_1 - \alpha_2 h_2}{\alpha_1 h_1 + \alpha_2 h_2}, \qquad (4.95)$$

and

$$\tan \theta_1 = \frac{\beta}{(k_1^2 - \beta^2)^{1/2}}, \qquad \tan \theta_2 = \frac{\beta}{(k_2^2 - \beta^2)^{1/2}}. \qquad (4.96)$$

For $h_1 > h_2$, is $k_1 < k_2$. Hence $\theta_1 > \theta_2$ and the wave number vector of the transmitted wave is directed more closely to the x axis than the incident wave vector. The quite opposite behaviour is observed when $h_1 < h_2$ (Fig. 4.6). Moreover, if we increase the angle of incidence θ_1, the β increases and if $\beta = k_2$, the angle $\theta_2 = \pi/2$. Thus, the transmitted wave propagates along the discontinuity ($\alpha_2 = 0$). Hence, eq. (4.96) gives:

$$(\theta_1)_{cr} = \arctan\left[\frac{k_2}{(k_1^2 - k_2^2)^{1/2}}\right], \tag{4.97}$$

and the value $\alpha_2 = 0$ corresponds to the total reflection.

When $\theta_1 > (\theta_1)_{cr}$, wave number α_2 becomes imaginary, and $\tan\theta_2$ loses its meaning, i.e.:

$$\zeta_{II} = \frac{1}{2}AK_T\exp(-\gamma_2 x)\exp[i(\beta y - \omega t)], \tag{4.98}$$

in which: $\gamma_2 = -i\alpha_2 = (\beta^2 - k_2^2)^{1/2}$.

The corresponding reflection and transmission coefficients take the form (Mei, 1983):

$$K_R = \frac{\alpha_1 h_1 - i\gamma_2 h_2}{\alpha_1 h_1 + i\gamma_2 h_2}, \qquad K_T = \frac{2\alpha_1 h_1}{\alpha_1 h_1 + i\gamma_2 h_2}, \tag{4.99}$$

and $\mid K_R \mid = 1.0$ because the reflection is perfect.

For slowly varying bottom, it is possible to derive the shallow–water wave equations describing wave propagation over a trench. For short waves this problem was discussed in Section 3.5. Hence, we consider a plane wave train arriving at an angle θ_1 with respect to the channel (Fig. 4.7). A coordinate system is established with z positive upwards and equal to zero at the free surface, and the y-axis extending along the trench axis. The depth outside the trench is constant and equal to h, while the depth in the trench is given by $h_2 = h_2(x)$. Following Selezov and Yakovlev (1978) we introduce the non–dimensional variables: $(x', y') = (x, y)/L$, $z' = z/h$, $t' = t/T$ and $\delta = kh \ll 1.0$ and $\nu \ll 1.0 (U \ll 75)$. The boundary value problem (in terms of the velocity potential ϕ) can be summarized as follows (for the convenience the primes were dropped):

$$\delta^2\left(\frac{\partial^2\phi}{\partial x^2} + \frac{\partial^2\phi}{\partial y^2}\right) + \frac{\partial^2\phi}{\partial z^2} = 0, \tag{4.100}$$

$$\delta^2\frac{\partial^2\phi}{\partial t^2} + \frac{\partial\phi}{\partial z} = 0 \qquad \text{at} \qquad z = 0, \tag{4.101}$$

$$\delta^2\frac{\partial\phi}{\partial x}\frac{\partial h}{\partial x} + \frac{\partial\phi}{\partial z} = 0 \qquad \text{at} \qquad z = -h(x). \tag{4.102}$$

For the velocity potential ϕ we introduce the series:

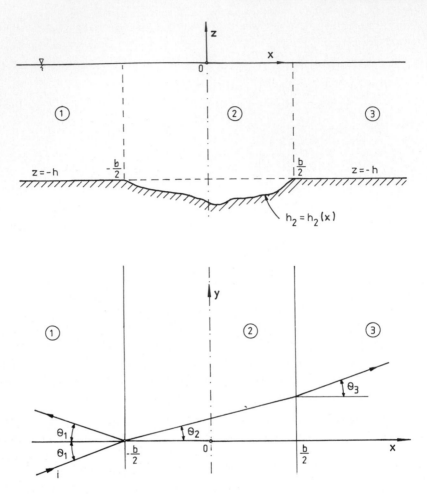

Figure 4.7: Underwater channel in the shallow water. Oblique incidence

$$\phi = \phi_0 + \delta^2 \phi_1 + \dots \tag{4.103}$$

After substituting (4.103) into (4.100) - (4.102) and involving the solution conditions we get (in dimensional variables):

$$\frac{\partial}{\partial x}\left(h\frac{\partial \phi}{\partial x}\right) - \frac{1}{g}\frac{\partial^2 \phi}{\partial t^2} = 0. \tag{4.104}$$

Eq. (4.104) can easily be extended to two–dimensional motion:

$$\nabla_h \cdot (h\nabla_h\phi) - \frac{1}{g}\frac{\partial^2 \phi}{\partial t^2} = 0. \tag{4.105}$$

If the linearized Bernoulli equation $g\zeta = \partial\phi/\partial t$ is used, eq. (4.104) leads to eq. (4.77).

In order to find the solution of eq. (4.105) in the particular regions of motion, let us to introduce the following scaling:

$$\left.\begin{array}{l} x' = x\left(\dfrac{b}{2}\right)^{-1}, \quad y' = y\left(\dfrac{b}{2}\right)^{-1}, \quad h' = \dfrac{h(x)}{h}, \quad \phi' = \phi\left(\dfrac{-igH_i}{2\omega}\right)^{-1} \\[3mm] s = l\sin\theta_1, \quad c = l\cos\theta_1, \quad l^2 = \dfrac{\omega^2}{gh}\left(\dfrac{b}{2}\right)^2, \quad \phi' = \varphi\exp(i\omega t) \end{array}\right\}. \qquad (4.106)$$

In region 1, the velocity potential should be of the following form:

$$\varphi_1 = \varphi_i + \varphi_r, \qquad \frac{\partial^2\varphi_1}{\partial x^2} + \frac{\partial^2\varphi_1}{\partial y^2} + l^2\varphi_1 = 0, \qquad (4.107)$$

and

$$\varphi_i = \exp[-i(cx + sy)], \qquad \varphi_r = K_R\exp[-i(-cx + sy)], \qquad (4.108)$$

On the transmitted side (region 3) we assume that:

$$\frac{\partial^2\varphi_3}{\partial x^2} + \frac{\partial^2\varphi_3}{\partial y^2} + l^2\varphi_3 = 0, \qquad \varphi_3 = K_T\exp[-i(cx + sy)]. \qquad (4.109)$$

In region 2, the velocity potential φ_2 should be solution of equation (4.105), i.e.:

$$h(x)\frac{\partial^2\varphi_2}{\partial x^2} + \frac{dh(x)}{dx}\frac{\partial\varphi_2}{\partial x} + h(x)\frac{\partial^2\varphi_2}{\partial y^2} + l^2\varphi_2 = 0. \qquad (4.110)$$

Let the function $\varphi_2(x, y)$ be denoted by:

$$\varphi_2(x, y) = \varphi_{2x}(x)\varphi_{2y}(y). \qquad (4.111)$$

Substituting (4.111) into (4.110) we obtain:

$$\left.\begin{array}{l} \dfrac{\partial^2\varphi_{2y}}{\partial y^2} + \chi^2\varphi_{2y} = 0 \\[4mm] h_2(x)\dfrac{\partial^2\varphi_{2x}}{\partial x^2} + \dfrac{dh_2(x)}{dx}\dfrac{\partial\varphi_{2x}}{\partial x} + [l^2 - \chi^2 h_2(x)]\varphi_{2x} = 0 \end{array}\right\}. \qquad (4.112)$$

The solution of the first equation of (4.112) simply is:

$$\varphi_{2y}(y) = \exp(-i\chi y).$$ (4.113)

The solutions to the full problem must satisfy certain matching conditions over the vertical planes separating the fluid regions, namely:

$$\left.\begin{array}{lll} \varphi_1 = \varphi_2, & \dfrac{\partial\varphi_1}{\partial x} = \dfrac{\partial\varphi_2}{\partial x} & \text{at} \quad x = -1 \\[3mm] \varphi_2 = \varphi_3, & \dfrac{\partial\varphi_2}{\partial x} = \dfrac{\partial\varphi_3}{\partial x} & \text{at} \quad x = +1 \end{array}\right\}.$$ (4.114)

Hence, $\chi = s = l\sin\theta_1$.
For the φ_{2x} we assume the following form:

$$\varphi_{2x}(x) = K_{2T}\,\psi_T(x) + K_{2R}\,\psi_R(x),$$ (4.115)

After substituting (4.115) into (4.112) we obtain:

$$h_2(x)\frac{d^2\psi}{dx^2} + \frac{dh_2(x)}{dx}\frac{d\psi}{dx} + (l^2 - h_2\chi^2)\psi = 0,$$ (4.116)

where ψ denotes ψ_R or ψ_T.
The general solution of eq. (4.116) does not exist. However, if the bottom profile can be approximated by the power series in x, i.e.:

$$h_2(x) = \sum_n a_n x^n,$$ (4.117)

the solutions ψ_T and ψ_R may be written as:

$$\psi_T(x)\,\{\psi_R(x)\} = \sum_m b_m x^m.$$ (4.118)

By combining eqs. (4.116) and (4.118) we get the recurence formulas for the coefficients b_m (Massel, 1986). When the coefficients b_m are known, the potentials, velocities and surface elevations in the arbitrary points can be found. In Fig. 4.8 the reflection coefficient $\mid K_R\mid$ is plotted against the nondimensional wave number l under the assumption of the trapezial cross–section of the trench and the incident angle $\theta_1 = 60^\circ$. The resonant behaviour of K_R is quite obvious. Fig. 4.8 is supplemented by Fig. 4.9 in which the normalized elevation $(\zeta^* = \zeta/h)$ and the horizontal velocity components $\left(u^*(v^*) = u(v)/\sqrt{gh}\right)$ are shown.

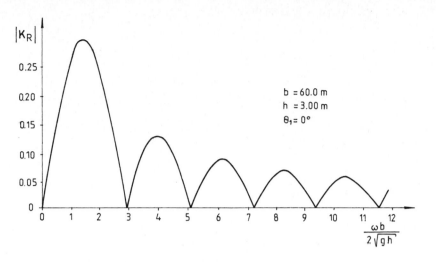

Figure 4.8: Reflection coefficient. Trapezoidal cross section

Let us now consider the interaction between sea bed topography and shallow–water waves. Using the linear wave theory, in Section 3.6 it was shown, that the reflected waves can be resonated by periodic sandbars if the wave length of seabed undulations is one half of the incident waves. Analysis of the typical scales of the wave motion on a gradually–sloping beach indicates the shallow–water approximation as a possible descriptive mode for the wave field. If the nonlinear and dispersive effects are of the same, small order of magnitude and bottom varies gradually, the resulting equations are the Boussinesq equations (4.7), which can be rewritten in the following non–dimensional form (Boczar–Karakiewicz et al., 1987):

$$
\left.
\begin{aligned}
\frac{\partial u_*}{\partial t_*} + \frac{\partial \zeta_*}{\partial x_*} + \epsilon u_* \frac{\partial u_*}{\partial x_*} &= \frac{1}{3} h_*^2 \delta^2 \frac{\partial^3 u_*}{\partial x_*^2 \partial t_*} \\
\frac{\partial \zeta_*}{\partial t_*} + \frac{\partial}{\partial x_*}[u_*(a_0 \zeta_* + h_*)] &= 0
\end{aligned}
\right\},
\qquad (4.119)
$$

where: $\epsilon = a_0/L_0$, $\delta = kh$ and:

$$
\left.
\begin{aligned}
x_* &= \frac{x}{L_0}, & t_* &= \frac{t}{L_0 \sqrt{gh_0}}, & u_* &= \frac{u}{\sqrt{gh_0}} \\
\zeta_* &= \frac{\zeta}{a_0}, & h_* &= \frac{h}{h_0}
\end{aligned}
\right\},
\qquad (4.120)
$$

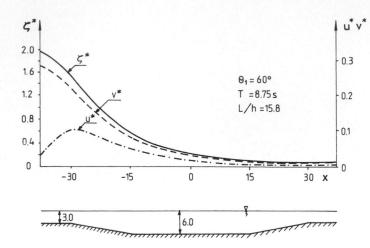

Figure 4.9: Normalized amplitude and horizontal velocity components for the channel as in Fig. 4.8

in which a_0, L_0, h_0 denote a typical amplitude, wave length and water depth, respectively.

The multiple scales analysis yields the following solution of eqs. (4.119) in which the stars " * " were omitted for convenience:

$$\left.\begin{array}{rcl} \zeta(x, X, t) & = & \sum_j \zeta_j(X)\exp[(i(k_j x - \omega_j t)] + c.c. \\[2mm] u(x, X, t) & = & \sum_j u_j(X)\exp[i(k_j x - \omega_j t)] + c.c. \end{array}\right\}, \qquad (4.121)$$

where: $X = \epsilon x$ and, the symbol $c.c.$ denotes the complex conjugate of the sum it follows.

On the basis of both field and laboratory experiments (Druet et al., 1972; Bendykowska, 1975; Boczar–Karakiewicz et al., 1981), attention was restricted to the first two harmonics, which obey the following system of two nonlinear equations (see also Lau and Barcilon, 1972):

$$\left.\begin{array}{l} \dfrac{d\zeta_1(X)}{dx} + w_1(X)\zeta_1(X) + w_2(X)\zeta_1^*(X)\zeta_2(X) = 0 \\[3mm] \dfrac{d\zeta_2(X)}{dx} + w_2(X)\zeta_2(X) + w_4(X)\zeta_1^2(X) = 0 \end{array}\right\}, \qquad (4.122)$$

where $w_n(n = 1, \ldots 4)$ are known functions of X.

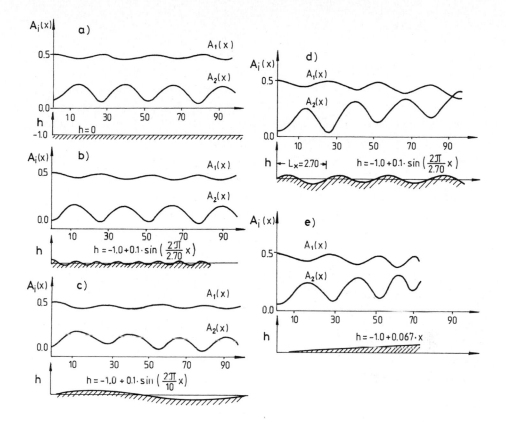

Figure 4.10: The waves response to the various bottom profiles

Figure 4.10 shows the waves response to the various bottom profiles ($A_j = | a_j |$ - harmonic amplitudes). In the first part of the Figure, the repetition of the amplitudes A_1 and A_2 for the horizontal bottom are displayed. The repetition length L_t is the mean distance between successive minima of A_2. Longuet–Higgins (1977) has demonstrated that for the shallow water, the distance L_t can be expressed as:

$$\frac{L_t}{L} = \frac{\left(\frac{L}{h}\right)^2}{4\pi^2(1 + rS_1)},$$

(4.123)

in which:

$$S_1 = \frac{H}{k^2 h^3} = \frac{U}{(2\pi)^2}, \qquad 3S_1 = \left[\frac{2}{\pi}m\mathbf{K}(m)\right]^{-2},$$

(4.124)

$$r = \frac{1}{m} \left\{ \left[\frac{2}{\pi} \mathbf{K}(m) \right]^{-2} + \left[(2 - m) - 3 \, \frac{\mathbf{E}(m)}{\mathbf{K}(m)} \right] \right\}. \tag{4.125}$$

In parts b) and c) of Fig. 4.10, the surface wave meets a wavy bed and the corrugation–length is quite different from the flat–bed repetition length L_t. If the wavy bed is "tuned" in L_t, a substantial increasing in the mean energy of the second harmonic is observed (part d). In part e) of the Figure, the development of the wave train over a slightly sloped bed is observed.

Recently, Yoon and Liu (1987) have derived the coupled nonlinear shallow–water wave equations describing wave propagation in a channel with corrugated boundaries. For the reflection of cnoidal waves a nonlinearity causes stronger reflection over the rippled patch than that predicted by linear theory. The reflection curve is a skewed one about the Bragg resonant condition. It should be noted that linear theory gives a symmetric reflection.

On the other hand, in the Keller (1988) paper it was shown that the amplitude equations for two resonant or nearly resonant gravity waves in water of nonuniform depth with variable frequency take the form:

$$\left. \begin{aligned}
\frac{d}{dt} \left(\frac{E_1}{\omega_1} \right) + \left(\frac{E_1}{\omega_1} \right) \nabla_h \cdot \left[(gh)^{1/2} \vec{k} \right] &= -\frac{3i}{h} E_1 a_2 \left(\frac{a_1^*}{a_1} \right) \exp \left(-\frac{i\Delta}{\epsilon} \right) \\
\frac{d}{dt} \left(\frac{E_2}{\omega_2} \right) + \left(\frac{E_2}{\omega_2} \right) \nabla_h \cdot \left[(gh)^{1/2} \vec{k} \right] &= -\frac{3i}{2h} E_1 a_2 \exp \left(\frac{i\Delta}{\epsilon} \right)
\end{aligned} \right\}, \tag{4.126}$$

in which: $E(a) = (1/2)\rho g a^2$, $\epsilon = O(a/L)$, $\Delta = 2\chi_1 - \chi_2$, χ_1 - phase function of one wave, χ_2 - phase function of the other wave. The eqs. (4.126) can be derived from the exact Euler equations or from the nonlinear shallow water theory or from the Boussinesq equations.

The interaction of shallow water waves and bottom sediment may led to the formation of equation of equilibrium bar and trough configurations. This problem, however, lies outside the scope of this book and the reader is referred to the papers by Boczar–Karakiewicz et al. (1983a 1983b), Benjamin et al. (1987).

4.6 Edge waves

The long–period (0.5 - 5 min) variations of the sea surface close to the shoreline were first observed by Munk (1949) and Tucker (1950). The amplitude of these "surf beats" was found to be approximately linearly related to the amplitude of incident wind–generated waves (Bychkov and Strekalov, 1971). Longuet–Higgins and Stewart (1964) and Gallagher (1971) attempted to explain surf beats in terms of nonlinear effects in the incident waves. Particularly, the wave groups of the incident wind waves would drive a long–period second–order wave travelling at the group velocity of the wind waves. This long wave could reflect from the beach as a free wave propagating into deeper water. In the simplest case, two incident waves of frequency ω_1 and ω_2 with longshore components of wave number k_1 and k_2 will generate a free wave if $k = k_1 - k_2$ and $\omega = \omega_1 - \omega_2$. If these relations are not obeyed, only a forced wave results.

Waves trapped to the coast with energy which decays asymptotically to zero at long distances from the shoreline are called *edge waves*. Ursell (1952) showed that at any given frequency such waves can only exist with longshore wave numbers greater than a critical "cut-off" wave number. Free waves with longshore wave numbers less than the cut–off are leaky, radiating energy seaward. Edge waves play an important role in generation of various longshore periodic coastal features (rip currents, crescentic bars, beach cusps etc.). In the following we consider a special case of the straight and long beach with a constant slope β, i.e.: $z = -h(x) = -\beta$; y axis coincides with the mean shoreline and water is in the region $x > 0$. For the wave propagating along the shoreline we try the solution:

$$\zeta(x, y, t) = \eta(x) \exp[i(\lambda y - \omega t)], \tag{4.127}$$

in which: λ - wave number in y direction.
Substituting (4.127) into two–dimensional version of the eq. (4.79) gives:

$$x\frac{\partial^2 \eta}{\partial x^2} + \frac{\partial \eta}{\partial x} + \left(\frac{\omega^2}{\beta g} - \lambda^2 x\right)\eta = 0. \tag{4.128}$$

Using the transformation (Mei, 1983):

$$\xi = 2\lambda x, \qquad \eta = \exp\left(-\frac{\xi}{2}\right) f(\xi), \tag{4.129}$$

we get:

$$\xi\frac{\partial^2 f}{\partial \xi^2} + (1 - \xi)\frac{\partial f}{\partial \xi} + \left[\frac{\omega^2}{2\lambda\beta g} - \frac{1}{2}\right] f = 0. \tag{4.130}$$

For the nontrivial solution we assume that the function f is finite at $\xi = 0$ and it is decreasing to zero at $\xi \to \infty$. Hence, the frequency ω should satisfy the following equation:

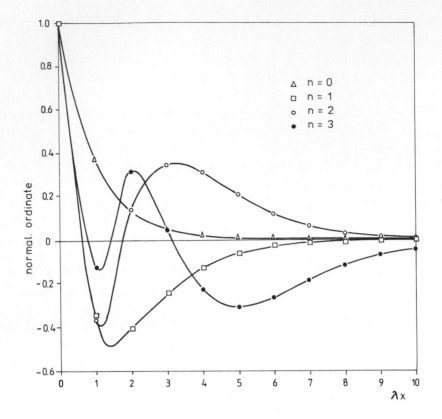

Figure 4.11: The first modes of the edge waves

$$\frac{\omega^2}{2\lambda\beta g} = n + \frac{1}{2}, \qquad n = 0, 1, 2, \ldots \tag{4.131}$$

The solution of eq. (4.130) is expressed now through the Laguerre polynomials (Gradshteyn and Ryzhik, 1980):

$$L_n(\xi) = \sum_{m=0}^{n} (-1)^m \begin{pmatrix} n \\ n-m \end{pmatrix} \frac{\xi^m}{m!}. \tag{4.132}$$

Therefore:

$$\zeta(x, y, t) = Re\{A_n L_n(2\lambda x) \exp(-\lambda x) \exp[i(\lambda y - \omega t)]\}. \tag{4.133}$$

The first few modes of edge waves are given in Fig. 4.11.

The frequency and longshore wave number are related by the dispersion relation:

$$\omega^2 = g\lambda(2n+1)\beta. \tag{4.134}$$

The eq. (4.128) can also be solved for standing waves on a plane beach. The simple analysis yields:

$$\zeta(x,y,t) = A_n L_n(2\lambda x)\exp(-\lambda x)\cos(\lambda y)\sin(\omega t). \tag{4.135}$$

The solutions (4.133) and (4.135) are valid under the shallow–water assumption. In the general case, the full linear Ursell (1952) solution should be used. Hence, for the plane beach we get:

$$\zeta(x,y,t) = A\{\exp(-\lambda x\cos\beta)+$$

$$+ \sum_{m=1}^{n} B_{mn}\left[\exp\left(-\lambda x\cos(2m-1)\beta\right)+\right.$$

$$+ \left.\exp\left(-\lambda x\cos(2m+1)\beta\right)\right]\}\cos(\lambda y)\cos(\omega t+\tau), \tag{4.136}$$

where:

$$B_{mn} = (-1)^m \prod_{r=1}^{n} \frac{\tan(n-r+1)\beta}{\tan(n+r)\beta}, \tag{4.137}$$

and

$$\omega^2 = g\lambda\sin\left[(2n+1)\beta\right], \qquad (2n+1)\beta < \frac{\pi}{2}. \tag{4.138}$$

When $(2n+1)\beta \ll 1$, the exact dispersion relation (4.138) and the shallow water equivalent (4.134) are in agreement, suggesting that the shallow water approximation is appropriate for gentle slopes and low mode numbers.

In cases of real beach topography the existence of underwater bars should be taken into account. The bars positions often remain reasonably fixed even under severe storm conditions. Such barred profile has the ability to modify the profiles of edge waves in a manner leading to self–maintenance. Kirby et al. (1981), using the wave equation of Berkhoff (1972), developed the numerical model to calculate edge wave profile and wave lengths for arbitrary bottom profile. The model was utilized to investigate the behavior of edge waves in the presence of longshore sandbars. The calculations indicate that sandbars can alter the resonant edge wave by trapping antinodes of the wave profiles at the bar locations. However, no effect of the bars on the dispersion relationship was found. Trapping of the edge wave antinodes at bars has the implication of longshore variability of wave height at the bars in the case of standing edge waves. This may control development and spacing of rip currents.

An interesting feature of the theory of edge waves is the prediction of cut–off frequency; waves with smaller frequency than cut–off, being non trapped, are radiating their energy away from the nearshore zone. The full linear theory yields a condition for trapped edge waves as (Ursell, 1952):

$$(2n+1)\beta \leq \frac{\pi}{2}. \tag{4.139}$$

If the beach profile has an exponential form:

$$h = h_\infty[1 - \exp(-\alpha x)], \tag{4.140}$$

in which h_∞ is the water depth at $x = \infty$, and α is a constant, the cut–off frequency is (Huntley, 1976):

$$\omega_{cr}^2 > n(n+1)\alpha^2 g h_\infty. \tag{4.141}$$

Thus, the edge waves are trapped only if $\omega > \omega_{cr}$.

On the other hand, for the beach of linear slope β out to x_0 and constant depth h_0 for $x > x_0$, where h_0 is still shallow enough for the shallow water equations to apply, the critical frequency ω_{cr} satisfies the condition:

$$\frac{\omega_{cr}^2 x_0}{g\beta} = 3.5n(n+1). \tag{4.142}$$

Physically, the cut–off frequency of an edge wave mode occurs when the long-shore wave number of the edge wave, k, becomes equal to the wave number of a free gravity wave, k_∞, at $x = \infty$. When $k > k_\infty$ only free waves which can exist are edge waves, which remain trapped against the shoreline. Spectral analysis of the experimental data taken in the shallow water zone of the South Baltic shows the presence of a set of discrete spectral peaks of low frequency (Massel and Musielak, 1980). Applying the Huntley suggestion that each of these peaks corresponds to a progressive edge wave mode at the cut–off frequency for the beach, they showed that the frequencies ω_{cr} for the consecutive modes obey approximately the law (Fig. 4.12):

$$\omega_{cr} \approx const\,[n(n+1)]^{1/2}, \tag{4.143}$$

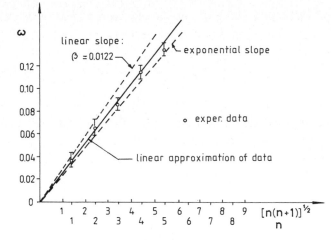

Figure 4.12: The cut–off frequency; measured versus computed

Vertical lines in Fig. 4.12 indicate the intervals of variability of ω_{cr} for consecutive peaks. However, the data support the conclusion that the lowest long–period peaks correspond to the cut–off frequencies of the lowest edge–wave modes.

It should be added that propagation of the edge waves in the coastal zone, in which water depth h is constant out to x_0 and then abruptly increases to $h = h_1(h_1 > h)$ for $x > x_0$, was discussed by Evans and Mc Iver (1984).

4.7 References

Abbott, M.B., Petersen, H.M., and Skovgaard, O., 1978. On the numerical modelling of short waves in shallow water. *Jour.Hydr. Res.*, 16: 173–204.

Abramowitz, M. and Stegun, I.A., 1975. *Handbook of mathematical functions.* Dover Publ., 1045 pp.

Badawi, E., 1981. Etude theorique de la reflexion des ondes de gravite de grande longeur d'onde relative sur les plages. L'Universite Scientifique et Medical de Grenoble, 129 pp (in French).

Bendykowska, G., 1975. Quelques aspects de la transformation du profil de la houle en eau peu profonde. *Arch.Hydrot.*, XXII: 97–103 (in French).

Benjamin, T.B., Boczar-Karakiewicz, B. and Pritchard, W.G., 1987. Reflection of water waves in a channel with corrugated beds. Univ. of Wisconsin, MRC Techn. Summary Rep., 2970, 56pp.

Berkhoff, J.C.W., 1972. Computation of combined refraction–diffraction. *Proc. 13th Coastal Eng. Conf.*, 1: 471–490.

Boczar-Karakiewicz, B., Paplinska, B. and Winiecki, J., 1981. Formation of sand bars by surface waves in shallow water. Laboratory experiments. *Rozpr.Hydrot.*, 43: 111–125.

Boczar–Karakiewicz, B., Long, .F. and Drapeau, G., 1983a. Formation and modification of a system of sand bars by progressive gravity waves in the presence of tides (on a example from the north coast of the St.Lawrence Gulf). Proc. Canadian Coastal Conf., Vancouver, pp. 37–51.

Boczar-Karakiewicz, B., Bona, J.L. and Chapalain, G., 1983b. On sediment transport rates and related time scales for sand bar formation in coastal zones. Proc. Canadian Coastal Conf., Vancouver, pp. 271–285.

Boczar-Karakiewicz, B., Bona, J.L. and Cohen, D.L., 1987. Interaction of shallow–water waves and bottom topography. In: J.L. Bona, C. Dafermos, J.L. Ericksen and D. Kinderlehrer (Editors), *Dynamical problems in continuum physics.* Springer–Verlag, pp. 131–176.

Bychkov, V.S., Strekalov, S.S., 1971. *Morskiye neregularnye volny.* Izd. Nauka, Moskva, pp. 132 (in Russian).

Crapper, G.D., 1978. *Introduction to water waves.* Ellis Horwood Ltd Publ., 224 pp.

Druet, Cz., Massel, S. and Zeidler, R., 1972. Statystyczne charakterystyki falowania wiatrowego w przybrzeznej strefie Zatoki Gdanskiej i otwartego Baltyku. *Rozpr. Hydrot.*, 30: 49–84(in Polish).

Evans, D.V. and McIver, P., 1984. Edge waves over a shelf: full linear theory. *Jour. Fluid Mech.*, 142: 79–95.

Fenton, J.D., 1972. A ninth–order solution for the solitary waves. *Jour. Fluid Mech.*, 53: 257–271.

Fenton, J.D., 1979. A higher–order cnoidal wave theory. *Jour. Fluid Mech.* 94: 129–161.

Fenton, J.D. and Gardiner–Garden, R.S., 1982. Rapidly–convergent methods for evaluating elliptic integrals and theta and elliptic functions. *Jour. Australian Math. Soc.*, B24: 47–58.

Gallagher, B., 1971. Generation of surf beat by non–linear wave interaction. *Jour. Fluid Mech.*, 49: 1–20.

Gradshteyn, I.S. and Ryzhik, I.M., 1980. *Tables of integrals, series and products.* Corrected and enlarged edition. Academic Press, 1160 pp.

Huntley, D.A., 1976. Long–period waves on a natural beach. *Jour. Geoph. Res.,* 81: 6441–6449.

Isobe, M. and Kraus, N.C., 1983. Derivation of a second-order cnoidal wave theory. Hydr. Lab., Dept. Civil Eng., Yokohama Nat. Univ., Tech. Rep. 83-2, 43 pp.

Keller, J.B., 1988. Resonantly interacting water waves. *Jour. Fluid Mech.,* 191: 529–534.

Kirby, J.T., Dalrymple, R.A. and Liu, P.L.F., 1981. Modification of edge waves by barred–beach topography. *Coastal Eng.,* 5: 35–49.

Laitone, E.V., 1962. Limiting conditions for cnoidal and Stokes waves. *Jour. Geoph. Res.,* 67: 1555–1564.

Lau, J. and Barcilon, A., 1972. Harmonic generation of shallow water waves over topography. *Jour. Phys. Oceanogr.,* 2: 405–410.

Longuet–Higgins, M.S., 1974. On the mass, momentum, energy and circulation of a solitary wave. *Proc. Roy. Soc. London,* A337: 1–13.

Longuet–Higgins, M.S., 1977. On the transformation of wave trains in shallow water. *Arch. Hydrot.,* XXIV: 445–457.

Longuet–Higgins, M.S. and Stewart, R.W., 1964. Radiation stresses in water waves: a physical discussion, with applications. *Deep Sea Res.,* 11: 529 562.

Longuet–Higgins, M.S. and Fenton, J.D., 1974. On mass, momentum, energy and calculation of a solitary wave. II. *Proc. Roy. Soc. London,* A340: 471–493.

Massel, S., 1983. Transformacja fali wodnej (generowanej mechanicznie) nad progiem podwodnym, w ujeciu nieliniowym. *Arch. Hydrot.,* XXX: 3–20 (in Polish).

Massel, S., 1985. Propagacja fal powierzchniowych nad kanalem podwodnym o przekroju prostokatnym. *Arch. Hydrot.,* XXXI: 3–29 (in Polish).

Massel, S., 1986. Dyfrakcja fal dlugich na kanale podwodnym o dowolnym przekroju poprzecznym. *Arch. Hydrot.,* XXXIII: 343-365 (in Polish).

Massel, S. and Musielak, S., 1980. Long–period oscillations in surf zone. *Rozpr. Hydrot.*, 41: 79–85.

Mei, C.C., 1983. *The applied dynamics of ocean surface waves.* A Wiley-Inter-Science Publication, New York, 734 pp.

Miles, J.W., 1979. On the Korteweg–de Vries equation for gradually varying channel. *Jour. Fluid Mech.*, 91: 181–190.

Miles, J.W., 1980. Solitary waves. *Ann. Rev. Fluid Mech.*, 12: 11–43.

Miles, J.W., 1981. The Korteweg–de Vries equation: a historical essay. *Jour. Fluid Mech.*, 106: 131–147.

Miles, J.W., 1982. On surface–wave diffraction by a trench. *Jour. Fluid Mech.*, 115: 315–325.

Munk, W.H., 1949. Surf beats. *Trans. Am. Geophys. Union*, 30: 849–854.

Osborne, A.R., Provenzale, A. and Bergamasco, L., 1985. A comparison between the cnoidal wave and an approximate periodic solution to the Korteweg–de Vries equation. *Il Nuovo Cimento*, 8C: 26–38.

Ostrovskiy, L.A. and Pelinovskiy, N.J., 1970. Transformatsiya voln na poverkhnosti peremennoy glubiny. *Fizika Atm. i Okeana*, 11: 935–939 (in Russian).

Ostrovskiy, L.A. and Pelinovskiy, J.N., 1975. Refraktsya nelineynykh morskikh voln v beregovoy zone. *Fizika Atm. i Okeana*, 11: 67–74 (in Russian).

Pelinovskiy, J.N., 1982. Nelineynaya dinamika voln tsunami. Inst. Prikl. Fiziki, Gorkiy, 222 pp (in Russian).

Peregrine, D.H., 1967. Long waves on a beach. *Jour. Fluid Mech.*, 27: 815–827.

Peregrine, D.H., 1972. Equations for water waves and the approximation behind them. In: R.E. Meyer (Editor), *Waves on beaches and resulting sediment transport.* Academic Press, pp. 95-121.

Sarpkaya, T. and Issacson, M.St.Q., 1981. *Mechanics of wave forces on offshore structures.* Van Nostrand Reinold, New York, 651 pp.

Schaper, L.W. and Zielke, W., 1984. A numerical solution of Boussinesq type wave equations. *Proc. 19th Coastal Eng. Conf.*, I: 1057–1072.

Selezov, I.T. and Yakovlev, V.V., 1978. *Difraktsiya voln na simmetrich-nykh neodnorodnostyakh*. Izd. Naukova Dumka, 144 pp (in Russian).

Selezow, I.T., Sidorchuk, W.N. and Yakovlev, V.V., 1983. *Transformatsiya voln v pribrezhnoy zone shelfa*. Izd. Naukowa Dumka, Kiyev, 205 pp (in Russian).

Sobey, R.J., Goodwin, P., Thieke, R.J. and Westberg, R.J., 1987. Application of Stokes' cnoidal and Fourier wave theories. *Proc. ASCE, Jour. Waterway, Port, Coastal and Ocean Eng.*, 113: 565–587.

Stiassnie, M. and Peregrine, D.H. , 1980. Shoaling of finite amplitude surface waves on water of slowly–varying depth. *Jour. Fluid Mech.*, 97: 783–805.

Svendsen, I.A. and Hansen, J.B., 1978. On the deformation of periodic long waves over a gently slopping bottom. *Jour. Fluid Mech.*, 87: 433–448.

Tucker, M.J., 1950. Surf beats: sea waves of 1 to 5 min. period. *Proc. Roy. Soc. London*, A202: 565–573.

Ursell, F., 1952. Edge waves on a slopping beach. *Proc. Roy. Soc. London*, A214: 79–97.

Vis, C. and Dingemans, M.W., 1978. An evolution of some wave theories. Delft Hydraulics Lab., 177 pp.

Wehausen, J.V. and Laitone, E.V., 1960. Surface waves. In: W.Flugge (Editor), *Handbuch der Physik*. Springer–Verlag, Berlin, pp. 446–778.

Whitham, G.B. , 1967. Variational methods and applications to water waves. *Proc. Roy. Soc. London*, A299: 6–25.

Yoon, S.B. and Liu, P.L.F., 1987. Resonant reflection of shallow–water waves due to corrugated boundaries. *Jour. Fluid Mech.*, 180: 451–469.

Chapter 5

WAVE MODULATION AND BREAKING

5.1 Nonlinear modulation of surface waves in space and time

Most practical predictions of wave propagation use approximations based on concepts of rays and group velocity. Although this is adequate in many instances, there are phenomena that can only be fully understood in terms of nonlinear effects. When waves are long compared with the depth of water, nonlinear effects are much more pronounced than in deep water. It is only for very small ratios of amplitude to depth, $\nu = A/h$, that nonlinear effects can be neglected. In Section 4.1 it was shown that the balance between some form of dispersion and weak nonlinearity, i.e.: $O(\nu) \approx 0(\delta^2) < 1$, can be described in the simpliest and most useful way by the KdV equation (4.31). Extensive numerical study of the KdV equation by Zabusky and Kruskal (1965) has shown that an initial hump disintegrates into a sequence of pulses, each of which has the properties of a solitary wave. Solitary waves also interact nonlinearly with the other waves and can be identified once more unchanged after the interaction, except for a spatial shift. Similar effects are observed for the solitary wave reflection (Renouard et al., 1985). These features are common in physics of electrons, protons, and so on. Zabusky and Kruskal adopted the word "soliton" for the solitary wave. The discovery of evolution of initial pulse into solitons led to development of so called *inverse–scattering procedure*. In this procedure, the initial nonlinear problem is replaced by a series of linear problems (Miles, 1981; Mei, 1983).

The asymptotic solution of the KdV equation for the arbitrary initial data $\eta(\xi, 0) \equiv \eta_0$ by inverse–scattering theory has been described in detail by Segur (1973). Here we briefly list, following to Hammack (1982), the most important features of the asymptotic solution to be illustrated in laboratory experiments:

a) initial disturbance evolves into a finite number of solitons ordered by their amplitude. When the solitons are well separated, the local shape of each is given by:

$$\eta = \alpha \cosh^{-2}\left[\left(\frac{\alpha}{2}\right)^{1/2}(\xi - \xi_0 - 2\alpha\tau)\right], \tag{5.1}$$

in which α and ξ_0 are constants;

b) the number N of solitons evolving from initial data of finite extent ($\eta_0 = 0$ for $\xi < \xi_1$ and $\xi > \xi_2$) is equivalent to the number of zeros of the function φ which satisfies the following equations:

$$\left.\begin{array}{ll} \dfrac{d^2\varphi}{d\xi^2} + \eta_0(\xi)\varphi & = 0 \\[2em] \varphi(\xi_1) = 1, \qquad \dfrac{d\varphi}{d\xi} & = 0 \end{array}\right\}, \tag{5.2}$$

c) when the net volume V in the initial wave is finite and positive, i.e.:

$$V = \int_{-\infty}^{\infty} \eta_0(\xi)d\xi > 0, \tag{5.3}$$

at least one soliton emerges,

d) when $\eta_0 \leq 0$ everywhere, no solitons emerge and the asymptotic solution consists of radiation components,

e) two other important classes of data are those for which $V < 0$ and those for which $V = 0$. No general statements regarding the asymptotic solution for these cases is provided theoretically.

In order to illustrate the application of the above features, the results of laboratory experiments are given in Fig. 5.1 where a leading negative wave is followed by a larger positive wave ($V = 30.5 cm^2 > 0$). After only twenty depths of propagation, the positive wave has separated into three separate crests. During subsequent propagation of three labelled crests of the positive wave appear to retain their integrity as they progress through the leading negative wave and emerge at the front of the wave train. At the last measurement station (after 400 depths of propagation), labelled waves 1 and 2 clearly resemble KdV solitons. The third wave is still interacting with the once–leading negative wave and can not be identified as a soliton. In Fig. 5.1, the normalized wave amplitudes are presented in the coordinate system which moves with the linear (nondispersive) speed $C = \sqrt{gh}$. This is illustrated by the motion of the point "*". Shifts of the waves to the left at succeeding stations indicate phase speed greater than C.

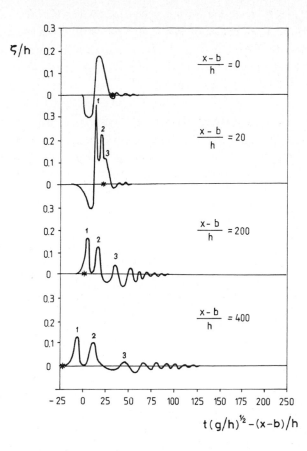

Figure 5.1: Surface wave evolution under the initial conditions where a leading negative wave is followed by a larger positive wave. (From Hammack, 1982)

The evolution of weakly nonlinear wave train on the deep, or moderate depth water is described by equations for the conservation of wave number, the conservation of wave action and the dispersion relation.

$$
\left.
\begin{aligned}
\frac{\partial k}{\partial t} &+ \frac{\partial \omega}{\partial x} = 0 \\[2mm]
\frac{\partial E}{\partial t} &+ \frac{\partial}{\partial x}(C_g E) = 0 \\[2mm]
\omega &= \sqrt{gk}\left[1 + \frac{1}{2}(ka)^2\right]
\end{aligned}
\right\}
\tag{5.4}
$$

These equations are derivable from the variational principles given in the Chapter 1. Using a perturbation approach, Benjamin and Feir (1967) demon-

strated that an initially uniform continuous wave train is unstable to modulational perturbations of its envelope if $kh > 1.363$. Further progress in the
subject was made by Chu and Mei (1970), who used multi–scale method to
derive the governing equations. Zakharov (1968) (and later Hasimoto and
Ono, 1972) showed that, if the wave number variation is small, all methods
lead to a single equation for the complex wave envelope $A(x,t)$, i.e. nonlinear
Schrodinger equation (Peregrine, 1983b, 1985):

$$\eta(x,t) = Re\left\{A\exp[i(k_0 x - \omega_0 t)]\right\}, \tag{5.5}$$

$$A(x,t) = a(x,t)\exp[i\theta(x,t)], \tag{5.6}$$

and

$$i\left(\frac{\partial A}{\partial t} + \frac{\omega_0}{2k_0}\frac{\partial A}{\partial x}\right) - \frac{\omega_0}{8k_0^2}\frac{\partial^2 A}{\partial x^2} - \frac{1}{2}\omega_0 k_0^2|\,A\,|^2 A = 0, \tag{5.7}$$

where: $\partial\theta/\partial t = \omega_0 - \omega, \quad \partial\theta/\partial x = k - k_0$.
Eq. (5.7) is a nonlinear parabolic equation and it arises in numerous circumstances where dispersive waves are weakly nonlinear. One of the most
important characteristics of the nonlinear Schrodinger equations is that it
can be solved exactly for initial conditions which decay sufficiently rapidly
as $|\,x\,|\rightarrow \infty$. Using the inversive–scattering technique Zakharov and Shabat
(1972) showed that any initial wave packet evolves into a number of "envelope
solitions" and a dispersive tail. This property of wave modulations has stimulated considerable interest in wave group occuring in the ocean (see Section
6.5). A wave train undergoing modulation instability has no steady end–state
but a series of modulation and demodulation cycles is observed. Such series is
known as the Fermi–Pasta–Ulam recurrence phenomenon (Miles, 1981; Yuen,
1982).

If the wave amplitude is sufficiently large, the uniform wave train is unique
but bifurcates into a train of waves with unequal crests or troughs. The critical
amplitudes at which bifurcation occurs lie in the range where the wave train is
stable to modulational perturbations and they coincide with points of neutral
stability. Longuet–Higgins (1985) has shown that the bifurcation points lie
around $k_0 a_0 \sim 0.4$. As was demonstrated by Zufiria (1987) recently, the
bifurcation from symmetric waves may generate also non–symmetric periodic
waves.

5.2   Wave breaking

Breaking of waves moving across a beach is the most significant phenomenon in the coastal dynamics. The physical significance of breaking arises from the fluid motion associated with breaking that absorbs most of the energy transmitted with the wave. The energy and momentum transmitted to the nearshore zone generate currents which may cause sediment transport both in the on–off shore and along shore directions (see Chapter 8). It should be noted, that breaking waves are not limited to the nearshore zone. Primarily during storms, waves may also break far at sea.

Considerable progress has been made in identification of different kinds of possible fluid motions in the surf zone. However, their interactions are only poorly understood at present and there is no unified model which is able to predict all spectrum of the flow. Partly it is due to difficulty of finding a precise mathematical description of a fluid that is in general nonlinear and time–dependent. The fluid accelerations can no longer be assumed to be small compared to gravity, nor is the particle velocity any longer small compared to the phase velocity. We begin the description of the breaking process with a brief discussion of the approach to wave–breaking.

5.2.1   Steep waves, approaching to breaking

As it was shown in Section 2.5, the increasing of the wave height (energy) is limited by some limiting wave steepness. The limiting steepness yields a distinction between nonbreaking and breaking waves (Fig. 2.2 and Table 2.3). One of the most widely quoted properties of the limiting wave is that the enclosed crest angle is 120 degree (Fig. 5.2). Stokes (1880) was the first to derive this value. Prior to give a proof of the limiting angle we define the

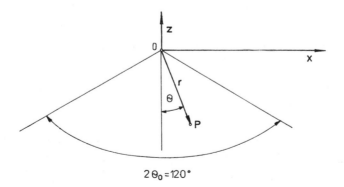

Figure 5.2: Enclosed crest angle for kinematically limited wave

kinematic breaking criterion. This criterion requires that the particle velocity at the wave crest exceeds the phase velocity. In the reference frame moving with the wave phase velocity, the wave motion is steady. Viewed from this reference frame, the kinematic criterion requires that the particle velocity at the crest be zero. Additionally we assume that the crest is formed by two intersecting straight lines which are the tangents to the real water surface curvature, and the velocity potential in the region of the crest in the polar coordinate system (r, θ) can be approximated by (Fig. 5.2):

$$\phi(r, \theta) = B r^n \sin(n\theta), \tag{5.8}$$

where B and n are coefficients to be evaluated, and r and θ are polar coordinates.

To evaluate n , the fact that the surface is a streamline was been considered. Thus:

$$u_\theta = -\frac{1}{r} \frac{\partial \phi}{\partial \theta} = 0, \tag{5.9}$$

or:

$$\frac{\partial \phi}{\partial \theta} = \cos(n\theta) = 0 \qquad \text{at} \qquad \theta = \theta_0, \tag{5.10}$$

and:

$$n\theta_0 = \frac{\pi}{2}. \tag{5.11}$$

Assuming that the pressure on the free surface is zero, the Bernoulli equation near the crest is:

$$\frac{1}{2} \left(u_r^2 + u_\theta^2 \right) + gz = 0. \tag{5.12}$$

Substituting velocities $u_r = -\partial\phi/\partial r$, $u_\theta = -\partial\phi/r\,\partial\theta$ and surface ordinate $z = -r \cos\theta$ into (5.12) yields:

$$\frac{1}{2} n^2 B^2 r^{2n-3} = g \cos\theta_0 = const. \tag{5.13}$$

Since the right–hand side of (5.13) is a constant, the power of r must be zero, i.e.:

$$2n - 3 = 0 \quad \text{or} \quad n = \frac{3}{2}. \tag{5.14}$$

Eq. (5.11) yields now:

$$\theta_0 = 60^o. \tag{5.15}$$

After calculating the B from (5.13), the velocity potential ϕ takes the form:

$$\phi(r, \theta) = \frac{2}{3} g^{1/2} r^{3/2} \sin\left(\frac{3}{2}\theta\right), \tag{5.16}$$

or in the complex form (Longuet–Higgins and Fox, 1977):

$$w = \frac{2}{3i} g^{1/2} z_1^{3/2}, \tag{5.17}$$

in which: $w = \phi + i\psi$ and $z_1 = r\exp(i\theta)$.
The particle velocity components are:

$$
\left.
\begin{aligned}
u_r &= -\frac{\partial \phi}{\partial r} = -g^{1/2} r^{1/2} \sin\left(\frac{3}{2}\theta\right) \\
u_\theta &= -\frac{1}{r}\frac{\partial \phi}{\partial \theta} = -g^{1/2} r^{1/2} \cos\left(\frac{3}{2}\theta\right)
\end{aligned}
\right\}. \tag{5.18}
$$

Both the velocity components tend to zero as r approaches zero, satisfying the kinematic breaking criterion. The second assumption, that the crest is formed by two intersecting straight lines, involves the identification of the free surface by the constant polar angle θ_0. The solution (5.15) suggests that the profile approaches the 120° corner–flow. This, however, refers only to the steepest possible waves. When $r > 0$ the crests are still rounded. To define the limiting flow we take radial coordinates (r, θ) with the origin 0 at the distance l (eq. (2.116) above the wave crest (Fig. 2.14). We require a solution which as $r/l \to \infty$, tends to the Stokes corner–flow (5.17). Longuet–Higgins and Fox (1977) have shown numerically that the free surface crosses its asymptote at about $r/l = 3.32$ and then approaches it in a very slowly damped oscillation. Between the two crossings of the asymptote at $r/l = 3.32$ and $r/l = 68.5$, the maximum angle of slope slightly exceeds 30° and the computed value is 30.37^o. Moreover, the vertical acceleration of a particle at the crest is 0.388 g but in the far field, as $r/l \to \infty$, the acceleration tends to the value $1/2g$ appropriate to the Stokes corner flow. In order to prove this value, we define the particle acceleration components in the moving coordinate system as:

$$\left.\begin{array}{rcl} a_r & = & u_r \dfrac{\partial u_r}{\partial r} + \dfrac{u_\theta}{r} \dfrac{\partial u_r}{\partial \theta} - \dfrac{u_\theta^2}{r} \\[2mm] a_\theta & = & u_r \dfrac{\partial u_\theta}{\partial r} + \dfrac{u_\theta}{r} \dfrac{\partial u_\theta}{\partial \theta} + \dfrac{u_r u_\theta}{r} \end{array}\right\}.$$

(5.19)

Using eqs. (5.18) to calculate the terms of eqs. (5.19) it is found that $a_\theta = 0$ and:

$$a_r = \frac{1}{2} g \sin^2 \left(\frac{3}{2}\theta\right) + \frac{3}{2} g \cos^2 \left(\frac{3}{2}\theta\right) - g \cos^2 \left(\frac{3}{2}\theta\right),$$

(5.20)

or:

$$a_r = \frac{1}{2} g.$$

(5.21)

Thus, the particle acceleration near the crest is directed radially downward from the crest with a magnitude $(1/2)g$.

For the deep water waves, the commonly quoted property of the "highest" wave is the wave steepness, i.e. wave height to length ratio of 0.142. Usually this is expressed as "the wave height which is one–seventh of the wave length". When examining the values obtained theoretically (see Table 5.1), this quote is observed to be very accurate (Gaughan et al., 1973; Longuet–Higgins, 1980; Williams, 1981). In the Table 5.1 the phase velocity is normalized to the velocity C_0, calculated by the small amplitude wave theory.

For the finite water depth waves, where the bottom is important to the breaking process, the local surface slope and the kinematic breaking criterion are the factors limiting shallow water wave growth. Beyond the limiting slope, i.e. when the surface is vertical, the wave is unstable, and water particles along the steep surface fall forward ahead of the wave. Miche (1944) obtained numerical estimations of $(H/L)_{max}$ in the form:

$$\left(\frac{H}{L}\right)_{max} = 0.142 \tanh(kh).$$

(5.22)

In practice, the kinematic criterion is usually used to estimate the maximum ratio of wave height to water depth $(H/h)_{max}$. These ratios for several water wave theories, are given in Table 2.3. The resulting values don't differ too much and in Fig. 2.2 the Cokelet calculations are used. In the special case of the solitary waves, the maximum values of $(H/h)_{max}$ suggested by the several Authors are listed for comparison in Table 5.2.

Table 5.1: Derived wave steepness for kinematically limited deep water waves

Author	Steepness	C/C_0
Mitchel (1893)	0.142	1.10
Havelock (1918)	0.1418	1.10
Davis (1951)	0.1443	1.08
Yamada (1957)	0.1412	1.09
Chappelear (1959)	0.1428	1.10
Dean (1968)	0.1723	-
Schwartz (1974)	0.1412	1.0922
Longuet–Higgins (1975)	0.1411	1.0923
Cokelet (1977)	0.1411	1.0921
Longuet–Higgins and Fox (1977)	0.14107	1.0923
Williams (1981)	0.141063	1.09228

Table 5.2: Maximum ratio of wave height to water depth for solitary waves

Author	H/h
Boussinesq (1871)	0.73
Rayleigh (1876)	1.0
Mc Cowan (1894)	0.78
Yamada (1957)	0.83
Lenau (1966)	0.827
Fenton 1972)	0.85
Longuet–Higgins and Fenton (1974)	0.827
Fox (1977)	0.8332
Williams (1981)	0.8332

5.2.2   Breaking wave mechanism

Waves break in different way, depending on wave height, wave period and beach slope. Laboratory studies (Galvin, 1972; Peregrine, 1983a; Wiegel, 1964) have shown that breakers can be classified into four principle types (Peregrine, 1983a):

- *Spilling*. White water appears at the wave crest and spills down the front face of the wave. The upper 25% of the front face may become vertical before breaking.
- *Plunging*. The whole front face of the wave steepens until vertical; the crest curls over the front face and falls into the base of the wave, sometimes preceeded by the projection of a small jet.
- *Collapsing*. The lower part of the front face of the wave steepens until vertical, and this front face curls over as an abreviated plunging wave. Minimal air pocket and usually no splash–up is observed.
- *Surging*. Wave slides up beach with little or no bubble production. Water surface remains almost plane except where ripples may be produced on the beach face.

Many properties of the surf zone appear to be governed by the parameter ξ_0 defined as (Battjes, 1974):

$$\xi_0 = \left(\frac{H_0}{L_0}\right)^{-1/2} \tan\beta, \tag{5.23}$$

where: β - bottom slope, H_0/L_0 - incident wave steepness.
The parameter ξ_0 can be used for classification of various breaker types into three main categories:

$$\begin{aligned}
surging \ and \ collapsing \ &if \ 3.3 < \xi_0 \\
plunging \ &if \ 0.5 < \xi_0 < 3.3 \\
spilling \ &if \quad\quad \xi_0 < 0.5
\end{aligned}$$

A transformation of ξ_0 into inshore parameter $\xi_b = (H_b/L_0)^{-1/2}\tan\beta$ yields:

$$\begin{aligned}
surging \ or \ collapsing \ &if \ 2.0 < \xi_b \\
plunging \ &if \ 0.4 < \xi_b < 2.0 \\
spilling \ &if \quad\quad \xi_b < 0.4
\end{aligned}$$

The parameters ξ_0 and ξ_b are equivalent to parameter $\chi_0 = (2\omega^2 a_0)/(g\tan^2\beta)$ also used in literature (Battjes, 1988).

In the following we restrict our attention to two first types of breaking. Plunging breakers derive its name from overturning wave front that curls forward, forming a jet that then plunges into the trough ahead of it (Basco, 1985; Sakai et al., 1986). Even symmetric waves tend to become unsteady and asymmetric long before their energy reaches the theoretical maximum.

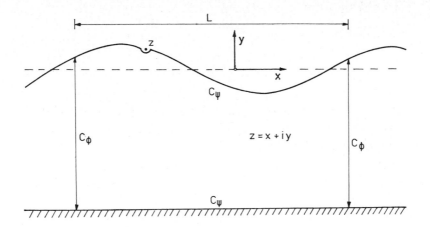

Figure 5.3: Integrating contour for breaking wave. (From Vinje and Brevik, 1981a)

Overturning looks very similar whatever its scale, suggesting that there may be some similarity solution that gives a local description of the overturning motion. This conclusion was partly supported by the numerical models, for the two–dimensional overturning wave. Longuet Higgins and Cokelet (1976) studied the evolution of waves on water of infinite depth mapping the physical flow domain onto the interior of a closed contour. If the initial free surface and potential ϕ on it are known, then a Cauchy–type integral equation may be formulated and approximated using discrete computational points on the surface to give subsequently the velocity of each point normal to the surface as the solution of a matrix equation. The tangential velocity can be found by numerical differentiation. The surface particle are allowed to move a finite distance in a small time step, giving a new surface location.

The Longuet–Higgins and Cokelet method has been modified by many authors (Vinje and Brevik, 1981a,b; Dold and Peregrine, 1984; New et al., 1985; Klopman, 1987). Particularly, Vinje and Brevik (1981a,b) have considered the finite water depth case (Fig. 5.3). As opposed to the above method, the computations were carried out for the physical contour C. Under the assumption that the contour C consists of subcontour C_ϕ (potential ϕ is known) and subcontour C_ψ (stream function ψ is known), the resulting set of the integral equations is (Vinje and Brevik, 1981a,b):

$$\left. \begin{array}{l} \pi\psi(x_0, y_0; t) \; + \; Re\left\{\oint_C \dfrac{\phi + i\psi}{z - z_0}dz\right\} = 0, \qquad \text{for} \quad z_0 \quad \text{at} \quad C_\phi \\[2em] \pi\phi(x_0, y_0; t) \; + \; Re\left\{i\oint_C \dfrac{\phi + i\psi}{z - z_0}dz\right\} = 0, \qquad \text{for} \quad z_0 \quad \text{at} \quad C_\psi \end{array} \right\}. \qquad (5.24)$$

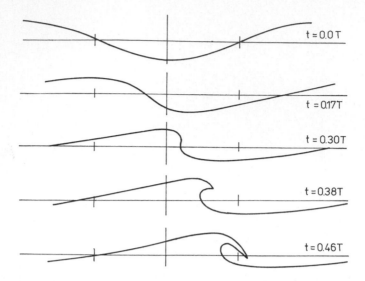

Figure 5.4: Evolution of sinusoidal wave profile of fairly large amplitude in time. (From Vinje and Brevik, 1981a)

In Figure 5.4, the evolution of sinusoidal wave of fairly large amplitude $(H/L) = 0.13$ and $(h/L) = 0.13$ is shown. At $t = 0.28T$, the surface becomes vertical. In Figure 5.5, the corresponding velocity and acceleration fields for time $t = 0.30\ T$ are given. The phase velocity C and acceleration of gravity g serve as the reference values.

Some discussion on the plunging breakers on the uneven bottom is given in New et al., (1985) and Klopman (1987) papers. There are also some analytic solutions that contribute to understanding aspects of an overturning wave. Longuet–Higgins (1982) considered a very simple analytic flow, with a time–dependent free surface. The surface profile is a parametric cubic curve which appears to correspond remarkably with the profile of the forward face of plunging breaker.

Consider now the plane wave approaching a slopping bottom at angle α_0. The breaking wave height H_b, water depth at breaking h_b, and the angle α_b are the unknown function of the initial parameters H_0, L_0 and α_0. Le Mehaute and Wang (1980) present a discussion of nonlinear shoaling and determine that no single theory can accurately account for wave shoaling from deep to shallow water. They found that limiting breaker height could be found for the parallel bottom contours from the empirical relationship:

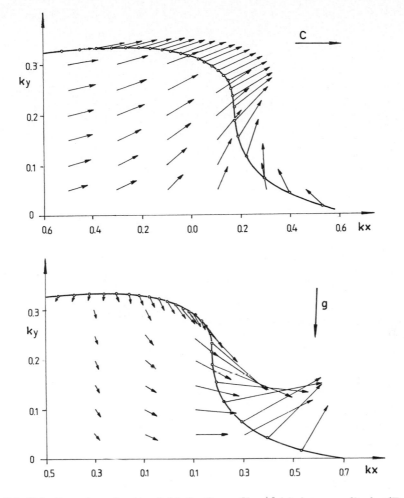

Figure 5.5: Velocity and acceleration fields for the profile of fairly large amplitude. (From Vinje and Brevik, 1981b)

$$\frac{H_b}{H_0} = 0.76\,(\cos\alpha_b)^{1/7}\left(\frac{H_0}{L_0}\,\beta^{4/7}\right)^{-1/4} K_R^{3/4}, \tag{5.25}$$

in which: β - bottom slope, K_R - refraction coefficient which relates the actual deep water wave height with the deep water wave height which is not influenced by the refraction.

The calculations and the hydraulic experiments yield the conclusion that the wave transformation (coefficient K_T) is described with sufficient accuracy by the cnoidal wave theory while the wave length is estimated in the best way by the linear theory. Using this fact Le Mehaute and Wang (1980) developed the hybrid prediction method for the parameters of shoaling wave. First of all, from the Snell's law (3.7) follows that:

$$\frac{L_b}{L_0} = \frac{\sin \alpha_b}{\sin \alpha_0},$$ (5.26)

and

$$K_R = \left(\frac{\cos \alpha_b}{\cos \alpha_0}\right)^{1/2}.$$ (5.27)

The breaker depth is calculated from the linear dispersion relation, i.e.:

$$(kh)_b = \frac{1}{2} \ln \left| \frac{1 + \dfrac{L_b}{L_0}}{1 - \dfrac{L_b}{L_0}} \right|.$$ (5.28)

5.2.3   Evolution of waves after breaking

The transformation of breaking and broken waves in the surf zone is the dominant factor in the hydrodynamics of many coastal processes, i.e.: nearshore circulation, run–up, and sediment transport etc. However, the development of rational models to describe breaking waves is just begining. In general, the models for propagation of breaking waves can be separated into three main categories (Izumiya and Horikawa, 1984):

- limiting breaker height concept,
- bore–propagation models,
- global description in terms of spatial variation of the integral quantities (wave energy, wave action).

Within the third category, the breaker height decaying is based on the energy balance equation:

$$\frac{\partial (EC_g)}{\partial x} = -\delta(x),$$ (5.29)

in which δ is the energy dissipation rate per unit surface area due to boundary shear, turbulence due to breaking, etc.

The laboratory data of Horikawa and Kuo (1966) and general observation indicate that the wave height is stabilizing at some value in a uniform depth following the initiation of wave breaking. Thus:

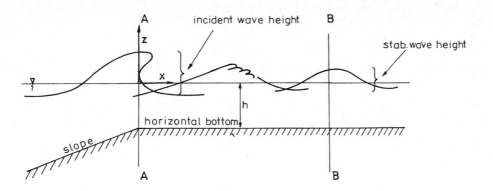

Figure 5.6: Shelf beach idealization of the surf zone

$$H_s = \Gamma h. \tag{5.30}$$

Consider a beach profile as in Fig. 5.6. The wave propagates onto this profile with characteristics such that breaking starts at the point where the bottom becomes horizontal (profile **AA**). The energy dissipated between profiles **AA** and **BB** is the difference between the local energy flux and the stable energy flux; hence eq. (5.29) yields:

$$\frac{\partial(EC_g)}{\partial x} = \frac{-K}{h}\left[EC_g - (EC_g)_s\right], \tag{5.31}$$

in which: $(EC_g)_s$ - energy flux associated with stable wave, K - dimensionless decay coefficient.
Because of eq. (5.30) and $C_g = \sqrt{gh}$ (shallow water is assumed) we have:

$$\frac{\partial G(x)}{\partial x} + \frac{K}{h(x)}G(x) = K\Gamma^2 h^{3/2}(x), \tag{5.32}$$

in which:

$$G(x) = H^2(x)h^{1/2}(x). \tag{5.33}$$

For simpler cases of breaking on beaches of idealized shapes, the closed form solutions exist (Dally et al., 1985):

uniform depth

$$\frac{H}{h} = \left\{ \left[\left(\frac{H}{h} \right)^2_b - \Gamma^2 \right] \exp\left(-K\frac{x}{h} \right) + \Gamma^2 \right\}^{1/2}. \tag{5.34}$$

At $x = 0$, $H/h = (H/h)_b$ - incipient of breaking.

uniform slope

If water depth is changing linearly with distance x, i.e.:

$$h(x) = h_b - \beta x, \tag{5.35}$$

we have:

$$\frac{H}{H_b} = \left[\left(\frac{h}{h_b} \right)^r (1 + \alpha) - \alpha \left(\frac{h}{h_b} \right)^2 \right]^{1/2}, \tag{5.36}$$

where:

$$\alpha = \frac{K\Gamma^2}{\beta \left(\frac{5}{2} - \frac{K}{\beta} \right)} \left(\frac{h}{H} \right)^2_b, \qquad r = \frac{K}{m} - \frac{1}{2}. \tag{5.37}$$

For the special case, if $K/\beta = 5/2$, eq. (5.36) takes the form:

$$\frac{H}{H_b} = \left(\frac{h}{h_b} \right) \left[1 - \alpha_1 \ln\left(\frac{h}{h_b} \right) \right]^{1/2}, \tag{5.38}$$

in which:

$$\alpha_1 = \frac{5}{2}\Gamma^2 \left(\frac{h}{H} \right)^2_b. \tag{5.39}$$

Note that if $K = 0$, eq. (5.36) reverts to the Green's law (3.17). The model depends strongly on the values of K and Γ factors. The determination of K and Γ coefficients, using a least squares procedure, yields the values presented in Table 5.3 (Dally et al., 1985).

Table 5.3: Coefficients Γ and K.

Slope	K	Γ
1/80	0.100	0.350
1/65	0.115	0.355
1/30	0.275	0.475

The transformation of the wave height over bottoms of arbitrary shape due to shoaling, breaking and reformation renders equations unsolvable analytically. Thus, a numerical procedure, possibly including the effects of set–up in mean water level, is needed. Usually, the slope of the mean water level $\bar{\zeta}$ is calculated by the "radiation stress" concept. From Longuet–Higgins and Stewart (1964), we have (see also Chapter 9):

$$\frac{\partial \bar{\zeta}}{\partial x} = \frac{-1}{\rho g (h + \bar{\zeta})} \frac{\partial S_{xx}}{\partial x}, \tag{5.40}$$

or for small water depth:

$$\frac{\partial \bar{\zeta}}{\partial x} = \frac{-3}{16} \frac{1}{(h + \bar{\zeta})} \frac{\partial H^2}{\partial x}, \tag{5.41}$$

in which S_{xx} is the onshore momentum flux S.

Fig. 5.7 illustrates the wave height decaying in the surf zone as a function of $(H/h)_b$ when $\Gamma = 0.5$ and $K/\beta = 7.5$. The faster decaying of the wave height is associated with the higher value of $(H/h)_b$. In general, the dependence of a breaker on beach slope appears in model explicitly, while the effect of wave steepness is contained implicitly through specification of the incipient conditions. A test run on the prototype and laboratory profile demonstrates the model ability to describe the wave height transformation observed in practice. The greatest assets of the model are its simplicity and ease of application.

The integral wave properties, i.e.: energy flux and radiation stress were determined by Svendsen (1984) using the crude approximation of the actual flow in surf zone. At the end we note that wave breaking can be included in the parabolic approximation to the mild–slope equation (Kirby and Dalrymple, 1986).

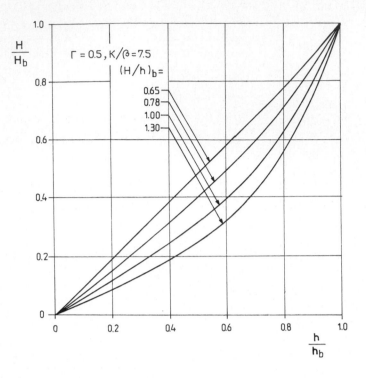

Figure 5.7: Wave height decaying as a function of $(H/h)_b$ for $\Gamma = 0.5$ and $K/\beta = 7.5$. (From Dally et al., 1985)

5.3   References

Basco, D.R., 1985. A qualitative description of wave breaking. *Proc. ASCE, Jour. Waterw., Port and Coastal and Ocean Div.*, 111: 171–188.

Battjes, J.A., 1974. Surf similarity. *Proc. 14th Coastal Eng. Conf.*, 1: 466–480.

Battjes, J.A., 1988. Surf–zone dynamics. *Ann. Rev. Fluid Mech.*, 20: 257–293.

Benjamin, T.B. and Feir, J.E., 1967. The disintegration of wave trains on deep water. *Jour. Fluid Mech.*, 27: 417–430.

Chu, V.C. and Mei, C.C., 1970. On slowly varying Stokes waves. *Jour. Fluid Mech.*, 41: 873–887.

Dally, W.R., Dean, R.G. and Dalrymple, R.A., 1985. Wave height variation across beaches of arbitrary profile. *Jour. Geoph. Res.*, 90: 11917–11927.

Dold, J.W. and Peregrine, D.H., 1984. Steep unsteady water waves: an efficient computational scheme. *Proc. 19th Coastal Conf.,* 1: 955–967.

Galvin, C.J., 1972. Wave breaking in shallow water. In: R.E. Meyer (Editor), *Waves on beaches and resulting sediment transport.* Academic Press, pp. 413–456.

Gaughan, M.K., Komar, P.D. and Nath, J.H., 1973. Breaking waves: a review of theory and measurements. School of Oceanogr., Oregon State Univ., Rep. 73-12, 145 pp.

Hammack, J.L., 1982. Small–scale ocean waves. *Proc. Inter. School of Physics "Enrico Fermi", Course LXXX-Topics in ocean physics,* Varenna, pp. 278–311.

Hasimoto, H. and Ono, H., 1972. Nonlinear modulation of gravity waves. *Jour. Phys. Soc. in Japan,* 33: 805–811.

Horikawa, K. and Kuo, C.T., 1966. A study of wave transformation inside surf zone. *Proc. 10th Coastal Eng. Conf.,* 1: 217–233.

Izumiya, T. and Horikawa, K., 1984. Wave energy equation applicable in and outside the surf zone. *Coastal Eng. in Japan,* 27: 119–137.

Kirby, J.T. and Dalrymple, R.A., 1986. Modeling waves in surfzones and around islands. *Proc. ASCE, Jour. Waterway, Port, Coastal and Ocean Eng.,* 112: 78–93.

Klopman, G., 1987. Numerical simulation of breaking waves on steep slopes. *Proc. ASCE Specialty Conf. on Coastal Hydrodynamics,* Delware, pp. 111-122.

Le Mehaute, B. and Wang, J.D., 1980. Transformation of monochromatic waves from deep to shallow water. Coastal Eng. Res. Center, Tech. Rep., 80-2, 43 pp.

Longuet–Higgins, M.S., 1980. The unsolved problem of breaking waves. *Proc. 17th Coastal Eng. Conf.,* 1: 1–28.

Longuet–Higgins, M.S., 1982. Parametric solutions for breaking waves. *Jour. Fluid Mech.,* 121: 403–424.

Longuet–Higgins, M.S., 1985a. Bifurcation in gravity waves. *Jour. Fluid Mech.,* 151: 457–475.

Longuet–Higgins, M.S. and Stewart, R.W., 1964. Radiation stresses in water waves: a physical discussion, with applications. *Deep Sea Res.*, 11: 529–562.

Longuet–Higgins, M.S. and Cokelet, E.D., 1976. The deformation of steep surface waves on water. I. Numerical method of computation. *Proc. Roy. Soc. London*, A350: 1–26.

Longuet–Higgins, M.S. and Fox, M.J.H., 1977. Theory of the almost–highest wave: the inner solution. *Jour. Fluid Mech.*, 80: 721–741.

Mei, C.C., 1983. *The applied dynamics of ocean surface waves.* A Wiley-Inter-Science Publication, New York, 734 pp.

Miche, R., 1944. Mouvements ondulatoires de la mer en profoundeur constante ou decroissante. *Ann. Ponts et Chaussees*, 121: 285–318 (in French).

Miles, J.W., 1981. The Korteweg–de Vries equation: a historical essay. *Jour. Fluid Mech.*, 106: 131–147.

New, A.L., McIver, P. and Peregrine, D.H., 1985. Computations of overturning waves. *Jour. Fluid Mech.*, 150: 233–251.

Peregrine, D.H., 1983a. Breaking waves on beaches. *Ann. Rev. Fluid Mech.*, 15: 149–178.

Peregrine, D.H., 1983b. Water waves, nonlinear Schrodinger equations and their solutions. *Jour. Austr. Math. Soc.*, B25: 16–43.

Peregrine, D.H., 1985. Water waves and their development in space and time. *Proc. Roy. Soc. London*, A400: 1–18.

Renouard, D.P., Seabra Santos, F.J. and Temperville, A.M., 1985. Experimental study of the generation, damping, and reflexion of a solitary wave. *Dynamics of Atm. and Oceans*, 9: 341–358.

Sakai, T., Mizutani, T., Tanaka, H. and Tada, Y., 1986. Vortex formation in plunging breaker. *Proc. Coastal Eng. Conf.*, 1: 711–723.

Segur, H., 1973. The Korteweg–de Vries equation and water waves. Solution of the equation: Part 1., *Jour. Fluid Mech.*, 59: 721–736.

Stokes, G.G., 1880. Considerations relative to the greatest height of oscillatory waves which can be propagated without change of form. *Mathematical and physical papers*, 1: 225–228.

Svendsen, I.A., 1984. Wave attenuation and set–up on a beach. *Proc. 19th Coastal Eng. Conf.,* 1: 54–69.

Vinje, T. and Brevik, P., 1981a. Numerical simulation of breaking waves. *Adv. Water Resources,* 4: 77–82.

Vinje, T. and Brevik, P., 1981b. Breaking waves on finite water depths. A numerical study. Mar. Techn. Centre, Trondheim, 77 pp.

Wiegel, R.L., 1964. *Oceanographical engineering.* Prentice–Hall, Englewood Cliffs, N.J.

Williams, J.M., 1981. Limiting gravity waves in water of finite depth. *Phil. Trans. Roy. Soc. London,* A302: 139–188.

Yuen, H.C., 1982. Nonlinear phenomena of waves on deep water. *Proc. Inter. School of Physics "Enrico Fermi", Course LXXX - Topics in ocean physics,* Varrena, pp. 205–234.

Zabusky, M.J. and Kruskal, M.D., 1965. Interaction of "solitons" in a collisionless plasma and the recurrence of initial states. *Phys. Rev. Letters,* 15: 240–243.

Zakharov, V.E., 1968. Ustoychivost periodicheskikh voln konechnoy amplitudy na poverkhnosti glubokoy zhidkosti. *Prikl. Mat. Teor. Fizika,* 2: 86–94 (in Russian).

Zakharov, V.E. and Shabat, A.B., 1972. Skhema integrirovaniya nelineynykh uravneniy matematicheskoy fiziki metodom obratnoy zadachi rasseyaniya. *Funkstyonalnyy analiz i ego prilozheniya,* 8: 43–53 (in Russian).

Zufiria, J.A., 1987. Weakly nonlinear non–symmetric gravity waves on water of finite depth. *Jour. Fluid Mech.,* 180: 371–385.

Chapter 6

WIND–INDUCED WAVE STATISTICS

6.1 General remarks

Of all the various types of wave motion that are possible in the ocean, wind waves are one of the most energetic and easily observed. Wind waves are generally considered to be surface gravity waves which are caused by the wind and propagate under the restoring force of gravity. Their wave lengths range from about 10 cm to about 1 km, with maximum energy density typically central at wave lengths of about 150 m. These water waves have maximum particle motion right at the air–sea interface, the particle motion decreasing rapidly with depth. The waves are dispersive over most of the ocean, with the long wave lengths travelling faster than the short wave lengths. Usually we distinguish the names "sea" for the rough, irregular waves in and near a storm area, and "swell" for the smooth, sinusoidal waves at some distance from the storm area.

Random seas as defined in this Chapter are those in which the sea surface is highly irregular and nonrepeatable in time and space. The energy transferred from winds to the sea surface spreads in space, and hence the waves propagate in different directions. Consequently, wave configuration is exceedingly complicated. Therefore, the properties of waves are not readily definable on a wave - by - wave basis. In general, the procedure of predicting random seas spans three domains: the time, frequency and probability domains (Ochi, 1982). In the time domain, the correlation functions are evaluated from the measured wave records. Once the autocorrelation function is evaluated, it is transferred, following the Wiener–Khintchine theorem, to the frequency domain by a Fourier transformation. The spectral density function is then completed with the probability functions for prediction of wave characteristics.

Shallow–water winds waves have many specific properties distinguishing them from deep–water waves. As waves approach to the shore, their dynamics is progressively more nonlinear and dissipative. Energy is transferred away from the peak of the spectrum to higher frequencies and also to low frequencies, in the form of surf beat. This is mainly due to water depth changes (Massel, 1985b). In the shallow water, a variety of processes composes very complicated picture of the sea surface and the wave spectrum. Among them there are nonlinear interactions between spectral components, wave transfer due to shoaling water, wave breaking, etc. Due to complexity of wave motion in the coastal zone, our present knowledge is based mainly on experiments in natural conditions and in laboratory and the application of statistical and spectral analysis. The primary objective of this Chapter is to characterize the properties of the wave motion within the shallow water area with modern statistical models.

6.2   Stress of wind on sea surface

The wind blowing over the sea exerts an effective tangential stress on the surface and causes the surface waves. To obtain the parameters characterized the wind–induced waves, it is assumed that the controlling parameters of wind field are:

- average wind speed V_w at some height (z) above still water surface (usually $z = 10\ m$),
- wind fetch X,
- wind duration t.

The mean velocity $V_w(z)$ at a fixed height is given by:

$$V_w(z) = \frac{u_*}{\kappa} \left| \ln\left(\frac{z}{z_0}\right) + f(R_i) \right|, \tag{6.1}$$

where u_* - friction velocity, defined as:

$$u_* = \sqrt{\frac{\tau}{\rho_a}}, \tag{6.2}$$

and τ is tangential wind stress, ρ_a - density of air, κ - Karman constant (approximately equal 0.42), z_0 - constant of integration giving the virtual origin of the profile, $f(R_i)$ - a function of *Richardson number* R_i, reflecting the influence of the atmosphere condition:

$$R_i = \frac{g}{\rho_a} \frac{d\rho_a}{dz} \left(\frac{dV_w}{dz}\right)^{-2}.$$ (6.3)

An important question concerns the determination of z_0 or equivalently of a drag coefficient C_D of the surface. This is defined in terms of the mean wind velocity at a convenient height z:

$$C_D = \frac{\tau}{\rho_a V_w^2(z)}.$$ (6.4)

For the aerodynamically rough flow, the length z_0 is approximately proportional to the friction velocity u_*^2, i.e.:

$$z_0 = m \frac{u_*^2}{g},$$ (6.5)

in which: m - empirical coefficient.
Eq. (6.5) represents the formula suggested originally by Charnock. Krylov et al. (1986) have pointed out that the experimental data on z_0 scatter widely (sometimes by factor 10). However, in view of the logaritmic form of (6.1), the scatter of C_D is considerably less. For the aerodynamically rough flow (wind velocity in the range 5 - 53 m/s), the experimental data in natural and laboratory conditions confirm the linear relationship between $C_D^{(10)}$ and $V_w^{\prime(10)} - V_s(10)$, i.e. (Krylov et al., 1986):

$$C_D^{(10)} = \left(0.71 + 0.071 V_w^{(10)}\right) 10^{-3},$$ (6.6)

or in general:

$$C_D^{(10)} = \left(a + b V_w^{(10)}\right) 10^{-3}.$$ (6.7)

In the Krylov et al. (1986) book, the experimental data on the drag coefficient $C_D^{(10)}$ and the values a and b are summarized. Also, a large number of formulas developed by various authors under conditions of neutral stability was reviewed by Garratt (1977).

When $V_w^{(10)} \leq 5\ m/s$, the drag coefficient is close to the value for turbulent, aerodynamically smooth flow and the Reynolds number $Re_x > (Re_x)_{cr}$, where $Re_x = u_* X / \nu_a$. The critical value $(Re_x)_{cr}$ corresponds the situation when the viscous sublayer is destroyed and the wind waves are going to influence on the wind profile. Then, we have:

$$\frac{gz_0}{u_*^2} = m_1 \left(\frac{u_*}{3\sqrt{g\nu_a}}\right)^{-3},$$

(6.8)

in which: ν_a - coefficient of the kinematic viscosity of air flow.

According to Krylov et al. (1986), the constant $m_1 = 43$. Therefore, the drag coefficient takes the form:

$$C_D = \left[2.5 \ln\left(\frac{u_* z}{\nu_a}\right) - 9.4\right]^{-2}.$$

(6.9)

In the coastal zone, the critical Reynolds number $(Re_x)_{cr}$ strongly depends on the initial coastal roughness; for example for the sandy dunes ($z_0 = 10^{-3} \div 10^{-2}$ m), we have $(Re_x)_{cr}, \approx 10^6$. Usually in the laboratory conditions, $Re_x < (Re_x)_{cr}$ and the typical roughness is smaller than the thickness of the viscous sublayer. For this regime, eq. (6.8) is still valid but the constant $m_1 = 0.11$, while in the natural conditions $m_1 = 0.46$ (Krylov et al., 1986).

In strong wind, i.e. near–neutral atmospheric conditions, eq. (6.1) gives:

$$\frac{V_w(z)}{V_w^{(10)}} = \frac{\sqrt{C_D}}{\kappa} \ln\left(\frac{z}{z_0}\right).$$

(6.10)

Assuming for example $V_w^{(10)} = 20$ m/s, eq. (6.6) yields $C_D = 2.13 \cdot 10^{-3}$. Thus, the roughness length z_0 is equal to about 1 mm.

Sometimes, it is very convenient to use power law profiles of the following type:

$$\frac{V_w(z)}{V_w^{(10)}} \approx \left(\frac{z}{10}\right)^\alpha.$$

(6.11)

Experience has shown that α is about 0.11.

For many practical purposes, the estimates of the wind conditions near the water surface are usually provided by the wind reports from meteorological stations or ships at sea. The estimates based on the isobaric pattern of synoptic weather charts are good suplement to the meteorological observations. The pressure gradient is in approximate equilibrium with the acceleration produced by the rotation of the earth. In the result of such equilibrium, the geostrophic wind is generated. The geostrophic wind blows parallel to the isobars and its speed is general higher than surface wind speed. The standard procedure to calculate the speed and direction of the geostrophic wind is published elsewhere (Shore Protection Manual, 1973; Krylov et al., 1986) and will not be repeated here. We only note that, the following instructions for obtaining estimates of the surface wind speed over the open sea from the geostrophic wind speed are recommended (Shore Protection Manual, 1973):

a) for moderately curved to straight isobars - no correction is applied,

b) for great anticyclonic curvature - add 10 percentage to the geostrophic wind speed,

c) for great cyclonic curvature - substract 10 percentage from the geostrophic wind speed.

To correct for air mass stability, the sea–air temperature difference must be computed.

6.3 Similarity laws for wind–generated waves

At the locations where no information on the wave climate is directly available, the wave characteristics may be estimated by application of available wind data, using the dimensional analysis approach. In general, the characteristic wave parameters - the significant wave height H_s and significant wave period T_s - are determined as functions of the constant wind speed V_w at the standard anemometer height of 10 m, the constant depth of water h, the distance of water (or fetch) X over which the constant wind speed $V_w^{(10)}$ is active, the duration t of activity of the constant wind speed V_w, and the gravitational acceleration g, i.e. (Sobey, 1986):

$$H_s, T_s = f\left(X, V_w, t, h, g\right). \tag{6.12}$$

The dimensional analysis applied to eq. (6.12) reduces the number of variables to three:

$$\frac{gH_s}{V_w^2}, \frac{gT_s}{V_w} = f\left(\frac{gX}{V_w^2}, \frac{gt}{V_w}, \frac{gh}{V_w^2}\right). \tag{6.13}$$

These results are commonly presented in terms of the *fetch–limited* or *duration–limited* graphs:

$$\left.\begin{aligned}
\frac{gH_s}{V_w^2} &= f_1\left(\frac{gX}{V_w^2}, \frac{gh}{V_w^2}\right) \\
\frac{gT_s}{V_w} &= f_2\left(\frac{gX}{V_w^2}, \frac{gh}{V_w^2}\right)
\end{aligned}\right\}, \tag{6.14}$$

and

$$
\left.
\begin{aligned}
\frac{gH_s}{V_w^2} &= f_3\left(\frac{gt}{V_w}, \frac{gh}{V_w^2}\right) \\[2mm]
\frac{gT_s}{V_w^2} &= f_4\left(\frac{gt}{V_w}, \frac{gh}{V_w^2}\right)
\end{aligned}
\right\}.
\tag{6.15}
$$

When both, the fetch and duration are sufficiently large for H_s and T_s to reach limiting values, these will become dependent only on the wind speed V_w and the condition of fully developed sea exists. In order to attain the *fetch–limited* condition, a certain time t_{min} is needed, where $gt_{min}/V_w = f\left(gX/V_w^2\right)$; for $t < t_{min}$, the *duration–limited* waves are observed. In the limit case, when the wind wave growth is fully controlled by water depth we have:

$$
\left.
\begin{aligned}
\frac{gH_s}{V_w^2} &= f_5\left(\frac{gh}{V_w^2}\right) \\[2mm]
\frac{gT_s}{V_w} &= f_6\left(\frac{gh}{V_w^2}\right)
\end{aligned}
\right\}.
\tag{6.16}
$$

The specific forms of the functions $f, f_1, \ldots, f_6$ for appropriate wave generation conditions will be given later.

In the alternative approach, the wave spectrum can be expressed directly in terms of the wind characteristics using the Kitaigorodskiy (1970) similarity law. Dependence was assumed on the constant wind–shear velocity u_*, the fetch X, the duration t, the constant water depth h, and the gravitational acceleration g:

$$
S_1(\omega, \theta) = g^2 \omega^{-5} F\left\{\frac{\omega u_*}{g}, \theta; \frac{gX}{u_*^2}, \frac{gt}{u_*}, \frac{gh}{u_*^2}\right\}.
\tag{6.17}
$$

By the same dimensional argument as employed by Kitaigorodskiy, the spectrum must be of the general form (6.17) with u_* replaced by $V_w = V_w^{(10)}$ which is easier to measure than u_*. Thus we have:

$$
S_1(\omega, \theta) = g^2 \omega^{-5} F\left\{\frac{\omega V_w}{g}, \theta; \frac{gX}{V_w^2}, \frac{gt}{V_w}, \frac{gh}{V_w^2}\right\}.
\tag{6.18}
$$

The Kitaigorodskiy representation was adopted in the presentation of the results of analysis of wave spectra in the North Atlantic (Pierson, Moskowitz, 1964) and specially of the results of the Joint North Sea Wave Project (JON-SWAP) - Hasselmann et al. (1973).

6.4 Statistical properties of the wind waves

As the sea waves are always random in the sense that the detailed configuration of the surface varies in an irregular manner in both space and time, the various statistical measures of the motion can be regarded as significant observationally or predictable theoretically (Fig. 6.1). In principle, in this Chapter we follow the standard nomenclature for the wave data analysis recomended by International Association for Hydraulic Research and Permanent International Association of Navigation Congresses (IAHR - PIANC, 1986) Bulletin.

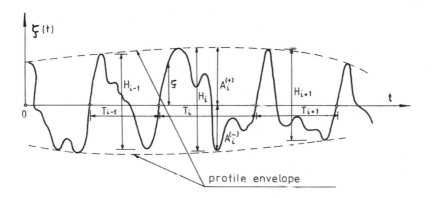

Figure 6.1: Random wave profile and its characteristics

6.4.1 Surface elevation

Surface displacement at a given point is frequently the resultant of many wave components that have been generated by the wind in different regions and have propagated to the point of observation. Under the assumption that the size of the generating area is large compared with the distance over which the surface displacements are significantly correlated and if the wave interactions can be neglected, the surface displacement at a given point can be regarded as a linear superposition of these harmonic waves. In further analysis, we assume considerable simplifications which are possible when the waves possess a narrow–band spectrum.

The assumption of the narrow–band spectra means that the resulting variation of the surface ordinate $\zeta(t)$ corresponds to a regular sinusoid with slowly varying envelope and phase. Therefore we have:

$$\zeta(t) = a(t) \cos\left[\omega_0 t + \epsilon(t)\right], \tag{6.19}$$

where: ω_0 - central frequency, $a(t)$ - amplitude of the envelope, and $\epsilon(t)$ - phase angle.

When we additionally assume that the phase $\epsilon(t)$ is uniformly distributed, $\zeta(t)$ has a *Gaussian distribution* (Phillips, 1977):

$$p_\zeta(\zeta) = \frac{1}{\sqrt{2\pi}\sigma_\zeta} \exp\left(\frac{-(\zeta - \bar{\zeta})^2}{2\sigma_\zeta^2}\right). \tag{6.20}$$

In above, the symbols $p(x)$, $\bar{\zeta}$ and σ_ζ denote the probability density function, mean value and standard deviation, respectively. The $\bar{\zeta}$ and σ_ζ are constants which are related to the probability density. Hence, the expected value of any function $g(x)$ of x is given as:

$$E\left[g(x)\right] = \int_{-\infty}^{\infty} g(x)p(x)dx, \tag{6.21}$$

provided that the above integral converges. Adopting this approach we may define the n-moment of x as:

$$E\left[x^n\right] = \overline{x^n} = \int_{-\infty}^{\infty} x^n p(x)dx; \tag{6.22}$$

thus:

$$E[x] = \bar{x} = \int_{-\infty}^{\infty} xp(x)dx. \tag{6.23}$$

It is often convenient to take moments about the mean of x rather than about $x = 0$; therefore we obtain:

$$\mu_n = E\left[(x - \bar{x})^n\right] = \int_{-\infty}^{\infty} (x - \bar{x})^n p(x)dx. \tag{6.24}$$

Particularly the second central moment:

$$E\left[(x - \bar{x})^2\right] = \sigma_x^2 = \int_{-\infty}^{\infty} (x - \bar{x})^2 p(x)dx = E\left[x^2\right] - \bar{x}^2, \tag{6.25}$$

defines the variance of x, denoted here σ_x^2. The standard deviation σ_x is the positive square root of the variance.

The assumption of (6.20), with the associated symmetry about the still water level, is expected to be realistic for small amplitude waves, especially in the deep water. The observations show that coastal waves have a definite excess of high crests and shallow troughs in contrast to those of waves in deep water. Moreover, in coastal areas the additional asymmetry with respect to a vertical line passed through the crest also exists. These facts yield the conclusion that the process is not *Gaussian* and the probability density function (6.20) is not valid. The non–Gaussian distribution was modelled by Longuet–Higgins (1963), using Gram–Charlier series in the form:

$$p_\zeta(\zeta) = p_n(\zeta) \left\{ 1 + \frac{\gamma_1}{3!} H_3 \left(\frac{\zeta - \bar{\zeta}}{\sigma_\zeta} \right) + \frac{\gamma_2}{4!} H_4 \left(\frac{\zeta - \bar{\zeta}}{\sigma_\zeta} \right) + \ldots \right\}, \qquad (6.26)$$

in which: $p_n(\zeta)$ - Gaussian distribution, H - Hermite polynomial of degree n, γ_1 - skewness coefficient and γ_2 - kurtosis coefficient; thus:

$$\gamma_1 = \frac{\mu_3}{\sigma_\zeta^3}; \qquad (6.27)$$

the γ_1 provides an indication of asymmetry of the probability distribution about the mean value and:

$$\gamma_2 = \frac{\mu_4}{\sigma_\zeta^4} - 3. \qquad (6.28)$$

Experiments in the shallow water zone (Bitner, 1980) demonstrate that the coefficients γ_1 are positive, in general. This is physically consistent with the vertical asymmetry of the nonlinear free surface, mentioned above. The coefficients γ_2 are rather small and their positive values correspond to the higher probability than normal one for the ordinates close to the mean.
This is clearly demonstrated in Fig. 6.2 which corresponds to the shallow water ($h \approx 2.8m$). In that case the Gram–Charlier distribution gives a significantly better representation of data than a simple Gaussian law did. Ochi and Wang (1984) have found that the parameter γ_1 is the dominant parameter affecting the non–Gaussian characteristics of coastal waves. It can be evaluated as a function of water depth and sea severity, i.e.:

$$\gamma_1 = 1.16 \exp^{-0.42h} H_s^r, \qquad (6.29)$$

where: $r = 0.74 \, h^{0.59}$, h and H_s are in meters.
According to Ochi and Wang (1984) for $\gamma_1 < 0.2$, the non–Gaussian presentation (in standardized form) of coastal wave profiles can be approximated by the Gaussian probability density function (also in standardized form).

In contrast to the approach considered above, Huang et al. (1983) obtained probability density function based on the approximation that waves can be expressed as a Stokes expansion carried to the third order. The amplitude and phase of the first–order component of the Stokes wave are assumed to

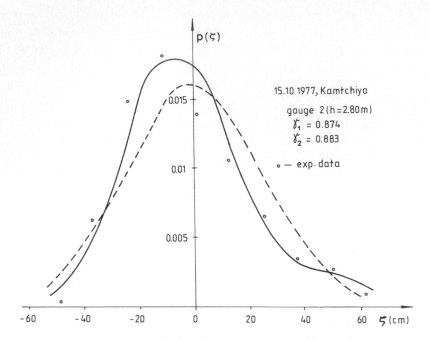

Figure 6.2: Probability density function for the shallow wave ordinates.

be Rayleigh and uniformly distributed and slowly varying, respectively. The probability density function depends on three parameters: the root–mean–square surface elevation, the significant slope and the non–dimensional depth. An important difference between this function and the Gram-Charlier representation is that the Huang et al. probability density is always non–negative. However, it should be noted that the Gram–Charlier expansion is more general in that it covers waves of any bandwidth, whereas the model based on the Stokes expansion is essentially one of narrow band and is therefore more restrictive as far as bandwidth is concerned.

It should be added that the Stokes expansion approach yields an expression for the skewness of the surface elevation of deep water waves, which is found to depend only on a "significant slope" - $\delta_s = \bar{\zeta}/L_p$ (L_p - wave length corresponding to the peak in the spectrum) - as the band–width of the first–order surface elevation approaches to zero (Srokosz and Longuet–Higgins, 1986; Langley, 1987). Thus, we have:

$$\gamma_1 = 6\pi \, \delta_s. \tag{6.30}$$

For the broader spectrum γ_1 may have a rather wide range of values, i.e.:

$$\gamma_1 = n\pi\,\delta_s, \tag{6.31}$$

in which: $n \approx 6 \div 9$.

6.4.2 Wave height

Under the assumption that waves are a narrow–band normal process (6.19) with zero mean and variance σ_ζ^2, it can be concluded that the wave amplitude (A) follows the Rayleigh distribution:

$$p_A(A) = \frac{A}{\sigma_\zeta^2}\exp\left(-\frac{A^2}{2\sigma_\zeta^2}\right), \tag{6.32}$$

or:

$$p_A(A) = \frac{2A}{\overline{A^2}}\exp\left(-\frac{A^2}{2\sigma_\zeta^2}\right), \tag{6.33}$$

in which: $\overline{A^2}$ - root–mean–square amplitude.

More general, the arbitrary moments for the wave amplitude can be expressed as:

$$\overline{A^m} = \frac{1}{\sigma_\zeta^2}\int_0^\infty A^{m+1}\exp\left(-\frac{A^2}{2\sigma_\zeta^2}\right)dA = \left(2\sigma_\zeta^2\right)^{m/2}\Gamma\left(1+\frac{m}{2}\right), \tag{6.34}$$

where: $\Gamma(\ )$ - gamma function.

In practice, the wave height (H) is preferred rather than amplitude (A), where possible. The assumption of a narrow–band spectrum leads to the very small probability that the maxima of the wave profile are located elsewhere than at the wave crests. Therefore, the wave envelope is regarded to represent the amplitude of individual waves themselves and the probabilities of wave crests and troughs are symmetric. Thus, if $H = 2A$, from (6.33) we have:

$$p_H(H) = \frac{H}{4\sigma_\zeta^2}\exp\left(-\frac{H^2}{8\sigma_\zeta^2}\right). \tag{6.35}$$

Because:

$$\bar{H} = \sqrt{2\pi}\,\sigma_\zeta, \tag{6.36}$$

eq. (6.35) yields:

$$p_H(H) = \frac{\pi}{2} \frac{H}{\overline{H}^2} \exp\left[-\frac{\pi}{4}\left(\frac{H}{\overline{H}}\right)^2\right],$$ (6.37)

and:

$$p_H(H) = \frac{H}{4\sigma_\zeta^2} \exp\left[-\frac{1}{8}\left(\frac{H}{\sigma_\zeta}\right)^2\right] = \frac{2H}{R} \exp\left(-\frac{H^2}{R}\right),$$ (6.38)

in which: $R = 8\sigma_\zeta^2$.

Following the common practice, it is convenient to define two characteristic wave heights. The first one is the *significant wave height H_s*, which is defined as the average of the highest one–third of wave heights. The development of reliable digital data recording and analysis techniques over the last 15 years has led to a fundamental change in the way significant wave height is estimated. Therefore, except H_s, the significant wave height is now commonly estimated from gauge records as:

$$H_{m_0} = 4\sigma_\zeta, \qquad \sigma_\zeta{}^2 = m_0 = \int_0^\infty S(\omega)\,d\omega,$$ (6.39)

in which: H_{m_0} - energy–based significant wave height determined as four times the square root of the area contained under the energy spectrum $S(\omega)$ (H_{m_0} is an internationally recommended notation; see for example IAHR-PIANC Bulletin, 1986). H_{m_0} is approximately equal to H_s except when water depth is very small or waves are very steep (Thompson and Vincent, 1985). Since H_{m_0} is based on energy, its applicability to estimating wave height statistics stems from its similarity to H_s. In situation where H_{m_0} is approximately equal to H_s either can be used to estimate various wave height statistics. However, when H_{m_0} differs from H_s, H_{m_0} cannot be used directly to estimate height statistics.

The second characteristic wave parameter: *root–mean–square wave height H_{rms}* - can be calculated as:

$$H_{rms} = \left[\frac{1}{N}\sum_{i=1}^{N} H_i^2\right]^{1/2},$$ (6.40)

or:

$$H_{rms}^2 = \int_0^\infty H^2 p(H)\,dH,$$ (6.41)

and

Table 6.1: Some wave height relations based on the Rayleigh distribution.

Characteristic height		$\dfrac{H}{H_{rms}}$	$\dfrac{H}{\sqrt{m_0}}$	$\dfrac{H}{H_s}$
Standard deviation of free surface,	$\sigma_\zeta = \sqrt{m_0}$	$1/2\sqrt{2}$	1.0	0.250
Root–mean–square height,	H_{rms}	1.0	$2\sqrt{2}$	0.706
Mode,	$\mu(H)$	$1/\sqrt{2}$	2.0	0.499
Median height,	$H(P = 1/2)$	$(\ln 2)^{1/2}$	$(8\ln 2)^{1/2}$	0.588
Mean height,	$\bar{H} = H_1$	$\sqrt{\pi}/2$	$\sqrt{2\pi}$	0.626
Significant height,	$H_s = H_{1/3}$	1.416	4.005	1.000
Average of tenth highest waves,	$H_{1/10}$	1.80	5.091	1.271
Average of hundreth highest waves,	$H_{1/100}$	2.359	6.672	1.666

$$H_{rms} = 2\sqrt{2}\,\sigma_\zeta = 0.706\,H_s. \tag{6.42}$$

Using the H_{rms} value, the wave height distribution $p(H)$ takes a particularly simple form:

$$p(H) = \frac{2H}{H_{rms}^2}\exp\left[-\left(\frac{H}{H_{rms}}\right)^2\right]. \tag{6.43}$$

Various other characteristic heights relating to the Rayleigh distribution are assembled in Table 6.1 (Sarpkaya and Issacson, 1981). As noted earlier, the profile of shallow–water waves is usually asymmetric with respect to the mean water surface with sharper, narrower crests and longer, shallower troughs than those of Gaussian waves. However, the experiments in shallow water zone indicate that the classical formula (6.36) which relates the mean wave height $\bar{H}$ to the standard deviation σ_ζ, is still applicable. In Fig. 6.3, the example taken from the paper by Massel and Robakiewicz (1986), gives:

$$\bar{H} = \alpha\sqrt{2\pi}\,\sigma_\zeta, \tag{6.44}$$

where: $\alpha \approx 0.95$.
Longuet–Higgins (1980) offers some explanation of the observed small discrepancy between eqs. (6.36) and (6.44). According to his idea, the Rayleigh distribution can still be used provided that the parameter R in eq. (6.38) should be rewritten in the form:

$$R = 8\sigma_\zeta^2\left[1 - \left(\frac{\pi^2}{8} - \frac{1}{2}\right)\nu^2\right] \approx 8\sigma_\zeta^2\left(1 - 0.734\,\nu^2\right), \tag{6.45}$$

where : ν - spectral width parameter.
Hence, for the typical value of the spectral width $\nu \approx 0.45$ is $R \approx 6.18\,\sigma_\zeta^2$. Note that for Rayleigh distribution it is:

$$R = H_{rms}^2 = \frac{4}{\pi}(\bar{H})^2; \tag{6.46}$$

therefore:

$$\frac{4}{\pi}(\bar{H})^2 = 6.18\,\sigma_\zeta^2 \qquad \text{or} \qquad \bar{H} \approx 0.88\sqrt{2\pi}\sigma_\zeta. \tag{6.47}$$

On the other hand, Naess (1983), using entirely different method, expressed the Rayleigh parameter R in the form:

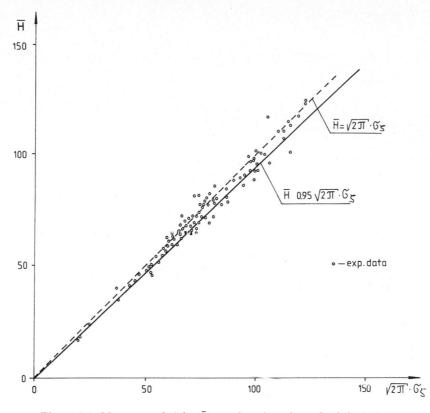

Figure 6.3: Mean wave height $\bar{H}$ as a function of standard deviation σ_ζ

$$R = 8\sigma_\zeta^2 \left\{ \frac{1}{2} \left[1 - r\left(\frac{T_p}{2}\right) \right] \right\}, \tag{6.48}$$

in which: $r\left(T_p/2\right) = \sigma_\zeta^{-2} K\left(T_p/2\right)$; $K(\tau)$ - autocorrelation function; T_p - wave period corresponding to the spectrum peak frequency ω_p.

The storm data from the North Sea indicates that $r\left(T_p/2\right) \approx 0.71$, what corresponds to $R = 6.84\,\sigma_\zeta^2$; thus $\bar{H} \approx 0.924\sqrt{2\pi}\,\sigma_\zeta$, what is close to observed value (6.44).

Except the Rayleigh distribution, other distributions for the probability density distributions for the wave height are also used. Glukhovskiy (1966) has developed some extension of the Rayleigh formula for limited water depth. His exceedence probability $P(\eta)$ takes the form:

$$P(\eta) = \exp\left\{ -\frac{\pi}{4} \frac{1}{1 + \eta^*/(2\pi)^{1/2}} \left[\eta^{2/(1-\eta^*)}\right] \right\}, \tag{6.49}$$

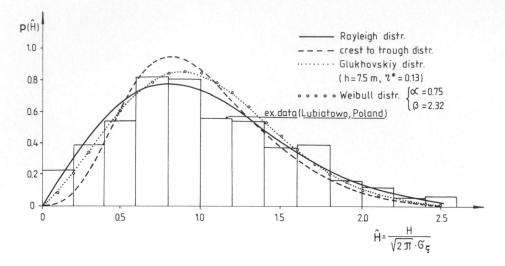

Figure 6.4: Comparison of the various probability distributions for the wave height

where: $\eta = H/\bar{H}, \quad \eta^* = \bar{H}/h$.

For the deep water, the *Glukhovskiy distribution* may be compared with the Rayleigh function. In the shallow water limit Glukhovskiy assumes that the wind waves break when $\bar{H}/h \approx 0.5$; therefore, the upper limit for η^* is 0.5. More information on the Glukhovskiy distribution can be found elsewhere (Glukhovskiy, 1966; Druet and Kowalik, 1970).

In Fig. 6.4 the comparison of the Rayleigh and Glukhovskiy distribution with experimental data is shown. It is observed that the Glukhovskiy distribution displays an excess of waves with heights near the midrange and a deficiency at the two extremes towards the Rayleigh distribution.

These shallow water data permit also the fitting to a *Weibull distribution* (Ochi, 1982):

$$p(\eta) = \alpha\beta\eta^{\beta-1}\exp\left(-\alpha\eta^\beta\right). \tag{6.50}$$

In Fig. 6.4 the values of $\alpha = 0.75$ and $\beta = 2.32$ were used.

In all definitions, given above, it was assumed that the magnitude of wave height is twice that of the wave amplitude. In practice, however, this is not always the case; it may be more appropriate to consider the sum of an upper–envelope value and a lower–envelope value that are separated by one–half of the average period. Using this definition recently, Tayfun (1981) has developed the probability function for crest–to–trough wave height in the form:

$$p(\eta) = \int_0^\eta p_2(\eta - u, u, \tau)du, \tag{6.51}$$

where:

$$\eta = \eta_1 + \eta_2; \qquad \eta_1 = \frac{A^{(+)}(t)}{\sigma_\zeta}; \qquad \eta_2 = \frac{A^{(-)}(t)}{\sigma_\zeta}, \tag{6.52}$$

and: $p_2(\ldots)$ - joint probability density function; $A^{(+)}$ - crest amplitude; $A^{(-)}$ - amplitude of the succeeding trough, separated by time τ, which must be defined separately.

For the function $p_2(x_1, x_2; \tau)$ we have (Middleton, 1960; Tayfun, 1981):

$$p_2(x_1, x_2; \tau) = \frac{x_1 x_2}{(1 - r_0^2)\,\sigma_\zeta^4}\, I_0 \left[\frac{r_0 x_1 x_2}{(1 - r_0^2)\,\sigma_\zeta^2} \right] \exp \left[-\frac{x_1^2 + x_2^2}{2\,(1 - r_0^2)\,\sigma_\zeta^2} \right], \tag{6.53}$$

$$r_0^2(\tau) = \rho_0^2(\tau) + \lambda_0^2(\tau) \le 1.0, \tag{6.54}$$

$$\left. \begin{aligned} \rho_0(\tau) &= \frac{1}{\sigma_\zeta^2} \int_0^\infty S(\omega) \cos(\omega - \omega_p)\tau d\omega \\[2mm] \lambda_0(\tau) &= \frac{-1}{\sigma_\zeta^2} \int_0^\infty S(\omega) \sin(\omega - \omega_p)\tau d\omega \end{aligned} \right\}, \tag{6.55}$$

ω_p - peak frequency, τ - time lag between amplitudes x_1 and x_2, r_0 - correlation coefficient, $I_0(\ )$ - modified Bessel function of zero order and first kind. Note that, as $\tau \to \infty$, r_0 approaches zero, and so:

$$p_2(x_1, x_2; \tau) = p(\eta_1)\,p(\eta_2). \tag{6.56}$$

Hence, the amplitudes η_1 and η_2 are uncorrelated. If $\tau \to 0$, then r_0 becomes unity and we get:

$$\lim_{\tau \to \infty} p_2(\eta_1, \eta_2; \tau) = p(\eta_1)\, \delta(\eta_2 - \eta_1), \tag{6.57}$$

in which $\delta(\ )$ is the delta function.

In order to estimate the parameter r_0, Forristall (1984) has assumed that $\tau = (1/2)\bar{T}_z$. The analysis of the measured wave height shows that the distribution of r_0 is rather narrow, with a mean of $r = 0.65$. However, in the shallow water the time lag τ is smaller than $\bar{T}_z/2$. Experiments in Kamtchiya (Black Sea) and in Lubiatowo (Baltic Sea) showed, that $\tau \approx 0.24\,\bar{T}_z$ and $0.68 < r_0(\tau) < 0.96$ (Massel, 1985a; Robakiewicz, 1988). Thus, in Fig. 6.4

the crest–to–trough distribution (with $r_0(\tau) = 0.75$) was compared with other distributions and experimental data. This distribution gives deficiencies of probability at both ends away from the mean, when comparing with the Rayleigh law.

The Rayleigh distribution does not have an upper bound. The probability density decreases exponentially as the independent variable (wave height or wave amplitude) becomes large. Therefore the largest wave height H_{max} is only a statistically defined quantity; it is the largest value that may occur in a sample of say, N, waves.

In particular, the probability density function for the largest wave height H_{max} is (Longuet–Higgins, 1952):

$$p\left(\frac{H_{max}}{\sigma_\zeta}\right) = \frac{N}{4}\left(\frac{H_{max}}{\sigma_\zeta}\right)\exp\left[-\frac{1}{8}\left(\frac{H_{max}}{\sigma_\zeta}\right)^2\right]\cdot$$

$$\cdot\left\{1 - \exp\left[-\frac{1}{8}\left(\frac{H_{max}}{\sigma_\zeta}\right)^2\right]\right\}^{N-1} \tag{6.58}$$

The expected value of H_{max} for the large values of N can be approximately expressed as:

$$\frac{\bar{H}_{max}}{\sigma_\zeta} = \frac{4}{\sqrt{2}}\left[\sqrt{\ln N} + \frac{\gamma}{2}(\ln N)^{-1/2}\right], \tag{6.59}$$

where: γ is Euler constant ($= 0.5772\ldots$).
For very large N, eq. (6.59) takes the simplified form:

$$\frac{\bar{H}_{max}}{\sigma_\zeta} \approx \frac{4}{\sqrt{2}}\sqrt{\ln N} = 0.707\sqrt{\ln N}. \tag{6.60}$$

The Glukhovskiy distribution (6.49) yields the following formula for the $\bar{H}_{max}$ (Matushevskiy and Karasev, 1977):

$$\frac{\bar{H}_{max}}{\bar{H}} = \left[\frac{4(1 + 0.4\eta)}{\pi}\right]^p\left[(\ln N)^p + \frac{\gamma}{2}(1 - \eta)(\ln N)^{-(p+\eta)}\right], \tag{6.61}$$

in which: $p = 1/2(1 - \eta)$.
Therefore, if $N = 500$, the mean of the highest wave height in deep sea is $\sim 2.94\,\bar{H}$ while in the breaker zone we have $\sim 1.81\bar{H}$. For particular purposes, however, it may be more meaningful to express the extreme wave height in

terms of time than as a function of number of observations. This can be done by using the formulation of the average period of zero up–crossings. Therefore, if we restrict ourselves to the first term in (6.59) we find:

$$\frac{\bar{H}_{max}}{\sigma_\zeta} = \frac{4}{\sqrt{2}} \left\{ \ln \left[\frac{(60)^2}{2\pi} T \sqrt{m_2/m_0} \right] \right\}^{1/2},$$ (6.62)

where: T - is the time in hours, m_n - moments of spectral function.
The number of studies (Forristall, 1978; Cavaleri, 1982; Massel, 1985b) in which the measured wave height distributions were compared with the Rayleigh distribution yield the conclusion that the Rayleigh distribution overpredicts the heights of larger waves. This may be partly due to the assumptions of a linear, Gaussian narrow band free surface. However, the difference between experimental and calculated values is rather small, and in most cases the Rayleigh distribution (original or modified) provides a good approximation to the distribution of individual wave heights defined by the zero up–crossing method.

6.4.3   Wave period

As was mentioned above, the wave period T is measured as the interval between successive zero up–crossing. A number of studies have been made on evaluation of the expected number of level crossing. Developing this approach, Longuet–Higgins (1975, 1983) has found that average period between successive zero up–crossing is:

$$\bar{T}_z = 2\pi \left[\frac{m_0}{m_2} \right]^{1/2},$$ (6.63)

where:

$$m_n = \int_0^\infty \omega^n S(\omega) d\omega, \qquad n = 0, 1 \ldots,$$ (6.64)

and: $S(\omega)$ - spectral density function.
The period $\bar{T}_z$ should be distinguished from the mean wave period $\bar{T}$ defined as:

$$\bar{T} = 2\pi \frac{m_0}{m_1},$$ (6.65)

and from the average period between successive crests $\bar{T}_c$

$$\bar{T}_c = 2\pi \left[\frac{m_2}{m_4}\right]^{1/2}. \tag{6.66}$$

The assumption that the surface oscillation $\zeta(t)$ is very similar to the regular sinusoid with slowly varying phase yields conclusion that the probability density for the wave period is narrower than that of wave heights, and the spread lies mainly in the range 0.5 to 2.0 times the mean wave period.

Usually in practice the probability density function of wave period is simply the marginal probability density function of the joint distribution in which the measure of the spectrum width is included. Longuet–Higgins (1975, 1983) presented a theory of the joint distribution of the heights and periods of waves with narrow–band spectra and discussed its applicability to actual sea waves. As a result, he found:

$$p\left(\tilde{H}, \tilde{T}\right) = \frac{\pi}{4} \frac{\sqrt{1 + \nu^2}}{\nu} \left(\frac{\tilde{H}}{\tilde{T}}\right)^2 \exp\left\{-\frac{\pi}{4}\tilde{H}^2 \left[1 + \frac{\left(1 - \frac{\sqrt{1+\nu^2}}{\tilde{T}}\right)^2}{\nu^2}\right]\right\} L(\nu), \tag{6.67}$$

in which: $\tilde{H} = H/\bar{H}$, $\tilde{T} = T/\bar{T}_z$, ν - spectral width parameter,

$$L = 2\left[1 + \left(1 + \nu^2\right)^{-1/2}\right]^{-1}. \tag{6.68}$$

under the assumption that $\nu^2 \leq 0.36$. The position of the maximum value of $p(\tilde{H}, \tilde{T})$ is given by the condition that $\partial p/\partial \tilde{H}$ and $\partial p/\partial \tilde{T}$ both vanish. Thus, we have:

$$\tilde{H} = \frac{2}{\sqrt{\pi}} \frac{1}{\sqrt{1 + \nu^2}} \qquad \text{and} \qquad \tilde{T} = \frac{1}{\sqrt{1 + \nu^2}}, \tag{6.69}$$

and

$$p_{max}\left(\tilde{H}, \tilde{T}\right) = \frac{\sqrt{1 + \nu^2}}{\nu} \frac{L}{e} \approx 0.368 \frac{\sqrt{1 + \nu^2}}{\nu} L. \tag{6.70}$$

The most probable value of $\tilde{T}$ at fixed values of the wave height $\tilde{H}$ is defined as:

$$\tilde{T}_{max.probl.} = 2\sqrt{1 + \nu^2} \left\{1 + \left(1 + \frac{16\nu^2}{\pi\tilde{H}^2}\right)^{1/2}\right\}^{-1}. \tag{6.71}$$

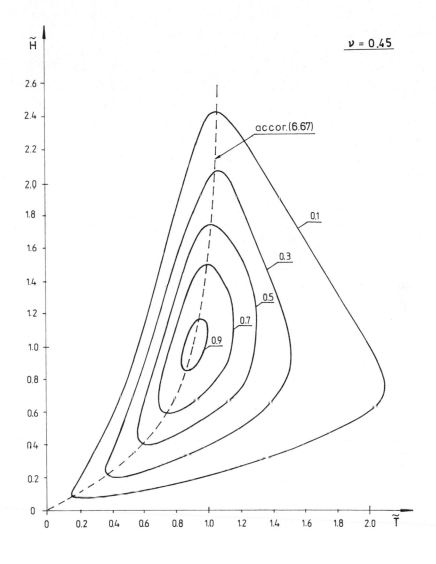

Figure 6.5: The function $p(\tilde{H}, \tilde{T})$ for $\nu = 0.45$

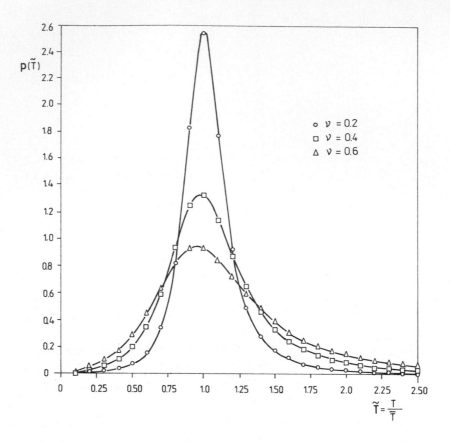

Figure 6.6: Probability density function for wave period ($\nu = 0.2; 0.4; 0.6$)

In Fig. 6.5, this curve is shown by the dashed line while the solid lines indicate the contours of $p(\tilde{H}, \tilde{T})$. It should be added that eq. (6.71) yields the following value of the most probable wave period $T \approx 1.004\,\bar{T}_z$.

After integration of eq. (6.67) against $\tilde{H}$ we find:

$$p\left(\tilde{T}\right) = \frac{1}{2}\frac{\sqrt{1+\nu^2}}{\nu}\,L\,\tilde{T}^{-2}\left\{1 + \nu^{-2}\left(1 - \frac{\sqrt{1+\nu^2}}{\tilde{T}}\right)^2\right\}^{-3/2}. \tag{6.72}$$

The density (6.72) shows some asymmetry with regard to $\tilde{T}$, what is observed in fact in the reality (Fig. 6.6). In practice, when evaluating the experimental distributions, the Weibull function (see eq. (6.50)) is frequently used. The particular values of α and β should be defined separately for the given wave field. Developing this approach, Davidan et al. (1985) have found that:

$$p(T) = \frac{3\alpha}{\bar{T}}\left(\frac{T}{\bar{T}}\right)^2 \exp\left[-\alpha\left(\frac{T}{\bar{T}}\right)^3\right],$$ (6.73)

in which: $\bar{T}$ - average period estimated from the wave record and

$$\alpha = \Gamma^3\left(1 + \frac{1}{3}\right) \approx 0.712.$$ (6.74)

In addition, the relations between various characteristical values of periods in the deep sea can be summarized as follows (Krylov, 1966; Longuet–Higgins, 1983):

$$\bar{T} = \left(1 + \nu^2\right)^{1/2}\bar{T}_z,$$ (6.75)

$$\omega_p \approx 0.8\,\bar{\omega}_z,$$ (6.76)

where: ω_p - peak frequency of the spectrum, ω_z - average zero crossing frequency.

However, the experiments taken in the coastal zone show that in the shallow water the relation (6.76) is not applicable. For example, the measurements in the South Baltic show that the ratio ω_p/ω_z remains practically constant troughout the coastal zone, ~ 0.62 on the average (Massel, 1980). On the other hand, in the mechanically generated, irregular wave trains propagating over shoaling water depth, the mean frequency $\bar{\omega}$ is increasing when the relative water depth is decreasing. The frequency rate $\omega_p/\bar{\omega}$ decreases with decreasing relative water depth for $k_{p_0}h < 1.2$ (k_{p_0} - wave number corresponding to the peak frequency in the deep water). When $k_{p_0}h > 1.2$, the rate $\omega_p/\bar{\omega} \approx const$ (Bendykowska,1986). Also, the experiments in the shallow water (Druet, 1978; Massel and Robakiewicz, 1986) suggest that the characteristic parameters of the Weibull distribution for the wave period usually are lying in the range:

$$\alpha \approx 2.86 \div 3.00 \qquad \text{and} \qquad \beta \approx 1.20 \div 2.00$$

The critical comparison of others joint distributions of wave height and period can be found in the paper by Srokosz and Challenor (1987).

6.5 Wave groups in the coastal zone

The observation of the sea surface shows that the heights of wind–generated waves are not uniform; they occur in successive groups in higher or lower

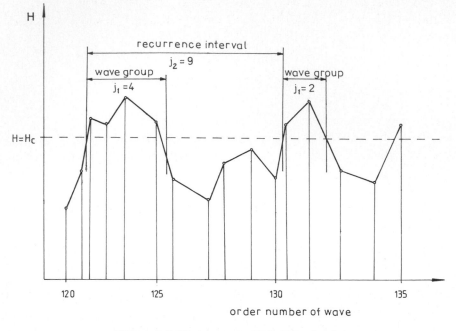

Figure 6.7: Wave groups. Definition sketch

waves. The existence of wave groups has been known to sea–farming men for a long time. Out of their experience has come the popular nation that "every seventh wave is large" or an old Icelandic saying that "large waves seldom come alone". The interest in the wave groups is stimulated by the suggestion that long–period oscillation of moored vessels, and irregular fluctuations of the mean water level near the shoreline with a period of several minutes or stability of the coastal defences may be caused by the occurence of runs of successive high waves (Goda, 1985).

The length of wave grouping can be described by counting the number of waves exceeding a specified value of the wave height H_c (e.q. significant wave height H_s, mean wave height $\bar{H}$, etc.). The succession of such high waves is called *a run* of high wave heights; the corresponding wave number we denote by j_1. In Fig. 6.7 we have two runs with $j_1 = 4$ and $j_1 = 2$, respectively. Similarly, we define a repetition length (or *total run*) over the threshold value such that a run begins at the first exceedance of wave height over this value, continuing through a sequence of the higher waves and then falling below the specific value, and ending at the first re–exceedance of threshold value. In the following we denote the total run by j_2. In Fig. 6.7, the total run is $j_2 = 9$.

Prior to discuss the j_1 and j_2 values, we may evaluate the group occurences by correlating succeeding wave heights; a correlation coefficient between wave heights takes the form (Goda, 1976):

$$\Phi_H(k) = \frac{1}{\Phi_H(0)} \frac{1}{N-k} \sum_{i=1}^{N-k} (H_i - \bar{H})(H_{i+k} - \bar{H}),$$ (6.77)

where:

$$\Phi_H(0) = \frac{1}{N} \sum_{i=1}^{N} (H_i - \bar{H})^2,$$ (6.78)

N - number of waves in the record, $\bar{H}$ - mean wave height, k - number of lags between waves in a sequence.
If the succeeding waves are uncorrelated, all of $\Phi_H(k)$ approach zero when $N \to \infty$. However, $\Phi_H(1)$ usually was found to be different from zero. For example, in the records taken in the coastal zone of the Baltic and the Black Sea (Cang, 1986), the average value of $\Phi_H(1)$ is $\approx 0.20 \div 0.40$, while $\Phi_H(k) \approx 0$ at $k > 1$. The positive sign suggests that large waves tend to be succeeded by large waves, while small tend to be succeeded by other small waves. If successive wave heights are uncorrelated, the probability of a run length j_1 is expressed as (Goda, 1976):

$$P(j_1) = p^{(j_1-1)}(1-p),$$ (6.79)

in which: p - occurrence probability that $H > H_c$. Thus, the mean value $\bar{j}_1$ and standard deviation σ_{j_1} are:

$$\bar{j}_1 = \frac{1}{q}, \qquad q = 1 - p,$$ (6.80)

and

$$\sigma_{j_1} = \frac{\sqrt{1-q}}{q},$$ (6.81)

in which: $p = $ prob.occur. $(H > H_c) = \exp\left[-(1/8)\eta_c^2\right]$ and $\eta_c = H_c/\sigma_\zeta$. The probability of a total run with the length j_2 can be derived by mathematical induction to give (Goda, 1976):

$$\bar{j}_2 = \frac{1}{p} + \frac{1}{q},$$ (6.82)

and

$$\sigma_{j_2} = \sqrt{\frac{p}{q^2} + \frac{q}{p^2}}. \tag{6.83}$$

The calculation by eqs. (6.80) - (6.83) and experiments in the coastal waters (Cang, 1986) indicate that at $H_c = H_s$, the mean values $\bar{j}_1$ and $\bar{j}_2$ are equal $1 \div 2$ and $7 \div 10$, respectively.

An alternative approach has been used by Kimura (1980) and Longuet–Higgins (1984), namely to consider the sequence of wave–heights as a Markov chain, with a non–zero correlation only between successive waves. This leads to expressions for $\bar{j}_1$ and $\bar{j}_2$ in terms of transition probabilities. However, the analysis by Longuet–Higgins (1984) shows that two approaches are roughly equivalent, to order ν.

6.6   Long–term statistics in the coastal zone

This Section deals with long–term wave statistics, which describe the severity of a sea in terms of the significant wave height over a long period of time. There are two methods generally used to estimate the impact of wave on the coastal zone: design wave and design-wave spectrum. For these two methods it is necessary to obtain information on the wave climate for the area of interest. Needed data comprise information on the frequency and duration of storms or sea states above or below a certain level, and return periods of storms above a certain H_s level and risk levels. The best data on wave action are obtained by instruments. However, because of the scarcity of instrumental wave data, visual observations of waves are extensively used in estimating the long–term distributions of wave height. Visual data from several ocean areas are reported by Bruun (1981), Davidan et al. (1985), Jardine (1979) and Massel (1978). A considerable scatter is observed when comparing visual observations with the results from a recording wave meter, for example (H_r - visual observed wave height, H_s - recorded wave height):

$H_s = 1.68 \, H_v^{0.75}$ (acc. Bruun, 1981),
$H_v = 0.22 \, H_s^2 + 0.78 \, H_s + 0.83$ (acc. Jardine, 1979),
with variance about the line $= 0.05 \, H_s^2 + 0.05 \, H_s + 0.75$.

With deep–ocean data, there is the possibility of combining data from a number of different locations to give an average for a particular area. However, as one approaches the shore, the wave activity can differ appreciably in relatively short distances. Therefore, requirements also may change in relatively short distances. Both these cases generate important problems of determining the

optimal distance between wave–measurement sites in such a way that the area
of interest is covered (Massel, 1978).

In an ideal situations, we should be able to use the data within such a
duration time that possible periodic climate variation are included. However,
normally only the time series of much shorter duration are available. Therefore, in order to estimate the parameters of extreme waves, the extrapolation
of the instrumental or visually observed data beyond the measurement range
is used. At present, various models for long–term prediction, first of all for
wave height, are taken into account. Usually, for long–term wave prediction,
two phases are considered. First is the long–term statistics of the severity of
sea, the second phase is the long–term statistics of all wave heights, irrespec-
tive of their magnitudes. The severity of a sea is usually expressed in terms of
significant wave height. For the probability density function applicable to the
significant wave height, the Weibull distribution and the log–normal distribution have often been considered. Some Authors (Bruun, 1981; Norwegian
Petroleum Directorate, 1976; Davidan et al., 1985) claim that the data of
significant wave can be represented very well by the three–parameter Weibull
distribution given by:

$$P(H_s) = 1 - \exp\left[-\left(\frac{H_s - H_0}{H_c - H_0}\right)^m\right]. \tag{6.84}$$

It should be noted, however, that the distribution (6.84) carries the minimum
nonzero value H_0 as one of its parameters. Hence, the Weibull distribution
does not satisfactorily represent the distribution, particularly for small significant wave height (Ochi, 1982). In contrast to the Weibull distribution,
sometimes the density function for the log–normal distribution is used:

$$f(H_s) = \frac{1}{2\pi\sigma H_s} \exp\left[-\frac{(\ln H_s - H_0)^2}{2\sigma^2}\right], \tag{6.85}$$

where: H_0 and σ are parameters to be determined from the observed data.
In Fig. 6.8 the comparison of the experimental distribution of the significant wave height with the Weibull and log–normal distributions is shown
(Ochi, 1982). The figure demonstrates that the log–normal probability density agrees well with the histogram over the entire range of wave height, while
the Weibull probability density agrees well for large significant wave heights,
but the agreement is rather poor for small wave heights. Moreover, some
additional argument in favour of the application of log–normal distribution
can be given. It is related to much easier development of the joint probability distribution of the significant wave height and wave period, under the

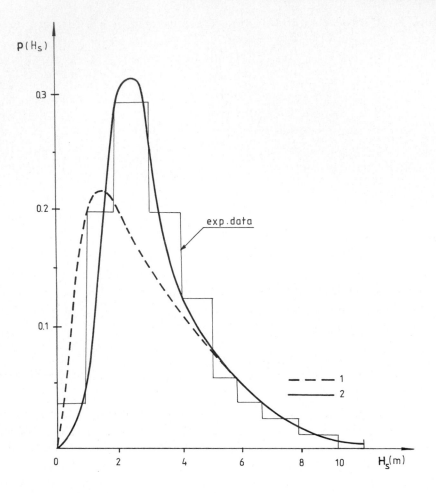

Figure 6.8: Comparison of experimental distribution of the significant wave height with the Weibull (1) and log–normal (2) distributions

assumption that the zero–crossing period follows the log–normal probability law. As Weibull distribution agrees well with the empirical distribution for large significant wave height and that it is no scientific basis for selecting any particular probability distribution to characterize the significant wave height in the remaining wave range, Davidan et al. (1978) have suggested to present the exceedance probability in the form:

$$
P_1\left(\tilde{H}'_s\right) =
\begin{cases}
\dfrac{s}{\sqrt{2\pi}} \displaystyle\int_{\tilde{H}'_s}^{\infty} \dfrac{1}{\tilde{H}_s} \exp\left[-\dfrac{1}{2}\ln^2\left(\tilde{H}_s\right)^s\right] d\tilde{H}_s & \text{for} \quad \tilde{H}_s \; \ge \; 1.0 \\[4mm]
\exp\left[-\alpha\left(\tilde{H}_s\right)^k\right] & \text{for} \quad \tilde{H}_s \; \le \; 1.0
\end{cases}
\tag{6.86}
$$

in which: $\tilde{H}_s = (H_s)/(H_{s_{50\%}})$, s - shape parameter equal $1/\sigma$, σ - parameter given as a function of $(H_s)/(H_{s_{50\%}})$ and s by Davidan et al. (1978).

The significant wave height is usually applied for modelling of the duration of sea states. The sea state is described by H_s and the duration of a storm above a particular level, H_s', is defined as the time when consequtive value of H_s equals or exceeds H_s' . The probability of the curve to cross the level H_s' during an interval of time is usually assumed to be described by the Poisson process. In the result, the following expression for the average duration of the storms (the amount of time the curve of H_s exceeds the level H_s' , divided by the expected number of storms) is given:

$$\bar{\tau}\left(H_s'\right) = \frac{\left[1 - P\left(H_s'\right)\right] L}{f\left(H_s'\right) L} = \frac{1 - P\left(H_s'\right)}{f\left(H_s'\right)}, \tag{6.87}$$

in which: L - total measurement period, and:

$$f\left(H_s'\right) = \int_0^\infty \dot{H}_s\, p\left(H_s', \dot{H}_s\right) d\dot{H}_s, \tag{6.88}$$

where: $p\left(H_s', \dot{H}_s\right)$ denotes the joint probability density function of H_s and time derivative $\dot{H}_s$.

When the forecasting of the statistical properties of wave heights in the long term (ranging 20 to 50 years) is needed, the long–term probability density should be considered. At present, a few prediction models are available, i.e. (Norwegian Petroleum Directorate, 1976):

- Det Norske Veritas model,
- Battjes model,
- storm model, and
- annual maximum model.

It is often necessary to estimate the extreme values in a specified longer period of time from the accumulation of daily, monthly, or yearly measured (or observed) largest values. Then, the estimation is based on the asymptotic distribution of the extreme values developed by Fisher and Tippett (Ochi, 1982). Moreover, in many practical applications the problem of the estimation of the probability of exceedance of the given wave height during the given time is a basic one. The suitable method for the calculation of the exceedance probability was developed recently by Van Heteren and Bruinsma (1981).

6.7   References

Bendykowska, G., 1986. Spectral characteristics of irregular waves over sloping bottom. *Proc. XX Convegno di Idraulica e Construzioni Idrauliche,* Padova, pp. 46–52.

Bitner, E.M., 1980. Nonlinear effects of the statistical model of shallow-water wind waves. *Appl. Ocean Eng.,* 2: 63–73.

Bruun, P., 1981. *Port engineering.* Gulf Publ. Comp., Houston, 800 pp.

Cang, L.T., 1986. Metodyka opisu grupowej struktury falowania wiatrowego. *Rozpr.Hydrot.,* 47: 65–96 (in Polish).

Cavaleri, L., 1982. Experimental characteristics of wind waves. *Proc. Inter. School of Physics "Enrico Fermi",* Course LXXX, Varenna, pp. 472–514.

Davidan, I.N., Lopatukhin, L.I. and Rozhkov, W.A., 1978. *Vetrovoye volneniye kak veroyatnostnyy gidrodinamicheskiy process.* Gidrometeoizdat, Leningrad, 287 pp. (in Russian).

Davidan, I.N., Lopatukhin, L.I. and Rozhkov,W.A., 1985. *Vetrovoye volneniya w mirovom okeane.* Gidrometeoizdat, Leningrad, 256 pp. (in Russian).

Druet, Cz., 1978. *Hydrodynamika morskich budowli i akwenow portowych.* Wydawnictwo Morskie, Gdansk, 390 pp. (in Polish).

Druet, Cz. and Kowalik, Z., 1970. *Dynamika morza.* Wydawnictwo Morskie, Gdansk, 428 pp. (in Polish).

Forristall, G.Z., 1978. On the statistical distribution of wave heights in a storm. *Jour. Geoph. Res.,* 83: 2353–2358.

Forristall, J.D.A., 1984. The distribution of measured and simulated wave heights as a function of spectral shape. *Jour. Geoph. Res.,* 89: 10547–10552.

Garratt, J.R., 1977. Review of drag coefficients over oceans and continents. *Monthy Weather Review,* 105: 915–929.

Glukhovskiy, B.H., 1966. *Issledovaniye morskogo vetrovogo volneniya.* Gidrometeoizdat, Leningrad, 284 pp. (in Russian).

Goda, Y., 1976. On wave groups. *Proc. Inter. Conf. BOSS'76,* 1: 115–128.

Goda, Y., 1985. *Random sea and design of maritime structures.* Univ. Tokyo Press, 323 pp.

Hasselmann, K., Barnett, T.P., Bouws, E., Carlson, H., Cartwright, D.E., Enke, K., Ewing, J.A., Gienapp, H., Hasselmann, D.E., Kruseman, P., Meerburg, A., Muller, P., Olbers, D.J., Richter, K., Sell, W. and Walden, H., 1973. Measurements of wind–wave growth and swell decay during the Joint North Sea Wave Project (JONSWAP). *Deutch. Hydr. Zeit.,* Reihe, A12, 95 pp.

Huang, N.E., Long, S.R., Tung, C.C., Yuan, Y. and Bliven, L.F. , 1983. A non-Gaussian statistical model for surface elevation of nonlinear random wave field. *Jour. Geoph. Res.,* 88: 7597–7606.

IAHR–PIANC, 1986. List of sea state parameters. Bulletin, 52, 24 pp.

Jardine, T.P., 1979. The reliability of visual observed wave heights. *Coastal Eng.,* 3: 33–39.

Kimura, A., 1980. Statistical properties of random wave groups. *Proc. 17th Coastal Eng. Conf.,* 3: 2955–2973.

Kitaigorodskiy, S.A., 1970. *Fizika vzaimodeistviya atmosfery i okeana.* Gidrometeoizdat, Leningrad, 284 pp. (in Russian).

Krylov, Y.M., 1966. *Spektralnye metody issledovaniya i rascheta vetrovykh voln.* Gidrometeoizdat, Leningrad, 254 pp. (in Russian).

Krylov, Y.M., and Galenin, B.G., Duginov, B.A., Krivitskiy, S.V., Podmogilnyy, J.A., Polyakov, J.P., Popkov, R.A., Strekalov, S.S., 1986. *Veter, volny i morskiye porty.* Gidrometeoizdat, Leningrad, 264 pp. (in Russian).

Langley, R.S., 1987. A statistical analysis of non–linear random waves. *Ocean Eng.,* 14: 389–407.

Longuet-Higgins, M.S., 1952. On the statistical distribution of the heights of sea waves. *Jour. Mar. Res.,* 11: 245–266.

Longuet-Higgins, M.S., 1963. The effect of non-linearities in statistical distributions in the theory of sea waves. *Jour. Fluid Mech.,* 17: 459–480.

Longuet-Higgins, M.S. 1975. On the joint distribution of the periods and amplitudes of sea waves. *Jour. Geoph. Res.,* 80: 2688–2694.

Longuet-Higgins, M.S., 1980. On the distribution of the heights of sea waves: some effects of nonlinearity and finite band width. *Jour. Geoph. Res.,* 85: 1519–1523.

Longuet-Higgins, M.S., 1983. On the joint distribution of wave period and amplitudes in a random wave field. *Proc. Roy. Soc. London,* A389: 241–258.

Longuet-Higgins, M.S., 1984. Statistical properties of wave groups in a random sea state. *Phil. Trans. Roy. Soc. London,* A310: 219–250.

Massel, S., 1978. Needs for oceanographic data to determine loads on coastal and offshore structures. IEEE, *Jour. Oceanic Eng.,* 3: 137–145.

Massel, S., 1980. Zwiazek dyspersyjny dla fal wiatrowych. *Arch. Hydrot.,* XXVII: 151–170, (in Polish).

Massel, S., 1985a. Rozklad prawdopodobienstwa wysokosci fal wzbudzanych wiatrem. *Arch. Hydrot.,* XXII: 343–360, (in Polish).

Massel, S., 1985b. Nonlinear statistical and spectral characteristics of wind-induced waves in coastal waters. Proc. Inter. Conf. " *Water waves research. Theory, laboratory and field* ", Hannover, pp. 285–318.

Massel, S. and Robakiewicz, M., 1986. Osobliwosci charakterystyk statystycznych falowania wiatrowego w strefie brzegowej morza. *Rozpr. Hydrot.,* 48: 21–51, (in Polish).

Massel, S. and Robakiewicz, M., 1986. Statystyczne i spektralne osobliwosci falowania wiatrowego w strefie brzegowej. *Rozpr. Hydrot.,* 48: 21–51, (in Polish).

Matushevskiy, G.V. and Karasev, N.J., 1977. O svyazi sredney i maksymalnoy vysot voln v pribrezhnoy zone morya. *Trudy Gos. Okean. Inst.,* 138: 72–75, (in Russian).

Middleton, D., 1960. *An introduction to statistical communication theory.* McGraw-Hill, New York.

Naess, A., 1983. On the statistical distribution of crest to trough wave heights. Publ. R-143.83, Norw. Hydr. Lab., Trondheim, 27 pp.

Norwegian Petroleum Directorate, 1976. Environmental conditions of Norwegian Continental Shelf. Special emphasis on engineering applications. Stavanger, 517 pp.

Ochi, M.K., 1982. Stochastic analysis and probabilistic prediction of random seas. *Adv. in Hydroscience*, 13: 217–375.

Ochi, M.K. and Wang, W.Ch., 1984. Non-Gaussian characteristics of coastal waves. *Proc. 19th Coastal Eng. Conf.*, 1: 516–531.

Phillips, O.M., 1977. *The dynamics of the upper ocean.* Cambridge Univ. Press, 336 pp.

Pierson, W.J. and Moskowitz, L.A., 1964. A proposed spectral form for fully developed sea based on the similarity theory of S.A. Kitaigorodskiy. *Jour. Geoph. Res.*, 69: 5181–5190.

Robakiewicz, M., 1988. Przydatnosc zmodyfikowanego rozkladu Rayleigha w statystyce fal wzbudzanych wiatrcm. *Studia i Mat. Oceanol.*, KBM, Gdansk, 52: 145–158 (in Polish).

Sarpkaya, T. and Isaacson, M., 1981. *Mechanics of wave forces on offshore structures.* Van Nostrand Reinhold Comp., New York, 651 pp.

Shore Protection Manual, 1973. U.S.Army, Coastal Engineering Research Center, Government Printing Office, Washington, D.C.

Sobey, R.J., 1986. Wind–wave prediction. *Ann. Rev. Fluid Mech.*, 18: 149–172.

Srokosz, M.A. and Longuet-Higgins, M.S., 1986. On the skewness of sea surface elevation. *Jour. Fluid Mech.*, 164: 487–497.

Srokosz, M.A. and Challenor, P.G., 1987. Joint distributions of wave height and period; A critical comparison. *Ocean Eng.*, 14: 295–311.

Tayfun, M.A., 1981. Distribution of crest-to-trough wave height. *Proc. ASCE, Jour. Waterway, Port, Coastal and Ocean Div.*, 107: 149–158.

Thompson, E.F. and Vincent, C.L., 1985. Significant wave height for shallow water design. *Proc. SCE, Jour. Waterway, Port, Coastal and Ocean Eng.*, 111: 828–842.

Van Heteren, J. and Bruinsma, J., 1981. A method to calculate the probability of exceedance of the design wave height. *Coastal Eng.*, 5: 83–91.

Chapter 7

SPECTRAL PROPERTIES OF WIND WAVES

7.1   General remarks

The properties of the random function $x(t)$ outlined so far do not depend upon this being a continous function of time. In fact, they are developed when x is a discrete rather than a continuous variable. In many problems, however, it is convenient to assume $x(t)$ as varying continuously with time. The simpliest analysis of such signal record needs the farther assumptions that $x(t)$ is both *stationary* and *ergodic*. For the stationary process $x(t)$, the statistical properties of $x(t)$ are independent of the origin of time measurement. In cases where $x(t)$ can be measured at different locations, an analogous assumption that $x(t)$ is *homogeneous* may also be made. In practice, $x(t)$ may be assumed stationary and homogeneous for a certain duration and region only. The statistical properties of random processes are usually established by a set of many simultaneous observations (*ensemble*), instead of by a single observation of the process. The assumption that $x(t)$ is ergodic, implies that the expected value of a function $g(x)$ can be interchanged with the temporal average of $g(x)$:

$$E[g(x)] = \lim_{T \to \infty} \left(\frac{1}{T} \int_{-T/2}^{T/2} g(x) dt \right). \tag{7.1}$$

Therefore, the acceptance of the ergodicity assumption permits derivation of all statistical information from a single record. It is well known that any periodic signal $x(t)$ taken from a real stationary random process having a zero mean value, that is $E[x(t)] = 0$, can be separated into its frequency components using a standard Fourier analysis. If this signal is represented only over the finite interval $-T/2 < t < T/2$, the Fourier series representation can be used, namely:

$$x(t) = \sum_{n=-\infty}^{\infty} a_n \exp\left(i\,n\,\omega_p t\right),\tag{7.2}$$

where:

$$a_n = \frac{1}{T}\int_{-T/2}^{T/2} x(t)\exp\left(-i\,n\,\omega_p t\right)dt,\tag{7.3}$$

and $\omega_p = 2\pi/T$.

Usually, the quantity of the greatest interest when analysing stationary random processes is the mean square value of $x(t)$ over the interval $-T/2 < t < T/2$; that is:

$$\sigma_x^2 = <x^2(t)> = \frac{1}{T}\int_{-T/2}^{T/2} x^2(t)dt = \sum_{n=-\infty}^{\infty} |\,a_n\,|^2.\tag{7.4}$$

If $x(t)$ is real, we must have $a_{-n} = a_n^*$ (the asterisk denotes the complex conjugate). Therefore, we have:

$$\sigma_x^2 = \sum_{n=1}^{\infty} \frac{1}{2}|\,A_n\,|^2 \qquad \text{with} \qquad A_n = 2a_n.\tag{7.5}$$

If T is now allowed to approach infinity, the summation in (7.5) becomes an integral:

$$\sigma_x^2 = \int_0^{\infty} S_x(\omega)d\omega,\tag{7.6}$$

where: $S_x(\omega)$ is a continuous function of frequency; $S_x(\omega)d\omega$ represents the contribution to the variance (or energy) of the signal $x(t)$ due to its content within the frequency range ω to $\omega + d\omega$:

$$S_x(\omega)\,d\omega \equiv \sum_{\omega_n}^{\omega_n+d\omega} \frac{1}{2}|\,A_n\,|^2.\tag{7.7}$$

The function $S_x(\omega)$ is known as the *spectral density* or the *power spectral density* or the *energy spectrum* of x .

The spectral density $S_x(\omega)$ and the autocorrelation function $R_x(\tau)$ are a Fourier transform pair by the Wiener–Khintchine theorem:

$$S(\omega) = \frac{2}{\pi}\int_0^{\infty} R(\tau)\cos\omega\tau\,d\tau,\tag{7.8}$$

and

$$R(\tau) = \int_0^\infty S(\tau) \cos \omega\tau \, d\omega. \tag{7.9}$$

The spectral density concept developed above may be extended to two dimensions. Thus, the wave profile in the particular point and at the particular time may be represented as an infinite number of component waves travelling with different wave number k, frequencies ω and directions Θ. The phase ϵ between components is assumed to be a random variable, distributed uniformly over the range $-\pi \le \epsilon \le \pi$. Thus, we have:

$$\zeta(x, y, t) = \sum_i A_i \cos \left[k_i \left(x \, \cos \Theta_i + y \, \sin \Theta_i \right) - \omega_i t + \epsilon_i \right], \tag{7.10}$$

in which the amplitudes A_i are treated also as random quantities.
For any frequency and directional interval $\Delta\omega\Delta\Theta$, the average wave energy becomes $(1/2)\rho g A_i^2$. That is:

$$S_1(\omega, \Theta)d\omega \, d\Theta = \sum_{\Delta\omega} \sum_{\Delta\Theta} \frac{1}{2} A_i^2, \tag{7.11}$$

in which: $S_1(\omega, \Theta)$ is the two–dimensional spectral density function. Therefore, the total wave energy can be represented as:

$$\sigma_\zeta^2 = \int_0^\infty \int_{-\pi}^{\pi} S_1(\omega, \Theta)d\omega \, d\Theta. \tag{7.12}$$

One–dimensional frequency spectrum may be obtained by integrating (7.12) over Θ:

$$S(\omega) = \int_{-\pi}^{\pi} S_1(\omega, \Theta)d\Theta. \tag{7.13}$$

If variation of surface ordinate $\zeta(t)$ corresponds to a regular sinusoid with slowly varying envelope and phase, the wave energy is concentrated around the central frequency ω_p within the narrow band. This is the case of the so called, *narrow spectrum.* However, the random sea waves exhibit a spectral form with the high frequency range attenuating as ω^{-n} $(n \approx 4 \div 5)$. Hence, the wave spectra can be treated as narrow and special parameters to measure the spectral bandwidth are needed.

7.2 Width of the energy spectra

In modern analysis the following measures of the spectral bandwidth are frequently used (Massel, 1985): $\epsilon, \epsilon_N, \nu, \nu_T, Q_p$ and Δ. The parameter ϵ was introduced by Cartwright and Longuet–Higgins (1956) in the form:

$$\epsilon^2 = 1 - \frac{m_2^2}{m_0 m_4}; \tag{7.14}$$

in which: ϵ takes values between 0 and 1; for narrow–band spectrum $\epsilon \rightarrow 0$. Alternatively, Longuet–Higgins (1975) has proposed two other parameters, namely ν and ν_T:

$$\nu^2 = \frac{m_0 m_2}{m_1^2} - 1, \tag{7.15}$$

and

$$\nu_T = \frac{\sqrt{3}}{2} IQR(\tau), \tag{7.16}$$

where: IQR is the interquartile range of the nondimensional wave period distribution.

For a narrow–band spectrum the ν value is approximately equal to one–half of ϵ. Tucker (1963) has proposed a simple method for estimating ϵ, i.e.:

$$\epsilon_N^2 = 1 - \left(\frac{\bar{T}_c}{\bar{T}_z}\right)^2 = 1 - \left(\frac{N_z}{N_c}\right)^2, \tag{7.17}$$

where: N_z and N_c denote the number of zero up-crossing and maxima within a wave record, respectively.

In some practical cases the parameter ϵ is not a criterion representing the bandwidth of a spectrum because it disregards the location of spectrum peaks. Thus, when evaluating the spectrum bandwidth for a given spectrum it is necessary to express the spectrum in normalized form. Developing this approach, Goda (1970) has proposed the following criterion, called the spectral peakedness parameter:

$$Q_p = \frac{1}{m_0^2} \int_0^\infty \omega [S(\omega)]^2 d\omega. \tag{7.18}$$

The value of parameter Q_p increases with increasing sharpness of the spectrum.

Finally, we can consider as a spectral width measure, the parameter Δ proposed by Belberov et al. (1983):

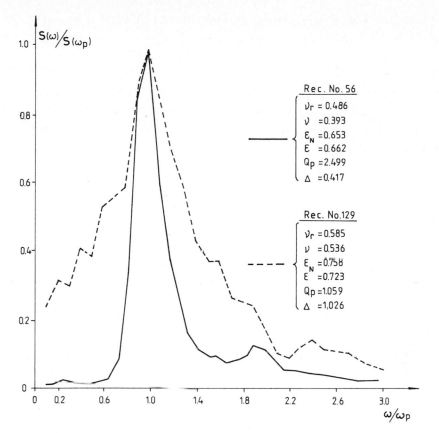

Figure 7.1: Narrow and broad spectra

$$\Delta = \frac{m_0}{\omega_p S(\omega_p)}. \tag{7.19}$$

In order to arrive at reliable values of probability distributions for wave height and period, it is important that the spectral width parameters are stable, i.e. not influenced by the high–frequency cut–off choice, when the moments of the spectrum are computed. This frequency represents the extent of small waves or ripples present in the wave field. Moreover, these small waves are generally of minor importance for the theoretical analysis and engineering practice. Thus, in order to compare the effectiveness of various spectral width parameters, the upper bound for the frequencies should be chosen (usually $\omega_{cf} \approx 3 \div 10\omega_p$). An accurate inspection of experimental data taken in shallow water regions shows that Q_p and Δ are really capable of distinguishing between narrow and broad spectra. That fact is demonstrated quite clearly in Fig. 7.1 (Massel, 1985). Two spectra normalized to their maximum values are different. However, the values of ϵ, ϵ_N, ν and ν_T are very similar to each other (difference is not greater than $\approx 20\%$). Meanwhile, Q_p and Δ values show the substantial difference for both spectra.

7.3   Spectral models of wind waves on limited water depth

The variety of wave spectra shapes is large. The shapes are strongly depending on the geographical location, duration and fetch wind, stage of growth and decay of a storm and existance of swell. However, the shape of wave spectra is not arbitrary and some fundamental properties associated with wave spectral energy can be distinguished. One of such properties is related to the upper limit of the spectral energy density for a given environmental condition. Once the spectrum reaches this saturated condition, the continuing energy input from the wind is lost by wave breaking, and by energy transfer from one wave frequency component to another. Phillips (1977) found that there exists an equilibrium range in the spectrum. The dimensional analysis yields the following formula for the spectral density over a range of wave frequencies higher than the spectral peak:

$$S(\omega) = \alpha g^2 \omega^{-5} \qquad \text{for} \qquad \omega \gg \omega_p, \tag{7.20}$$

where: α is a dimensionless constant ($\alpha = 7.4 \cdot 10^{-3}$). According to Kitaigorodskiy (1970) the constant α is in fact a function of dimensionless fetch. Recent measurements of wave spectra indicate that Phillips conception of an upper–limit asymptote to the spectrum is no longer tenable. Phillips (1985) has reexamined the nature of the equilibrium range, using dynamic insights into wave–wave interactions, energy input from the wind and wave breaking. As the result he obtained:

$$S(\omega) = \beta g u_* \omega^{-4}, \tag{7.21}$$

in which: u_* - shear velocity of the wind over the water surface, β - constant. This form has been found also by Toba (1973), Kitaigorodskiy (1983), Zaslavskiy and Zakharov (1982). A reanalysis of the JONSWAP data made recently by Battjes et al. (1987) indicates that the high frequency part of the wave spectrum can be better approximated with an ω^{-4} tail than with an ω^{-5} tail. However, it should be noted that the above results are valid for deep water waves only. Therefore, in the following we will keep the formulation (7.20).

In general, the wave spectrum over all frequency range takes the form:

$$S(\omega) = g^2 \omega^{-5} f\left(\frac{\omega}{\omega_p}\right). \tag{7.22}$$

If $\omega/\omega_p \gg 1.0$, then $f \to \alpha$. The explicit forms for the function f are usually developed on the basis of experimental studies. From the results of analysis of wave spectra in the North Atlantic, Pierson and Moskowitz (1964) have developed a spectral formulation representing fully developed wind generated seas in the form:

$$S(\omega) = \alpha g^2 \omega^{-5} \exp\left[-\frac{5}{4}\left(\frac{\omega}{\omega_p}\right)^{-4}\right], \tag{7.23}$$

with: $\alpha = 8.1 \cdot 10^{-3}$.

An extensive wave measurement program - JONSWAP - was carried out in 1968 and 1969 into the North Sea (Hasselmann et al., 1973). From the analysis of the measured spectra, the following spectrum, appropriate for the wind–generated seas with a fetch limitation, was developed:

$$S(\omega) = \alpha g^2 \omega^{-5} \exp\left[-\frac{5}{4}\left(\frac{\omega}{\omega_p}\right)^{-4}\right] \gamma^r, \tag{7.24}$$

where:

$$r = \exp\left[-\frac{(\omega - \omega_p)^2}{2\sigma_0^2 \omega_p^2}\right]. \tag{7.25}$$

This form contains five parameters, i.e., α, γ, ω_p and $\sigma_0 = \sigma_0'$ for $\omega \leq \omega_p$ and $\sigma_0 = \sigma_0''$ for $\omega > \omega_p$. From the mean JONSWAP spectrum, $\gamma = 3.3$, $\sigma_0' = 0.07$ and $\sigma_0'' = 0.09$. For the constant α and peak-frequency ω_p, the experiment yields the forms:

$$\alpha = 0.076\left(\frac{gX}{V_w^2}\right)^{-0.22}, \tag{7.26}$$

and

$$\omega_p = 22.0\left(\frac{g}{V_w}\right)\left(\frac{gX}{V_w^2}\right)^{-0.33}, \tag{7.27}$$

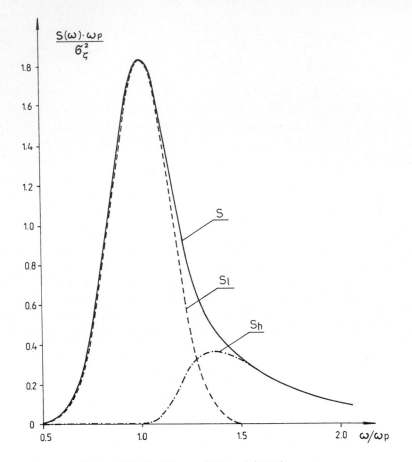

Figure 7.2: Strekalov and Massel (1971) spectrum

in which: X - wind fetch, and V_w - wind velocity.

Many wave spectra observed in the ocean have double peaks; one in the lower frequency part, the other in the higher frequency part. Sometimes three peaks exist in a spectrum, specially in shallow water conditions. Even for spectra with a single peak, it is often difficult to express the entire spectral shape in a simple mathematical formula. One way to represent the entire spectral shape in a mathematical formulation is to decompose the spectrum into two parts, the low-frequency component and the high–frequency component. Strekalov and Massel (1971) proposed the decomposition in the following form (Fig. 7.2):

$$\frac{S(\omega)\omega_p}{\sigma_\zeta^2} = S_l\left(\frac{\omega}{\omega_p}\right) + S_h\left(\frac{\omega}{\omega_p}\right), \qquad (7.28)$$

where:

$$S_l\left(\frac{\omega}{\omega_p}\right) = \frac{A}{\sqrt{2\pi}\, r}\left(\frac{\omega_p}{\bar{\omega}}\right)\exp\left[-\left(\frac{\omega_p}{\bar{\omega}}\right)^2\frac{\left(\frac{\omega}{\omega_p}-1\right)^2}{2r^2}\right],\qquad(7.29)$$

$$S_h\left(\frac{\omega}{\omega_p}\right) = B\left(\frac{\omega_p}{\bar{\omega}}\right)^{1-n}\left(\frac{\omega}{\omega_p}\right)^{-n}\exp\left[-q\left(\frac{\omega_p}{\bar{\omega}}\right)^{-m}\left(\frac{\omega}{\omega_p}\right)^{-m}\right],\qquad(7.30)$$

in which: $A = 0.69$; $B = 1.38$; $n = 5$; $m = 8$; $r = 0.12$; $q = 1.34$.
In that representation, the ratio $\omega_p/\bar{\omega} \approx 0.8$. However, in general this ratio
is a function of the non–dimensional fetch and duration. The high–frequency
part of the spectrum (7.30) reflects the presence of the second peak close to the
frequency $\omega/\omega_p \approx 1.35$. Moreover, according to eq. (7.28) , the characteristic
value $S(\omega_p)\omega_p/m_0 \approx 1.84$, what is in good agreement with experimental value
≈ 1.80 reported by Krylov et al. (1986). These authors have combined
data from about 200 different spectra published by Barnett, Clark, Davidan,
Ewing, Hasselmann, Kostichkova, Krylov, Mitsuyasu, Polakov and others and
have presented them in a non–dimensional form. Ochi and Hubble (1976)
have developed other type of multimodal spectral function. Each part of the
spectrum is expressed by the following formula:

$$S(\omega) = \frac{1}{4}\frac{\left[(4\lambda+1)\frac{\omega_m^4}{4}\right]^\lambda}{\Gamma(\lambda)}\frac{H_s^2}{\omega^{4\lambda+1}}\exp\left[-\left(\frac{4\lambda+1}{4}\right)\left(\frac{\omega_m}{\omega}\right)\right],\qquad(7.31)$$

where: H_s - significant wave height; ω_m - modal frequency and λ - shape
parameter. By combining two sets of three–parameter spectra, one representing the low–frequency components and the other the high–frequency components, we can derive the six–parameter spectral representation.
For the "old waves", when $gX/V_w^2 \geq 10^4$, Davidan et al. (1985) have formulated three–modal spectrum.

The water depth h does not appear explicitly in any mentioned spectrum.
One the other hand, the knowledge of spectral densities of wind–induced
waves in shallow water is of vital importance in many oceanographical and engineering applications. The similarity principles applied to the wind–induced
waves yield the saturation range of the wave spectrum in deep water as in
eq. (7.20). For waves in water of finite depth, Kitaigorodskiy et al. (1975)
extended the Phillips equilibrium range concept on the whole depth range.
Thus, we have:

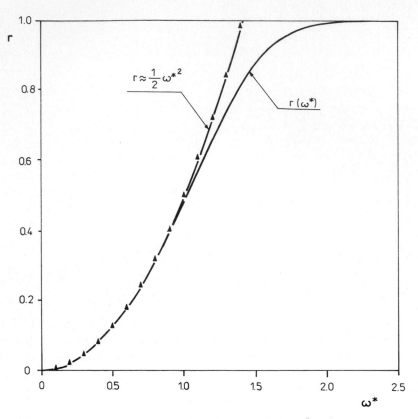

Figure 7.3: The function $r(\omega^*)$.

$$S(\omega) = \alpha g^2 \omega^{-5} r\left(\omega^*\right), \tag{7.32}$$

where:

$$r(\omega^*) = \frac{1}{f^2\left(\omega^*\right)} \left\{ 1 + \frac{2\omega^{*2} f\left(\omega^*\right)}{\sinh\left[2\omega^{*2} f\left(\omega^*\right)\right]} \right\}^{-1}, \tag{7.33}$$

$$\omega^* = \omega \sqrt{\frac{h}{g}}; \qquad f(\omega^*) = \tanh^{-1}\left[k\left(\omega^*\right) h\right]. \tag{7.34}$$

The function $r(\omega^*)$ - is illustrated in Fig. 7.3. It is easily verified that $r(\omega^*) \rightarrow 1$ when $h \rightarrow \infty$ and eq. (7.32) is equivalent to eq. (7.20) . In the other extreme case, when $h \rightarrow 0$, the function $r(\omega^*) \rightarrow (1/2)\omega^{*2}$ and eq. (7.32) gives:

$$S(\omega) = \frac{1}{2}\,\alpha\,gh\,\omega^{-3}. \tag{7.35}$$

The experimental data collected in the shallow water zone support the conclusion that the power of frequency should be in the range $(-5, -3)$. Bouws et al. (1985) assumed that a first approximation for a self–similar, finite water depth spectrum might be achieved by substitution of the factor $r(\omega^*)$ into JONSWAP spectrum $\dot{S}_J(\omega)$:

$$S(\omega, h) = S_J(\omega)r(\omega^*). \tag{7.36}$$

This spectral expression is called the "TMA–spectrum" after the three data sets including the wide range of depth, bottom materials, and wind–generation conditions. The hypothesis of a similarity appears to hold everywhere.

Huang et al. (1983) have proposed alternative spectral form - "Wallops spectrum" for the finite water depth, which is based on the Pierson–Moskowitz formula, i.e.:

$$S(\omega) = \frac{\alpha g^2}{\omega_p^5}\left(\frac{\omega}{\omega_p}\right)^{-s}\exp\left[-\frac{s}{4}\left(\frac{\omega}{\omega_p}\right)^{-4}\right], \tag{7.37}$$

in which:

$$s = \left| \frac{\ln\left\{\sqrt{2}\pi\delta\coth\left(k_p h\right)\left[1 + \frac{3}{\left(2\sinh^2\left(k_p h\right)\right)}\right]\right\}}{\ln\sqrt{2}} \right|, \tag{7.38}$$

and

$$\alpha = \frac{\left(4\pi\delta\right)^2 s^{s_1}}{4^{s_1}}\,\frac{\tanh^2\left(k_p h\right)}{\Gamma\left(s_1\right)}, \tag{7.39}$$

where: $s_1 = (s-1)/4$, $\quad \delta = \sigma_\zeta/L_p$, $\quad L_p$ - wave length corresponding to the frequency $\omega = \omega_p$, $\quad \Gamma$ - gamma function.
When $k_p h < 0.75$ (very shallow water), the power s is calculated from the solitary waves theory:

$$s = \left| \frac{\ln\left(\cosh\sigma_1\right)}{\ln\left(\sqrt{2}\right)} \right|, \tag{7.40}$$

in which:

$$\sigma_1 = \frac{\pi}{(7.5 U_r)^{1/2}}; \qquad U_r = \frac{2\pi\delta}{(k_p h)^3},$$

(7.41)

$$\alpha = (4\pi\delta)^2 \left(\frac{s}{4}\right)^{s_1} \left(\frac{C^2 k_p}{g}\right)^2 \Gamma^{-1}(s_1).$$

(7.42)

For the same depth range, Liu (1985) suggests that the spectrum takes the form ("GLERL spectrum"):

$$S(\omega) = C_1 \left(\frac{\sigma_\zeta}{\omega_p}\right)^2 \left(\frac{\omega}{\omega_p}\right)^{-C_2} \exp\left[-C_3 \left(\frac{\omega}{\omega_p}\right)^{\frac{-C_2}{C_3}}\right].$$

(7.43)

In the paper by Liu (1987), the detailed comparisons of spectral models given above, with Lake Erie surf zone wave measurements and measurements from a research tower offshore from the surf zone in 14 m water depth were made. In order to compare each spectrum against an objective assessment, so called deviation index (D.I.) was introduced as a measure of fitness of the models, i.e.:

$$D.I. = \sum \left[|S(f_i) - S_m(f_i)| S(f_i)^{-1} 100\right] \left[S(f_i) \Delta f E^{-1}\right],$$

(7.44)

in which: $S(f_i)$ and $S_m(f_i)$ are spectral densities calculated from the measurement and from the models, respectively, and Δf is the frequency interval used in calculating the spectrum.

The zero value of D.I. indicates a perfect representation; a smaller D.I. implies a better fit. The results of averaging all the deviation indexes calculated from the model comparison yield the conclusion that none of the models is clearly superior, but they can all provide crude to reasonable estimations of wave spectra. As these models are either purely empirical or semi–theoretical, a choice of which model to use depends upon the availability of required inputs. However, all the models failed to depict a fairly strong secondary peak on the high frequency side of the spectrum which are very clearly seen in the experimental spectra (see e.g. Druet et al., 1972; Banach, 1985). This can be done by the decomposition of the spectral shape into a few components (see for example eqs. (7.28) and (7.31). Massel et al. (1980) using the spectral representation as in eq. (7.28), presented the shallow water spectrum in the form:

$$
\frac{S(\omega)\,\omega_p}{(\bar{H})^2} = 1.12\left(\frac{\omega_p^2\,\bar{H}}{g}\right)^{-0.71}\left(\frac{\bar{H}}{h}\right)^{1.13}\exp\left[-20.0\left(\frac{\omega}{\omega_p}-1\right)^2\right] +
$$

$$
+ \left[2.52 - 6.98\left(\frac{\omega_p^2\,\bar{H}}{g}\right)^{-0.71}\left(\frac{\bar{H}}{h}\right)^{1.13}\right]\frac{\omega_p^2 h}{g}\,\Psi\left(\frac{\omega^2 h}{g}\right)\cdot
$$

$$
\cdot\left(\frac{\omega}{\omega_p}\right)^{-3}\exp\left[-12.21\left(\frac{\omega}{\omega_p}\right)^{-8}\right], \tag{7.45}
$$

in which:

$$
\Psi\left(\frac{\omega^2 h}{g}\right) = \frac{\tanh(kh)}{kh}; \tag{7.46}
$$

wave number k follows from the dispersion relation:

$$
\omega^2 = gk\tanh(kh). \tag{7.47}
$$

Authors suggest to use the spectrum (7.45) within the range $1.80 \le \omega_p^2 h/g \le 3.00$.

The spectrum is better expressed in wave number space because depth is included explicitly through wave number. Therefore, instead (7.36) we have (Vincent, 1984):

$$
F(k) = \frac{\alpha}{2}k^{-3}\,\Psi(\gamma, \omega_p, \omega, \sigma_0), \tag{7.48}
$$

where: Ψ is dimensionless function.

Moreover, we introduce a nondimensional wave number for the peak frequency of the spectrum:

$$
k = \frac{V_w^2 k_p}{g}. \tag{7.49}
$$

In that case, the prognostic equations for α and γ take the form:

$$
\left.\begin{array}{rcl}
\alpha &=& 0.0078 k^{0.49} \\[2mm]
\gamma &=& 2.47 k^{0.39}
\end{array}\right\}. \tag{7.50}
$$

Under the assumption, that the Kitaigorodskiy et al. (1975) expression:

$$F(k) = \frac{\alpha}{2} k^{-3} \qquad (7.51)$$

is a good scaling of the spectrum, Vincent obtained the following relationship between α and steepness parameter $\delta = \sigma_\zeta / L_p$ where $L_p = 2\pi/k_p$:

$$\alpha = 16\pi^2 \delta^2. \qquad (7.52)$$

The comparison of (7.52) with the experimental data (water depth $h = 17\ m$ and $h = 2\ m$) indicates an excellent fit at high steepness.

In order to predict an upper limit on wave energy as a function of frequency and water depth, we define a depth–controlled wave height as (Vincent, 1985):

$$H_l = 4(E)^{0.5}. \qquad (7.53)$$

According to Kitaigorodskiy et al. (1975), the Phillips equilibrium spectrum in finite depth water is expressed by (7.32), which can be rewritten in slightly different form:

$$S(f,h) = \alpha g^2 (2\pi)^{-4} f^{-5} r\left(\omega^*\right). \qquad (7.54)$$

When $h \to 0$, $r\left(\omega^*\right) \to \frac{1}{2}\omega^*$ (see Fig. 7.3); thus:

$$S(f,h) \approx \frac{\alpha g h}{8\pi^2} f^{-3}. \qquad (7.55)$$

The upper bound on the depth–controlled wave energy, E_h, can be estimated if low frequency cut–off value, f_c, is known, i.e.:

$$E_h = \int_{f_c}^{\infty} S(f,h)df \qquad \text{and} \qquad H_l = 4(E_h)^{0.5}. \qquad (7.56)$$

Please note, that in shallow water, H_l is expected to be different from the significant wave height H_s. Examination of storm spectra indicates that the spectral peak is quite sharp. Consequently, a choice for $f_c \approx 0.9\,f_p$ allows for a conservative estimate of H_l. The parameter α can be obtained by fitting eq. (7.54) to observed data if they are available. If not, α can be estimated by knowledge of the peak frequency f_p, and wind speed V_w through the relations (7.26) and (7.27). In order to give an estimate of H_l, we integrate eq. (7.55) analytically, i.e.:

$$H_l = \frac{1}{\pi}(\alpha gh)^{1/2} f_p^{-1},$$ (7.57)

or:

$$H_l = \frac{1.1}{\pi}(\alpha gh)^{1/2} T_p.$$ (7.58)

Eq. (7.57) suggests that H_l varies with the square root of depth (if α and f_c are held constant) when the primary spectral components are depth controlled. With the assumption that $\alpha \approx (V_w \omega_p/g)^{2/3}$ (see eqs. (7.26) and (7.27)), eq. (7.57) yields:

$$\frac{gH_l}{V_w^2} = \frac{1.1 g\alpha^{1/2}}{V_w f_p}\left(\frac{gh}{V_w^2}\right)^{0.5},$$ (7.59)

and

$$\frac{gH_l}{V_w^2} = C_4 \alpha^{-1}\left(\frac{gh}{V_w^2}\right)^{0.5}.$$ (7.60)

The predictive capability of eqs. (7.59) and (7.60) was checked against data collected in Lake Okeechobee. These data were used by Bretschneider (1958) in the development of shallow water forecasting curves. Expressions (7.59) and (7.60) are not algebraically equivalent to the Bretschneider equation; however, the results are in good agreement with the observed data.

We consider now the case of fully saturated (or developed) wind sea in shallow water. At a given depth, all waves of frequency f^* or less have the same speed $C = (gh)^{1/2}$, if f^* is the highest frequency where the shallow water dispersion relation is valid. Vincent and Hughes (1985) propose that the frequency where shallow water wave growth would stop is determined by:

$$2\pi f_p\left(\frac{h}{g}\right)^{1/2} = 0.9 \quad \text{or} \quad \frac{gT_p}{V_w} = 6.98\left(\frac{gh}{V_w^2}\right)^{0.5}.$$ (7.61)

If we assume that the peak frequency of the fully developed shallow water spectrum at a given depth, h, coincides with the highest frequency where the linear dispersion relation for shallow water is valid, eqs. (7.58) and (7.60) yield:

$$\frac{gH_l}{V_w^2} = 0.210\left(\frac{gh}{V_w^2}\right)^{0.75}.$$ (7.62)

In the above formula, the relation:

$$\alpha = 0.0086 \left(\frac{k}{1.23} \right)^{0.49} \tag{7.63}$$

was assumed (Bouws et al., 1985). This result is in agreement with Krylov et al. (1976). Their analysis of the experimental data yields the following forecasting equations:

$$\frac{g\bar{H}}{V_w^2} = 0.16 \left\{ 1 - \left[1 + 6.0 \cdot 10^{-3} \left(\frac{gX}{V_w^2} \right)^{0.5} \right]^{-2} \right\} \cdot$$

$$\cdot \ \tanh \left\{ 0.625 \left(\frac{gh}{V_w^2} \right)^{0.8} \left[1 - \left(1 + 6.0 \cdot 10^{-3} \left(\frac{gX}{V_w^2} \right)^{0.5} \right)^{-2} \right]^{-1} \right\}, \tag{7.64}$$

and

$$\frac{g\bar{T}_z}{V_w} = 6.2\pi \left(\frac{g\bar{H}}{V_w^2} \right)^{0.625}. \tag{7.65}$$

In the shallow water limit and when the fetch effect is disregarded, eq. (7.64) yields:

$$\frac{g\bar{H}}{V_w^2} = 0.10 \left(\frac{gh}{V_w^2} \right)^{0.8}. \tag{7.66}$$

If we assume that the relation $H_s \approx 1.6\,\bar{H}$ is valid also in the shallow water we get:

$$\frac{gH_s}{V_w^2} \approx 0.16 \left(\frac{gh}{V_w^2} \right)^{0.8}, \tag{7.67}$$

what is in close agreement with the Vincent estimation (see (7.62)). Moreover, it was shown that period T_p satisfies the following relationship (Krylov et al., 1976):

$$T_p \sqrt{\frac{g}{h}} \approx 5.5, \tag{7.68}$$

or:

$$\frac{gT_p}{V_w} \approx 5.5 \left(\frac{gh}{V_w^2}\right)^{0.5}, \tag{7.69}$$

while for the mean wave period $\bar{T}_z$ is:

$$\bar{T}_z = 4.6\sqrt{\frac{h}{g}}. \tag{7.70}$$

Again, (7.68) can be compared with the Vincent formula (7.61).

7.4 Evolution of spectral function over shoaling water depth

As surface gravity waves approach a beach, their shapes change dramatically until, in most cases, they break. In this Section we describe the transformation that occurs as a spectrum of surface gravity waves propagates shoreward over a mildly sloping bottom. We concentrate first on the "shoaling" regions, outside and specifically exluding the break zone. Linear theory has often been used as the basis for shoaling wave models. For the physically interesting case of small bottom slope, the solution has been obtained on the assumption of no reflected energy. Moreover, these solutions locally satisfy flat–bottom equation.

7.4.1 Transformation of wave spectra and statistical characteristics of waves

We limit our attention to steady–state situations in which the time dependency which may result from wind generation is negligible; also, the energy dissipation due to wave–seafloor interactions, white caps, and breakers.

Prior to demonstrate the transformation procedures, we note that the directional energy density spectrum can be defined by a multiplicity of mathematical forms depending upon the choice of parameters. Therefore, we have (Le Mehaute and Wang, 1982):

$$S_1(k_x, k_y) = \frac{1}{k} S_1(k, \Theta) = \frac{C_g C}{\omega} S_1(\omega, \Theta) = \frac{C_g}{2\pi k} S_1(f, \Theta), \tag{7.71}$$

in which: C - phase velocity, C_g - group velocity, $f = \omega/2\pi$, $k_x = k\cos\Theta$, $k_y = k\sin\Theta$.

Following the theoretical approach, we consider a spectrum defined in terms of wave number space, k_x, k_y. When the conservation of energy is applied we get:

$$\frac{\partial S_1}{\partial t} + \frac{\partial S_1}{\partial x}\frac{dx}{dt} + \frac{\partial S_1}{\partial y}\frac{dy}{dt} + \frac{\partial S_1}{\partial k_x}\frac{dk_x}{dt} + \frac{\partial S_1}{\partial k_y}\frac{dk_y}{dt} = 0. \qquad (7.72)$$

The last two terms on the left–hand side of eq. (7.72) represent combined refraction and shoaling which are not present in the balance equation for the deep water. Eq. (7.72) can be rewritten in the simple form as:

$$\frac{d\,S_1(k_x,k_y)}{d\,t} = 0. \qquad (7.73)$$

when the conservation of energy is applied.
Inserting the particular relationships into eq. (7.72) we get:

$$\frac{C_g}{2\pi\omega}\frac{d}{ds}[CC_gS_1(f,\Theta)] = 0, \qquad (7.74)$$

i.e.:

$$CC_gS_1(f,\Theta) = const \qquad \text{or} \qquad \frac{C_g}{k}S_1(\omega,\Theta,x,y) = const, \qquad (7.75)$$

and

$$S_1(\omega,\Theta) = \frac{k}{k_0}\frac{C_{g_0}}{C_g}S_{1_0}(\omega,\Theta_0); \qquad (7.76)$$

subscript (0) denotes the initial value at the point (x_0,y_0).
Let us now consider the simple case of a periodic wave arriving at an angle with a parallel bottom contour, $h = h(x)$. The Snell's law is then simply: $k\sin\Theta = const$, i.e.:

$$\frac{\sin\Theta}{C} = \frac{\sin\Theta_0}{C_0}. \qquad (7.77)$$

Thus:

$$\Theta_0 = \arcsin\left(\frac{k}{k_0}\sin\Theta\right). \qquad (7.78)$$

Inserting (7.78) into (7.76) finally yields (Krasitskiy, 1974):

$$S_1(\omega, \Theta) = \frac{k}{k_0} \frac{C_{g_0}}{C_g} S_{1_0} \left[\omega, \arcsin \left(\frac{k}{k_0} \sin \Theta \right) \right]. \tag{7.79}$$

One notes that to be valid, eq. (7.79) implies that:

$$\frac{k(\omega, x)}{k_0(\omega, x)} \sin \Theta \le 1. \tag{7.80}$$

If eq. (7.80) is no longer valid, the corresponding spectral component (ω, Θ) could not have come from deep water. Considering the simplicity of this formula, as compared to the procedure for 3–dimensional bathymetry, it appears that it is cost effective to consider near–shore bathymetry as sufficiently well–defined by a plane bathymetry.

In general, for 3–dimensional bathymetry the following system of ordinary differential equations should be solved:

$$\frac{dS_1(k_x, k_y)}{dt} = 0 = \frac{C_g}{2\pi\omega} \left\{ \cos\Theta \, \frac{\partial\left(CC_g S_1(f, \Theta)\right)}{\partial x} + \sin\Theta \, \frac{\partial\left(CC_g S_1(f, \Theta)\right)}{\partial y} + \right.$$

$$\left. + \, \frac{1}{C} \left(\sin\Theta \, \frac{\partial C}{\partial x} - \cos\Theta \, \frac{\partial C}{\partial y} \right) \frac{\partial\left(CC_g S_1(f, \Theta)\right)}{\partial \Theta} \right\}, \tag{7.81}$$

and

$$\frac{dx}{ds} = \cos\Theta, \qquad \frac{dy}{ds} = \sin\Theta, \qquad \frac{d\Theta}{ds} = \frac{1}{C} \left(\sin\Theta \, \frac{\partial C}{\partial x} - \cos\Theta \, \frac{\partial C}{\partial y} \right), \tag{7.82}$$

in which: s - the distance along a ray.

At present, some numerical schemes for solving these equations are available; for example, those by Collins (1972) and Shiau and Wang (1977). In the first step we should find the rays, i.e., solve the system of eq. (7.82) for each frequency. Next, the variation of energy level along the ray should be determined by application of the formula, $CC_g S(f, \Theta) = const$, which gives the spectral change along a ray for a given frequency.

Generally, the linear approach to the transformation of the wave spectrum is based on the superposition of elementary spectral components. For each frequency, the energy level (proportional to the square of the surface elevation $\zeta^2(\omega)$) is treated as an invariant during the transformation. Therefore, the transformation of each spectral component can be treated as identical to the transformation of the monochromatic wave of the same amplitude and frequency, and that the energy contained in each frequency band and direction travels along its corresponding wave ray at group velocity. The shallow water

spectrum is then determined from the deep water spectrum by simply applying the transformation coefficient square, K_H^2, to each frequency component,
i.e. (Krylov, 1966; Krylov et al., 1976):

$$S_1(\omega, \Theta) = S_{1_0}(\omega, \Theta_0) K_H^2(\omega, \Theta_0, h), \tag{7.83}$$

in which:

$$K_H^2 = \frac{b_0}{b} \frac{C_{g_0}}{C_g}, \tag{7.84}$$

b_0 - distance between two selected adjacent wave orthogonals, b - spacing
between orthogonals in the shallower water and $S_{1_0}(\omega, \Theta_0)$ - incident wave
spectrum in the deep water.
It should be noted that:

$$K_H^2 = \left(\frac{b_0}{b}\right)\left[\frac{1}{2}\frac{g}{\omega}\left(\frac{d\omega}{dk}\right)^{-1}\right] = K_R^2 K_T^2, \tag{7.85}$$

K_R - refraction coefficient, K_T - shoaling coefficient.
Changing the variables in (7.83), gives:

$$S_1(\omega, \Theta)d\omega d\Theta = S_{1_0}(\omega, \Theta_0) K_H^2(\omega, \Theta_0, h)\frac{\partial \Theta_0}{\partial \Theta} d\omega d\Theta, \tag{7.86}$$

and

$$S_2(k, \Theta) = S_1[\omega(k), \Theta]\frac{1}{k}\frac{d\omega}{dk}, \tag{7.87}$$

in which Θ_0 is a function of ω, Θ and h.
Hence:

$$S_2(k, \Theta) = S_1[\omega(k), \Theta]\frac{1}{k}\frac{d\omega}{dk} = S_{1_0}(\omega, \Theta_0) K_H^2\frac{\partial \Theta_0}{\partial \Theta}\frac{1}{k}\frac{d\omega}{dk}, \tag{7.88}$$

or:

$$S_2(k, \Theta) = \frac{1}{2} S_{1_0}(\omega, \Theta_0)\frac{b_0}{b}\frac{g}{\omega}\frac{1}{k}\frac{\partial \Theta_0}{\partial \Theta}. \tag{7.89}$$

In the simple case of a wave arriving at an angle with a parallel bottom
contour, eq. (7.89) implies that:

$$S_2(k, \Theta) = \frac{1}{2} g^2 \omega^{-3} S_{1_0}(\omega, \Theta_0).$$
(7.90)

Krylov et al. (1976) applies the above formula to predict the evolution of mean wave height in the shallow water zone with a parallel bottom contour. The rate of $(\bar{H}/\bar{H}_0)^2$ is (when the relation (6.36) is assuming):

$$\left(\frac{\bar{H}}{\bar{H}_0}\right)^2 = \frac{\int_0^\infty \left[\int_{\Theta_1}^{\Theta_2} f_0(\omega_n, \Theta_0)\, d\Theta\right] \omega_n^{-3} \kappa\, d\kappa}{\int_0^\infty \left[\int_{\Theta_3}^{\pi/2} f_0(\omega_n, \Theta_0)\, d\Theta\right] \omega_n^{-3} \kappa_0\, d\kappa_0},$$
(7.91)

in which:

$$\omega_n = \frac{\omega}{\bar{\omega}_{z_0}}, \qquad \Theta_3 = \Theta_0^* - \frac{\pi}{2},$$
(7.92)

$\bar{\omega}_{z_0}$ - mean frequency corresponding to $\bar{T}_{z_0}$ in deep water,

$$\kappa = \frac{k}{k_0}, \qquad \bar{k}_0 = \frac{\bar{\omega}_{z_0}^2}{g}, \qquad f_0 = \frac{\bar{\omega}_{z_0}}{\bar{H}_0^2} S_2\left[\kappa_0(\kappa), \Theta_0(\kappa, \Theta)\right],$$
(7.93)

and

$$\omega_n^2 = \kappa \tanh(2\pi\kappa h^*), \qquad \omega_n^2 = \kappa_0, \qquad h^* = \frac{h}{\bar{L}_0^*}, \qquad \bar{L}_0^* = \frac{g\bar{T}_{z_0}^2}{2\pi},$$
(7.94)

$$\Theta_0 = \arcsin\left[\frac{\sin\Theta}{\tanh(2\pi\kappa h^*)}\right], \qquad \Theta_0^* - \frac{\pi}{2} = \arcsin\left[\frac{\sin\Theta_1}{\tanh(2\pi\kappa h^*)}\right],$$
(7.95)

$$\frac{\pi}{2} = \arcsin\left[\frac{\sin\Theta_2}{\tanh(2\pi\kappa h^*)}\right].$$
(7.96)

where: Θ_0^* - angle between mean wave direction in the deep water and normal to shoreline.

Similar formulas have been derived for the wave length and wave period. In Table 7.1 the values $K_{\bar{H}}^{(s)} = \bar{H}/\bar{H}_0$, $K_{\bar{L}}^{(s)} = \bar{L}/\bar{L}_0$ and $K_{\bar{T}}^{(s)} = \bar{T}_z/\bar{T}_{z_0}$ are listed for different values of $h/\bar{L}_0^*$ and Θ_0^*. The spectrum in the deep water was assumed in the form (Krylov et al., 1976):

$$\frac{S_1(\omega, \Theta)\bar{\omega}_{z_0}}{\bar{H}_0^2} = \frac{1}{2}\left(\frac{\omega}{\bar{\omega}_{z_0}}\right)^{-7} \exp\left[-0.785\left(\frac{\omega}{\bar{\omega}_{z_0}}\right)^{-4}\right] D(\Theta, \omega_{z_0}),$$
(7.97)

Table 7.1: Mean transformation coefficients $K_H^{(s)}$, $K_L^{(s)}$ and $K_T^{(s)}$ for the shallow water zone with a parallel bottom contour (accor. Krylov et al., 1976).

$\frac{h}{L_0'}$	$K_H^{(s)}$				$K_T^{(s)}$	$K_L^{(s)}$
	0°	30°	60°	90°		
0.04	1.05	1.00	0.90	0.78	1.03	0.55
0.10	0.91	0.88	0.80	0.72	1.00	0.75
0.20	0.90	0.87	0.82	0.76	0.97	0.86
0.30	0.96	0.96	0.94	0.92	0.98	0.95
0.60	0.98	0.97	0.96	0.95	0.99	0.97

and

$$D\left(\Theta, \omega_{z_0}\right) = 2^r \, \frac{\Gamma[2(1+r)]}{\Gamma^2(1+r)} \cos^r \Theta, \tag{7.98}$$

in which: $r = 1.8/\bar{\omega}_{z_0}$.

The values in Table 7.1 can be compared with the transformation coefficients for the regular waves K_H and K_L listed in Table 7.2 (the coefficient $K_T \equiv 1.0$).

Table 7.2: Transformation coefficients K_H and K_L for regular waves in the shallow water zone with a parallel bottom contour.

$\frac{h}{L_0}$	K_H				K_L
	0°	30°	60°	80°	
0.04	1.06	1.00	0.79	0.47	0.48
0.10	0.93	0.90	0.74	0.46	0.71
0.20	0.92	0.91	0.81	0.55	0.89
0.30	0.95	0.94	0.90	0.70	0.96
0.50	0.99	0.99	0.98	0.95	1.00
0.60	1.00	1.00	1.00	0.99	1.00

The comparison indicates that the evolution of the mean length of irregular waves in the coastal zone is very close to that of regular waves. On the other hand, the change of the irregular wave height is substantial. For $\Theta_0^* = 60$, this change is smaller, if compare with the regular waves. When $\Theta_0^* < 30^o$, the trend of change is opposite. As the result, for the various wave directions Θ_0^*, the range of the mean height changing in the shallow water zone is quite small. Thus, the shoaling and refraction have no substantial effect on the wave heights within the shallow water zone, in contrast with the regular waves.

In order to verify the equations previously presented, measurements of the directional function in the Black Sea and the Kaspian Sea can be used (Krylov et al., 1976). The measurements were done at the locations where the bottom contours can be considered straight and parallel (bottom slope $0.005 \div 0.01$) and the wind velocities were in the range 10 - 20 m/s; the angle Θ_0^* lies in the range $0^o \div 90^o$. The comparison of the experimental data with the theoretical curves is illustrated in Fig. 7.4. From the inspection of the Figure, it is seen that the linear transformation theory of wave spectrum is quite well verified. Particularly, the constancy of the wave period is clearly shown. In general, the methods to analyse the problem of wave spectrum change on a sloped beach, outlined above, can be summarized as follows:

a) For the parallel contours, we simply apply the eq. (7.79).

b) In the case of 3–dimensional bathymetry, if one–dimensional shallow water spectrum is needed only, then apply:

$$S(\omega) = K_T^2(\omega, h) \int_{-\pi/2}^{\pi/2} K_R^2(\omega, \Theta_0, h) S_{1_0}(\omega, \Theta_0) d\Theta_0. \qquad (7.99)$$

c) For the 3–dimensional bathymetry, the shallow water directional spectrum is calculated by eqs. (7.81) - (7.82) or (7.83).

d) In the case of 3–dimensional bathymetry, and if the mean wave height is needed only, apply (7.91).

Slowly varying, linear, finite–depth theory is roughly consistent with observations of shoaling wave heights, but some spectral features are apparently due to nonlinear effects. Specially, the processes immediately proceeding wave breaking are essentially nonlinear. Considerable effect has been made in attempts to use Stokes–type perturbation expansions on the equations of motion for waves propagating over a sloping bottom (Le Mehaute and Webb, 1964; Chu and Mei, 1970). Most solutions are based on expansion of the dependent variables in a small parameter equivalent to the Ursell number $U = (H/h)(L/h)^2$. The typical parameters of wind waves in the shoaling region lead to Ursell number of $O(75)$.

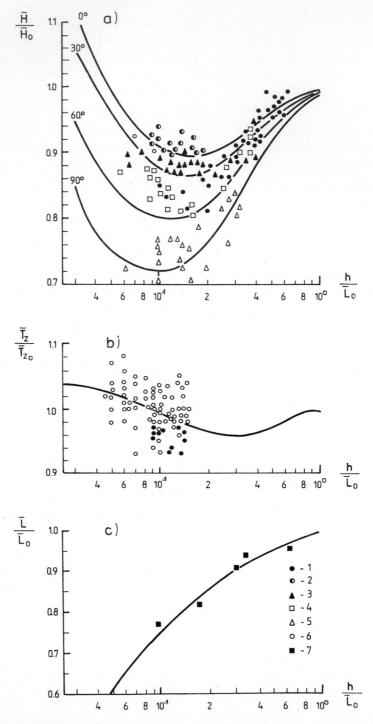

Figure 7.4: Comparison of the experimental data with the theoretical curves for $\bar{H}/\bar{H}_0$, $\bar{L}/\bar{L}_0$ and $\bar{T}_z/\bar{T}_{z_0}$.

On the other hand, this region corresponds to the evoluation equations developed by Boussinesq and Korteweg–de Vries. Both of these equations are containing terms accounting for weak dispersion due to finite depth, and weak nonlinearity due to finite amplitude. Boussinesq–type equations were used by Freilich and Guza (1984) as a basis of two nonlinear models for the evaluation of the wave field in the shoaling region (a consistent shoaling model and dispersive shoaling model). According to Authors, the basic mechanism for the transformation is related to the nonlinear triad interactions across the entire wind wave frequency band.

The theoretical results were compared with experimental data in the shoaling region between depths of 10 and 3 m; the beach slope was 0.022. Power spectral comparison, as well as the coherences between predictions of all models and the data, indicate that the nonlinear models accurately predict Fourier coefficients of the wave field through the shoaling region. On the other hand, the linear theory was considerably less accurate except under broad banded, low energy conditions. Directional effects appear to play little role in the nonlinear evolution of the analysed data sets. Moreover, the influence of the sloping bottom is rather small in determining the spectral evolution of the wave field through the shoaling region, under the consideration.

Elgar and Guza (1985) have extended comparisons of the observed statistics of non–breaking waves and nonlinear models, mentioned above, for the shoaling region between 4 and 1 m depths (beach slope 0.05). Again, the nonlinear model was shown to be generally superior to linear finite–depth theory.

Recently a new concept of the wind wave transformation in a shallow water zone was developed. It was based on the so called parabolic approximation to the mild–slope equation (Isobe, 1987; Izumiya and Horikawa, 1987).

Let us now consider the transformation of significant wave heights. This transformation can be achieved by the linear theory under the assumption of the wave power conservation between locations when the wave reflection and refraction are negligible. The resultant transformed significant wave height in the shallow water depth usually turns out to be larger than the corresponding wave height in deeper water. The linear procedure gives reasonable estimates for the narrow band spectrum. However, Hughes and Miller (1987) have demonstrated that for banded wind sea, the significant wave height is decreasing with decreasing depth. Assuming the validity of TMA spectrum (7.36) and applying Kitaigorodskiy (1983) spectral scaling they have found:

$$\frac{H_{m_0}}{L_p^{3/4}} = const, \qquad (7.100)$$

where: L_p - wave length associated with the spectral peak frequency from the linear dispersion relation.

For any two depths we can write:

$$\frac{(H_{m_0})_1}{(H_{m_0})_2} = \frac{(L_p)_1^{3/4}}{(L_p)_2^{3/4}}. \tag{7.101}$$

It should be noted that the linear transformation can be written as:

$$\frac{(H_{m_0})_1}{(H_{m_0})_2} = \frac{(C_g)_2^{1/2}}{(C_g)_1^{1/2}}. \tag{7.102}$$

The verification of the eqs. (7.101) and (7.102) against the experimental data yields the conclusion that, qualitatively, the eq. (7.101) performs a better (than the linear theory) transformation of the significant wave height. However, the transformation shows a tendency to underpredict H_{m_0} at the longer wave periods ($h/L_0 < 0.1$). Practical usage of eq. (7.101) should be limited to cases conforming to the criteria mentioned above.

When we are approaching near the wave breaking, the wave height H_s can exceed H_{m_0} by 40% . Thompson and Vincent (1985) developed a simple empirical method for interrelating H_s and H_{m_0} with the CERC laboratory data, two samples of field data, and stream function wave theory.

7.4.2   Wind waves in surf zone

In previous analysis, the dissipation effects were totally neglected. Outside the surf zone, the wave decay due to dissipation mechanism (i.e. boundary layer near the bed, percolation in a porous bottom, etc.) is rather weak. However, inside the surf zone, the dissipation of wave energy in the breaking process is dominant. Large storm waves build to larger heights as they approach shore until they crash over, and throwing up a great wall of water which rushes up the beach. The flow in breaking waves is highly complex and does not lead itself to a detailed deterministic treatment. However, in the first approximation the relationship between various characteristic parameters can be assumed to be the same as for regular waves. In particular, many properties of the surf appear to be governed by the parameter ξ_0 defined by Battjes (1974) - see Section 5.2. Usually, we consider the ratio of the wave height to water depth at breaking, which is also an important parameter of the surf zone; thus:

$$\gamma_b = \frac{H_b}{h_b}.$$ (7.103)

According to Galvin (1968), γ_b of 0.8, 1.1 and 1.2 are typical for spilling, plunging and collapsing breakers. Bowen et al. (1968) suggest that γ_b may be a function of ξ_0 only. However, even values presented by various Authors for the same values of ξ_0 show considerable scatter. It may reflect the difficulties and ambiguities inherent in defining experimentally and measuring breaker characteristics. Another factor contributing to the scatter may be appearance of secondary waves. They affect the breaking process in a manner depending on the phase difference with the primary waves.

However, in many applications it is usefull to present the transformation of the characteristic wave height as a function of the initial wave steepness and the beach slope. Let us consider the prediction scheme for the nearshore significant wave height H_s. This scheme is based on the procedure developed by Goda (1985). Goda theory describes the wave heights using a modified Rayleigh distribution in which the tail of the distribution is shortened to represent the decrease in wave height due to breaking. Finally, the breaker height is expressed by:

$$\frac{H_b}{H_0} = A \frac{L_0}{H_0} \left\{ 1 - \exp\left[-1.5 \frac{h}{L_0} (1 + K\beta^s) \right] \right\},$$ (7.104)

in which: β - beach slope, $A = (0.12 \div 0.18)$, $K = 15$ and $s = 4/3$. Seelig (1980) used Goda theory to calculate the maximum wave heights and critical water depth. He defined the $H_{s_{max}}$ value which is the peak value of significant wave height and $H_{1_{max}}$ value which is average of the highest 1% of the waves; it is the wave height with an exceedance probability of approximately 1/260. Goda model predicts that the peak value of H_1 occurs just seaward of $H_{s_{max}}$. Fig. 7.5 shows the design curves for $H_{s_{max}}$ and $H_{1_{max}}$ as a function of deep water wave steepness and beach slope. These curves show that the peak wave heights decrease as the wave steepness increases and the beach slope becomes flatter. Thornton et al. (1984) compared breaking wave heights measured in the field and in random wave experiments in the laboratory with the Goda (1975) model and with Shore Protection Manual (1977). The Goda model reasonably predicts H_s and H_{max} for the higher wave steepness laboratory data. For initially low steepness waves the Goda model over–predicts H_s, but more reasonably predicts H_{max}. Moreover, the depth at breaking corresponding to the breaking wave height compares favorably with the Goda model for all wave steepness values.

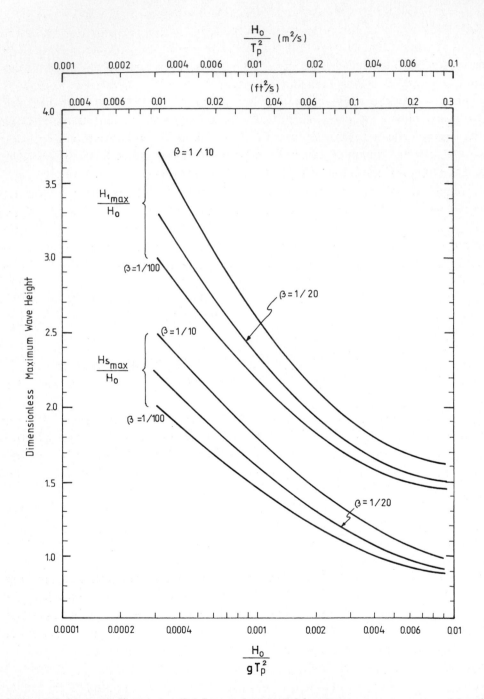

Figure 7.5: Max. significant wave height as a function of deepwater wave steepness and beach slope. (From Seelig, 1980)

As waves start to break, the turbulent dissipation of the wave energy is the dominant dissipative mechanism, and breaking processes dominate the wave transformation. However, in contrast to monochromatic waves, there is no well–defined breakpoint for random waves. Because of the randomness of wind generated waves, the occurence of breaking at a fixed location is itself a random process. There are two classes of random wave shoaling and breaking models. The earlier models describe shoaling as only dependent on the "local" water depth. The second type of model is based on the integration of the energy–balance equation with wave height dependent on the shoaling processes along the integral path starting in deep water.

Applying this approach, Battjes and Janssen (1978) presented a method, in which the local mean rate of energy dissipation is modelled, based on that occuring in a bore and on the local probability of wave breaking. Under the assumption of a steady–state, one–dimensional propagation, and no interaction with the mean motion, the wave energy–balance equation reduces to:

$$\frac{dP_x}{dx} + D = 0, \tag{7.105}$$

where: P_x is x-component of the time–mean energy flux per unit length, x - horizontal coordinate, perpendicular to the shore line, and D - time–mean dissipated power per unit area.

The advantage of this approach compared with earlier models are that other sources or sinks than those due to breaking can be accommodated without difficulty for the arbitrary bottom changes. The energy dissipation in a breaking wave is modelled after that in a bore connecting two regions of uniform flow (see also (4.24)); hence (Battjes and Janssen, 1978):

$$D = \frac{1}{4}\beta f_p \rho g \frac{H^3}{h}. \tag{7.106}$$

The factor β incorporates the effect of the approximations involved and is of order one. The key element in the model is the estimation of the fraction (Q) of waves which at any point are breaking or broken. This is done by assuming that the cumulative probability distribution is of the Rayleigh type, cut off dicontinuously at $H = H_m$. The depth–limited height is given by a Miche type expression (see eq. (5.22)):

$$H_m = 0.88k_p^{-1} \tanh\left(\gamma k_p h / 0.88\right), \tag{7.107}$$

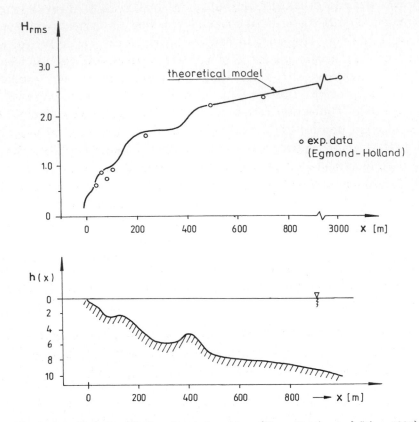

Figure 7.6: Variation of H_{rms} due to breaking. (From Battjes and Stive, 1985)

where: γ - parameter to account for influence of bottom slope and incident wave steepness.

The relation between Q and H_{rms}/H_m, is:

$$\frac{1-Q}{-\ln Q} = \left(\frac{H_{rms}}{H_m}\right)^2. \tag{7.108}$$

Substituting the approximations mentioned above in the average relation (7.106), gives:

$$\tilde{D} = \frac{1}{4}\beta Q f_p \rho g H_m^2, \tag{7.109}$$

where: f_p - peak frequency.

For known depth profile $h(x)$, incident wave values for f_p and H_{rms}, and estimates of the coefficients β and γ, eq. (7.105) with $D = \tilde{D}$ can be integrated numerically to obtain the variation of H_{rms}. It should be noted that the mean depth (h) is the sum of the bottom depth below still water level and a wave–induced set–up. The verification of the model against the empirical data indicates that the model predicts the r.m.s. wave height variation due to breaking very well in a wide range of conditions as far as bottom profile and incident waves are concerned. This is illustrated in Fig. 7.6 for the field data (Battjes and Stive, 1985). The applicability of the model was also supported by Thornton and Guza (1983) who presented some refinement of the model and found that r.m.s. relative error was about 9% . Therefore, the model can be used with confidence for purposes of prediction, using a parametrization of its single adjustable coefficient in terms of wave steepness.

7.5   References

Banach, L., 1985. Parametryzacja widma falowania wiatrowego w strefie brzegowej morza na przykladzie rejonu Lubiatowa. *Rozpr. Hydrot.*, 47: 97–115 (in Polish).

Battjes, J.A., 1974. Computation of set-up, long-shore currents, run-up and overtopping due to wind-generated waves. Communications on Hydraulics, Dept. Civil Eng., Delft Univ. of Technology, Rep. 74–2, 244 pp.

Battjes, J.A. and Janssen, J.P.F.M., 1978. Energy loss and set-up due to breaking of random waves. *Proc. 16th Coastal Eng. Conf.*, 1: 569–587.

Battjes, J.A. and Stive, M.J.F., 1985. Calibration and verification of a dissipation model for random breaking waves. *Jour. Geoph. Res.*, 90: 9159–9167.

Battjes, J.A., Zitman, T.J. and Holthuijsen, L.H., 1987. A reanalysis of the spectra observed in JONSWAP. *Jour. Phys. Oceanogr.*, 17: 1288–1295.

Belberov, Z.K., Zhurbas, V.M., Zaslavskiy, M.M. and Lobysheva, G., 1983. Integralnye kharakteristiki chastotnykh spektrov vetrovykh voln. In: *Vzaimodeystviye atmosfery, gidrosfery i litosfery v pribrezhnoy zone morya. Rezultaty mezhdunarodnogo eksperimenta "Kamchiya 79 ".* Publ. House of Bulg. Acad. Sciences, Sofia, pp. 143–166 (in Russian).

Berkhoff, J.C., 1972. Computation of combined refraction–diffraction. *Proc. 13th Int. Conf. Coastal Eng.*, 2: 471–490.

Bouws, F., Gunther, H., Rosenthal, W. and Vincent, C.L., 1985. Similarity of the wind wave spectrum in finite depth water: 1. Spectral form. *Jour. Geoph. Res.*, C1-90: 975–986.

Bowen, A.J., Inman, D.L. and Simons, V.P., 1968. Wave "set-down" and "set-up". *Jour. Geoph. Res.*, 73: 2569–2577.

Bretschneider, C.L., 1958. Revisions in wave forecasting: deep and shallow water. *Proc. 6th Coastal Eng. Conf.*, 1: 30–67.

Cartwright, D.E. and Longuet–Higgins, .S., 1956. The statistical distribution of the maxima of a random function. *Proc. Roy. Soc. London*, A.237: 212–232.

Chu, V.H., and Mei, C.C., 1970. On slowly-varying Stokes waves. *Jour. Fluid Mech.*, 41: 873–887.

Collins, I.L., 1972. Prediction of shallow water spectra. *Jour. Geoph. Res.* 17: 2693–2707.

Davidan, I.N., Lopatukhin, L.I. and Rozhkov,W.A., 1985. *Vetrovoye volneniye w mirovom okeane*. Gidrometeoizdat, Leningrad, 256 pp. (in Russian).

Druet, Cz., Massel, S. and Zeidler, R., 1972. Statystyczne charakterystyki falowania wiatrowego w przybrzeznej strefie Zatoki Gdanskiej i otwartego Baltyku. *Rozpr. Hydrot.*, 30: 49–84 (in Polish).

Elgar, S. and Guza, R.T., 1985. Shoaling gravity waves: comparison between field observations, linear theory, and a nonlinear model. *Jour. Fluid Mech.*, 158: 47–70.

Freilich, M.H. and Guza, R.T., 1984. Nonlinear effects on shoaling surface gravity waves. *Phil. Trans. Roy. Soc. London*, A311: 1–41.

Galvin, C.J., 1968. Breaker type classification on three laboratory beaches. *Jour. Geoph. Res.*, 73: 3651–3659.

Goda, Y., 1970. Numerical experiments on wave statistics with spectral simulation. Port. Harbour Res. Inst. Rep., 9: 3–57.

Goda, Y., 1975. Irregular wave deformation in the surf zone. *Coastal Eng. in Japan*, 18: 13–26.

Goda, Y., 1985. *Random sea and design of maritime structures*. Univ. Tokyo Press, 323 pp.

Hasselmann, K., Barnett, T.P., Bouws, E., Carlson, H., Cartwright, D.E., Enke, K., Ewing, J.A., Gienapp, H., Hasselmann, D.E., Kruseman, P., Meerburg, A., Muller, P., Olbers, D.J., Richter, K., Sell, W. and Walden, H., 1973. Measurements of wind–wave growth and swell decay during the Joint North Sea Wave Project (JONSWAP). *Deutch. Hydr. Zeit.,* Reihe, A12, 95 pp.

Huang, N.E., Wang, H., Long, S.R. and Bliven, L.F., 1983. A study on the spectral models for waves in finite water depth. *Jour. Geoph. Res.,* 88: 9579–9587.

Hughes, S.A. and Miller, H.C., 1987. Transformation of significant wave heights. *Proc. ASCE, Jour. Waterway, Port, Coastal and Ocean Eng.,* 113: 588–605.

Isobe, M., 1987. A parabolic equation model for transformation of irregular waves due to refraction, difraction and breaking. *Coastal Eng. in Japan,* 30: 121–135.

Izumiya, T., and Horikawa, K., 1987. On the transformation of directional random waves under combined refraction and difraction. *Coastal Eng. in Japan,* 30: 49–65.

Kitaigorodskiy, S.A., 1970. *Fizika vzaimodeystviya atmosfery i okeana.* Gidrometeoidat, Leningrad, 284 pp. (in Russian).

Kitaigorodskiy, S.A., 1983. On the theory of the equilibrum range in the spectrum of wind-generated gravity waves. *Jour. Phys. Oceanogr.,* 13: 816–827.

Kitaigorodskiy, S.A., Krasitskiy, V.P. and Zaslavskiy, M.M., 1975. On Phillips' theory of equilibrum range in the spectra of wind-generated gravity waves. *Jour. Phys. Ocean.,* 5: 410–420.

Krasitskiy, W.P., 1974. K teorii transformatsii spektra pri refraktsii vetrovykh voln. *Fizika Atm. i Okeana,* X: 72–82 (in Russian).

Krylov, Y.M., 1966. *Spektralnye metody issledovaniya i raschyeta vetrovykh voln.* Gidrometeoizdat, Leningrad, 254 pp. (in Russian).

Krylov, Y.M., Strekalov, S.S. and Tsyplukhin, W.F., 1976. *Vetrovye volny i ikh vozdeystviye na sooruzheniya.* Gidrometeoizdat, Leningrad, 255 pp. (in Russian).

Krylov, Y.M. (Editor), 1986. *Veter, volny i morskiye porty.* Gidrometeoizdat, Leningrad, 264 pp. (in Russian).

Le Mehaute, B. and Webb, L.M., 1964. Periodic gravity waves over a gentle slope at a third order of approximation. *Proc. 9th Coastal Eng. Conf.*, I: 23–40.

Le Mehaute, B. and Wang, J.D., 1982. Wave spectrum changes on sloped beach. *Proc. ASCE, Jour. Waterway, Port, Coastal and Ocean Eng.*, 108: 33–47.

Liu, P.C., 1985. Representation frequency spectra for shallow water waves. *Ocean Eng.*, 12: 151–160.

Liu, P.C., 1987. Assessing wind wave spectrum representations in a shallow lake. *Ocean Eng.*, 14: 39–50.

Longuet–Higgins, M.S. 1975. On the joint distribution of the periods and amplitudes of sea waves. *Jour. Geoph. Res.*, 80: 2688–2694.

Massel, S., 1985. Nonlinear statistical and spectral characteristics of wind-induced waves in coastal waters. *Proc. Inter. Conf." Water waves research. Theory, laboratory and field "*, Hannover, pp. 285–318.

Massel, S., Kostichkova, D. and Cherneva, Z., 1980. Parametrizatsya spektrov vetrovogo volneniya na vkhode v beregovuyu zonu. In: *Vzaimodeystviye atmosfery, gidrosfery i litosfery v pribrezhnoy zone morya. Rezultaty mezhdunarodnogo eksperimenta "Kamchiya 77"*. Izd. Bulg. Akad. Nauk, pp. 173–179, (in Russian).

Ochi, M.K. and Hubble. E.N., 1976. On six-parameter wave spectra. *Proc. 15th Coastal Eng. Conf.*, 1: 301–328.

Pierson, W.J. and Moskowitz, L.A., 1964. A proposed spectral form for fully developed sea based on the similarity theory of S.A. Kitaigorodskii. *Jour. Geoph. Res.*, 69: 5181–5190.

Phillips, O.M., 1977. *The dynamics of the upper ocean.* Cambridge Univ. Press, 336 pp.

Phillips, O.M., 1985. Spectral and statistical properties of the equilibrium range in wind-generated gravity waves. *Jour. Fluid Mech.*, 156: 505–531.

Seelig, W.N., 1980. Maximum wave heights and critical water depths for irregular waves in the surf zone. U.S. Army Corps of Eng., Coastal Eng. Research Center, 80–1, 11 pp.

Shiau, I.C. and Wang, H., 1977. Wave energy transformation over irregular bottom. *Proc. ASCE, Jour. Waterway, Port, Coastal and Ocean Eng.*, 103: 57–68.

Shore Protection Manual, 1977. U.S. Army Coastal Eng. Res. Center, Washington, I–III.

Strekalov, S.S. and Massel, S.R., 1971. Niektore zagadnienia widmowej analizy falowania wiatrowego. *Arch. Hydrot.,* XVIII: 457–485 (in Polish).

Thompson, E.F. and Vincent, .L., 1985. Significant wave height for shallow water design. *Proc. ASCE, Jour. Waterway, Port, Coastal and Ocean Eng.,* 111: 828–842.

Thornton, E.B. and Guza, R.T., 1983. Transformation of wave height distribution. *Jour. Geoph. Res.,* 88: 5925–5938.

Thornton, E.B., Wu, C.S. and Guza, R.T., 1984. Breaking wave design criteria. *Proc. 19th Coastal Eng. Conf.,* 1: 31–41.

Toba, Y., 1973. Local balance in the air–sea boundary process. III: On the spectrum of wind waves. *Jour. Oceanogr. Soc. Japan,* 29: 209–220.

Tucker, M.J., 1963. Analysis of records of sea waves. *Proc. Inst. Civil. Eng.,* 26: 305–316.

Vincent, C.L., 1984. Shallow water waves: a spectral approach. *Proc. 19th Coastal Eng. Conf.,* I: 370–382.

Vincent, C.L., 1985. Depth-controlled wave height. *Proc. ASCE, Jour. Waterway, Port, Coastal and Ocean Eng.,* 111: 459–475.

Vincent, C.L. and Hughes, S.A., 1985. Wind wave growth in shallow water. *Proc. ASCE, Jour. Waterway, Port, Coastal and Ocean Eng.,* 111: 765–770.

Zaslavskiy, M.M. and Zakharov, W.E., 1982. K teorii prognoza vetro-vykh voln. *Doklady Akademii Nauk SSSR. Geofizika,* 265: 567–571, (in Russian).

Chapter 8

CURRENTS IN COASTAL ZONE

8.1 Currents classification

In all water wave problems, discussed in the previous Chapters, some approximations were made to find mathematical solutions in order to gain physical understanding. Almost always the current speed was supposed to be equal to zero or uniform with depth and not varying with distance. Moreover, the discussion of the currents generation mechanisms has been omitted. However, the obvious effect of wind stress on the sea is that the wind also produces a steady movement of the surface layer of water in the same general direction as the wind. This *wind–driven current* constitutes of one component of the oceanic circulation. In the proximity of a coast line, the movement of water will be restricted, leading to a rising of water and the development of surface slopes which may extend away from the coast. The surface slopes give rise to horizontal pressure gradient in the water and this in turn generates currents, so called *gradient currents*. Moreover, the wave–induced current that flows alongshore and is generally confined between the first breaker and the shoreline is called the *longshore current*. The typical values of the longshore current velocity are well under one meter per second and the highest velocity is observed at the breaker line. Field experiments have generally recognized the variability of longshore currents across the surf zone and down the coast. It should be noted that the longshore current plays an important role in the generation of the sediment transport.

Another wave induced current motion, which also exists over a horizontal bed is the so–called *mass transport*. This is a second–order effect, so the current velocity increases rapidly with wave height. The flow across the surf zone produces a tilt of the mean water surface from the breakers to the shore. Also longshore variations in the breakers height create gradients of the mean water level within the surf zone that generate currents flowing from positions of highest breaker height to positions of lowest breakers. Here, the longshore currents converge and turn seaward as *rip currents*. Rip currents are narrow

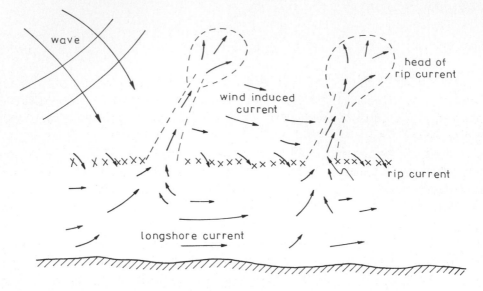

Figure 8.1: Nearshore classification system.

strong return flows directed through the surf to sea. Together with longshore currents they can create a two–dimensional coastal current system within and beyond the surf zone, i.e.: *near–shore circulation* (Basco, 1982). A series of rip currents is usually found along the coast with longshore currents feeding the rips and forming independent circulation cells. Spacings have been reported as 30 to 400 m. The crude estimation of the width of rip currents gives the values as 10 to 30 m, in the region of rip head (Fig. 8.1).

The gravitational attraction of the moon and the sun generate the tides. In the case ot tides in coastal seas, the influence of these forces is largely indirect. The currents associated with the rising or falling of the sea surface are termed the *tidal currents*. The relation between elevation and current is straightforward in a gulf or estuary and this relation is less clear off an open coast or well away from land.

Let us consider now the particular constituents of the near–shore circulation in detail.

8.2 Wind currents in coastal zone

The wind generated currents observed in the coastal zone are driven first of all by oceanic currents. At present, the current fields are modelling by the numerical methods. In some cases the numerical procedure is simplifying considerably. This is particularly true for the shallow sea for which we have

(Felzenbaum, 1976):

$$(ah)^2 \ll 1.0, \tag{8.1}$$

where: $a = (\rho f/2A_z)^{1/2}$, $f = 2\Omega_z \sin \varphi$ - the Coriolis parameter, Ω - earth angular rate of rotation ($= 7.29 \cdot 10^{-5}$ radians per second), φ - latitude, positive to the north of the equator, A_z - coefficient of eddy diffusion in the z direction.

The depth h follows from the bottom bathymetry in the case of uniform, well–mixed sea. However, a seasonal thermocline in summer months occurs in temperate and higher latitudes. Whether or not a thermocline develops in a particular area it may be expected to depend on the strength of the tidal currents and the depth of water. Where a thermocline develops, the vertical flux of heat is restricted because of the reduction in turbulent mixing. Whereas in vertically mixed water the whole depth is available, in stratified water, the effective depth h in eq. (8.1) is only that of the surface layer (Bowden, 1983). From eq. (8.1) follows that for the wind speed higher than 5 m/s, the sea can be treated as shallow till water depth $h = 25$ m. For example, this value corresponds to the mean thickness of the mixing layer in the Baltic Sea. After averaging of the current velocity in depth and time, we get (Felzenbaum, 1976):

$$\left. \begin{aligned} U_1 &= a\,\tau_x + b\,\frac{\partial \bar{\zeta}}{\partial x} \\[2mm] U_2 &= a\,\tau_y + b\,\frac{\partial \bar{\zeta}}{\partial y} \end{aligned} \right\}, \tag{8.2}$$

in which: a, b - coefficients depending on water depth h and shearing stress vector $\vec{\tau}$.

The water depth is assumed to be constant; therefore in this model the coastal zone is not taken into account. The numerical solution of eqs. (8.2) is given elsewhere (Felzenbaum, 1976; Druet, 1978) and therefore will be omitted here.

In the coastal zone we use equations similar to eqs. (8.2). When the izobats are parallel to the y - axis, $h = h(x)$, the simplified equations can be presented in the form (Shadrin, 1972):

$$\left. \begin{aligned} A_z\,\frac{\partial^2 U_2}{\partial z^2} &= 0 \\[2mm] A_z\,\frac{\partial^2 U_1}{\partial z^2} &= \frac{\partial p}{\partial x} \end{aligned} \right\}. \tag{8.3}$$

Eq. (8.3) is completed with the continuity equation:

$$\frac{\partial U_1}{\partial x} + \frac{\partial W}{\partial z} = 0, \tag{8.4}$$

and hydrostatic approximation:

$$p = \rho g \int_z^\zeta dz = \rho g (\zeta - z). \tag{8.5}$$

If τ_x, τ_y are the components of shearing stress, and they are related to the velocity components U_1, U_2 at the sea surface by:

$$\tau_x = -A_z \frac{\partial U_1}{\partial z}, \qquad \tau_y = -A_z \frac{\partial U_2}{\partial z} \qquad \text{at} \qquad z = 0. \tag{8.6}$$

In the shallow water, the steady state can be achieved when the energy dissipated at the bottom is equal to the energy supplied by the wind (energy dissipation in the fluid layer is neglected). Under the assumption that stress at the sea bottom is represented by the square of the velocity, for x - direction we get:

$$\tau_x = c_f \rho U_1^2 \qquad \text{at} \qquad z = -h, \tag{8.7}$$

in which: c_f is a drag coefficient.
The currents within the shallow water zone are related to the flow outside of the coastal region. Thus, the conditions at the seaward boundary are:

$$\left.\begin{array}{llll} U_1 = U_1^{(max)}, & U_2 = U_2^{(max)} & \text{at} & h = h_g \\[2mm] U_1 = 0, & U_2 = 0 & \text{at} & h = 0 \end{array}\right\}. \tag{8.8}$$

We consider now the right–handed rectangular coordinate system with the origin at the water line and the x and y axes horizontal and the z axis vertically upwards. Moreover, the water depth is assumed to be:

$$h(x) = \beta x, \tag{8.9}$$

where: β - bottom slope.
According to Shadrin (1972), the solution for current velocities can be expressed as:

$$U_1(z) = \frac{\tau_x}{4A_z}\left(h + 4z + \frac{3z^2}{h}\right) - \frac{1}{2}\sqrt{\frac{\tau_x}{\rho c_f}}\left(1 - 3\frac{z^2}{h^2}\right), \tag{8.10}$$

$$U_2(z) = \frac{\tau_y}{A_z}(h + z) + \sqrt{\frac{\tau_y}{\rho c_f}}, \tag{8.11}$$

$$W(z) = z\beta\left[\frac{1}{4}\frac{\tau_x}{A_z}\left(1 - \frac{z^2}{h^2}\right) + \sqrt{\frac{\tau_x}{\rho c_f}\frac{z^2}{h^2}}\right]. \tag{8.12}$$

8.3 Longshore currents induced by waves

8.3.1 Radiation stress concept

The longshore currents play a very important role in the hydrodynamics of the shallow water zone. In the theory of longshore currents, two different approaches have been developed. The first one is related to the radiation stress theory (Longuet–Higgins and Stewart, 1962) while the other is based on the Boussinesq equation. In both the approaches the mass and momentum conservation laws are used; the velocity profile is assumed to be uniform over the water depth. The term "radiation stress" comes from EM wave theory where a radiation pressure impinges on a surface. It should be noted that the radiation stresses resulting from time–averaging wave orbital motions are completely analogous to Reynolds stresses resulting from time–averaging turbulent flow motions. In this Chapter, the radiation stress principle will be used to develop the expressions for the various coastal phenomena:

- longshore current profiles,
- wave set–down and set–up,
- nearshore circulation systems and rip currents.

Consider wave propagating in X - axis direction (Fig. 8.2). The time–average value of the total flux of horizontal momentum across a vertical plane minus the force resulting from the still water hydrostatic pressure can be expressed as:

$$S_{XX} = \overline{\int_{-h}^{\zeta}(p + \rho u^2)\,dz} - \int_{-h}^{0} p_0 dz, \tag{8.13}$$

where p_0 is a still water hydrostatic pressure.

Eq. (8.13) can be rearranged as follows (Basco, 1982):

$$S_{XX} = \int_{-h}^{0}\overline{\rho\left(u^2 - w^2\right)}\,dz + \frac{1}{2}\rho g\overline{\zeta^2}. \tag{8.14}$$

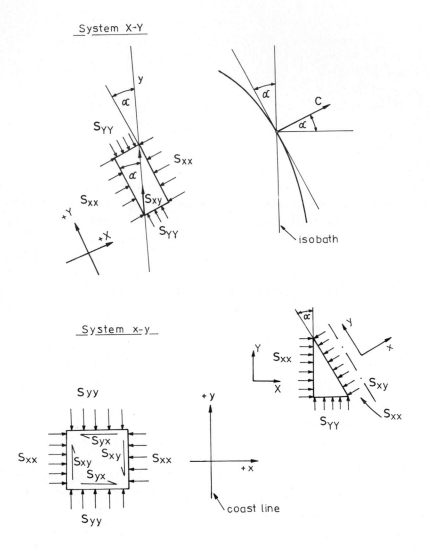

Figure 8.2: Definitions for radiation stress components

Analogous to eq. (8.14), we define the transverse stress component as:

$$S_{YY} = \int_{-h}^{0} \overline{\rho w^2} \, dz + \frac{1}{2} \rho g \overline{\zeta^2}. \tag{8.15}$$

Here, the orbital velocity v in the YY - direction, parallel to the wave crests equals zero. Therefore, the shear component is identically zero:

$$S_{XY} \equiv \int_{-h}^{\zeta} \overline{\rho u v} \, dz = 0. \tag{8.16}$$

In the coordinate system (O, X, Y), the S_{XX} and S_{YY} are principal stresses and they represent the horizontal forces per unit width acting normal to and parallel to the wave crest, respectively.
It will be more convenient to rewrite the radiation stresses components in the coordinate system oriented in the alongshore (y - coordinate) and shore–normal (x - coordinate) directions. Therefore, the coordinate transformation gives:

$$S_{xx} = \frac{1}{2}(S_{XX} + S_{YY}) + \frac{1}{2}(S_{XX} - S_{YY})\cos 2\alpha, \tag{8.17}$$

$$S_{yy} = \frac{1}{2}(S_{XX} - S_{YY}) - \frac{1}{2}(S_{XX} - S_{YY})\cos 2\alpha, \tag{8.18}$$

$$S_{xy} = \frac{1}{2}(S_{XX} - S_{YY})\sin 2\alpha, \tag{8.19}$$

where α is the angle between the wave crest and the shoreline. The S_{xy} is the shear stress component in the longshore direction due to the excess momentum flux of oblique wave incidence.
After substituting the particle orbital velocities u and w and surface elevation ζ as defined by the linear wave theory we get (Basco, 1982):

$$S_{XX} = E\left(2m - \frac{1}{2}\right), \qquad S_{YY} = E\left(m - \frac{1}{2}\right). \tag{8.20}$$

Thus:

$$
\left.
\begin{aligned}
S_{xx} &= E\left(\frac{3}{2}m - \frac{1}{2}\right) + \frac{1}{2}Em\cos 2\alpha \\[2mm]
S_{yy} &= E\left(\frac{3}{2}m - \frac{1}{2}\right) - \frac{1}{2}Em\cos 2\alpha \\[2mm]
S_{xy} &= Em\sin\alpha\cos\alpha
\end{aligned}
\right\},
\tag{8.21}
$$

where: E - total wave energy, m - parameter given by (2.57).
The computation of the higher order radiation stresses using the third–order Stokes theory or cnoidal wave theory indicates that the higher order theories give similar values as the linear theory (James, 1974).

Waves approaching a sloping coastline will undergo modifications in the wave parameters. Therefore, the spatial changes in the radiation stress components must result. For the steady–state situation, the equilibrium will be established for the time–averaged water level and time–averaged currents present. We restrict our analysis to the plain sloping beach with bottom contours parallel to the wave crests (Fig. 8.3). The momentum balance between a rate of change of the radiation stress and a rate of mean water level change is now expressed in the form:

$$
\frac{dS_{xx}}{dx} + \rho g\left(h + \bar{\zeta}\right)\frac{d\bar{\zeta}}{dx} = 0,
\tag{8.22}
$$

where: $\bar{\zeta}$ - change of the mean water level against the still water level.
A more detail information on the mean water level changes will be given in the next Chapter. At present we only note that over a plain sloping beach, seaward of the breakers $\bar{\zeta}$ is (Longuet–Higgins and Stewart, 1962):

$$
\bar{\zeta} = -\frac{1}{8}\frac{kH^2}{\sinh(2kh)}.
\tag{8.23}
$$

Thus a lowering of the mean water level below SWL takes place since $\bar{\zeta}$ is negative. Shoreward of the breakers, the reduction of wave height produces a negative gradient of S_{xx}, which should be balanced by a positive gradient of $\bar{\zeta}$ in the x - direction. This is called wave *set–up* (see Fig. 8.3). However, we leave the detailed discussion on mean water level changes to the next Chapter.

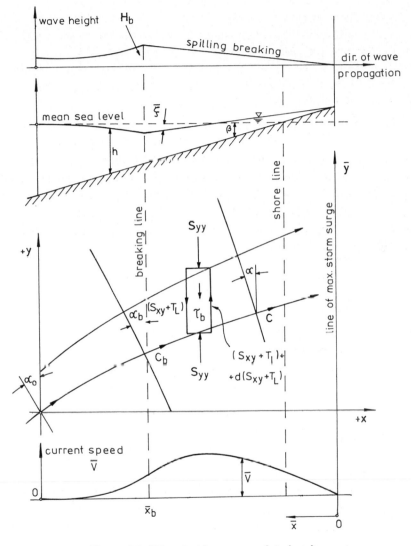

Figure 8.3: Wave incidence on a plain beach

8.3.2   Longshore currents

There is a large number of factors influencing longshore currents. In order to
allow theoretical treatment, the complexity of the forcing field, geometry, and
fluid must be reduced. Therefore, we assume that the wave field is represented
by a steady monochromatic gravity wave train. These waves are long–crested
and obliquely incident on the beach. Within the surf zone the breaker ratio
is assumed constant. Moreover, the assumptions on the beach conditions can
be summarized as follows:

- infinite length, straight and parallel contours,
- plane bottom slope,
- gentle slope,
- impermeable bottom.

The resulting longshore currents are depth–integrated, and time–averaged.
Additional, simplifying assumptions are related to neglecting the following
mechanisms:

- surface wind stresses,
- Coriolis acceleration,
- tides,
- bed shear stresses outside the surf zone, and
- wave–current interaction stresses.

Under the above assumptions there is a balance of forces such that:

driving force + bottom friction + lateral friction = 0.

For a plane, infinite beach there is no gradient in the radiation stress component S_{yy} in the y - direction. However, the mean fluxes of y - momentum due
to the waves, across two parallel boundaries of a differential element are not
the same. This is because all factors $(E, m$ and $\alpha)$ in eq. (8.21) for S_{xy} vary
in the x - direction. It follows that the waves produce a net flux of momentum given by the gradient dS_{xy}/dx. This driving stress in the y - direction is
resisted by the bed shear stress, $\bar{\tau}_b$, and the lateral shear force over the total
depth, T_L, due to turbulent mixing (i.e. wave orbital velocity interactions in
the x - and y - directions). Hence, the momentum balance in y - direction
becomes (Longuet–Higgins, 1970):

$$\frac{dS_{xy}}{dx} - \bar{\tau}_b + \frac{dT_L}{dx} = 0. \tag{8.24}$$

The longshore current velocity, U_2, apears in the bed shear–stress term $\bar{\tau}_b$,
and in the lateral shear force term T_L. After substitution of the appropriate

expressions for these quantities and for S_{xy}, it is possible to integrate eq. (8.24) and derive an expression for the distribution of longshore current $U_2(x)$ across the nearshore zone.

A uniform longshore current profile was first presented by Bowen (1969), Thornton (1970) and Longuet–Higgins (1970), all working independently. The original theory essentially neglect the cross–shore effects such as set–up. Moreover, the longshore current and wave incident angle are assumed small to linearize the bed shear stress term. At present, many modifications of the original theory are known. In general, they are attempting to improve the generality and accuracy of the first proposed models by the inclusion of the arbitrary beach profile, and wave set–up and set–down, and various lateral mixing shear stress formulations. In the following we present the modified model developed by Kraus and Sasaki (1979). Prior to summarize the main results, we introduce a new coordinate system where the origin is taken at the maximum set–up line and a positive $\bar{x}$ facing seaward (see Fig. 8.3). This simply gives the longshore current $U_2 = 0$ at $\bar{x} = 0$. The breaker position is $\bar{x}_b$. In the system $0(\bar{x}, \bar{y})$, the driving stresses take the form:

outside surf zone

$$\frac{dS_{xy}}{d\bar{x}} = 0, \tag{8.25}$$

inside surf zone

$$\frac{dS_{xy}}{d\bar{x}} = \frac{5}{16} \rho g \gamma_b^2 \left(\frac{h^3}{h_b}\right)^{1/2} \left(1 + \frac{3}{8} \gamma_b^2\right)^{-1} \tan \beta \sin \alpha_b \cdot$$

$$\cdot \left[\left(1 - \frac{h}{h_b} \sin^2 \alpha_b\right)^{1/2} - \frac{\sin^2 \alpha_b}{5} \frac{h}{h_b} \left(1 - \frac{h}{h_b} \sin^2 \alpha_b\right)^{-1/2}\right], \tag{8.26}$$

in which: α_b - angle between the crest and izobat lines at the breaking point. It should be noted that after neglecting wave set–up, wave refraction and assuming small α_b, eq. (8.26) reduces to that developed by Longuet–Higgins (1970), i.e.:

$$\frac{dS_{xy}}{d\bar{x}} = \frac{5}{16} \rho \, \gamma_b^2 (gh)^{3/2} \left(\frac{\sin \alpha}{C}\right) \frac{dh}{d\bar{x}}. \tag{8.27}$$

in which: $\gamma_b = (H/h)_b$.

In order to define the time–average bottom shear stress, $\bar{\tau}_b$, Longuet–Higgins (1970) made an assumption that instantaneous bottom shear stress τ_b can be presented as:

$$\vec{\tau}_b = \rho\, c_f\, |\vec{u}_t|\, \vec{u}_t, \tag{8.28}$$

where: $\vec{u}_t = \vec{u}_b + \vec{U}$.
The instantaneous bottom shear stress $\vec{\tau}_b$ is assumed to be in the direction of the resultant velocity $\vec{u}_t$ of the vector sum of longshore current velocity $\vec{U}(0, U_2, 0)$ and bottom wave orbital velocity $\vec{u}_b$. We assume that the long-shore current velocity U_2 is small in comparison with the wave orbital velocity, u_b. Hence, the wave incident angle α_b is also small, so that $\vec{U}$ is roughly nor-mal to $\vec{u}_b$. Moreover, using linear theory to obtain u_b in shallow water, the time–averaged bottom shear stress becomes:

$$\bar{\tau}_b = \frac{1}{\pi}\rho c_f \gamma_b (g\bar{x}\tan\beta)^{1/2} U_2. \tag{8.29}$$

As observed in the laboratory and field for the very gentle slope, it is the occurence of relatively large wave incidence angle α_b that drives the longshore current. Taking this fact Kraus and Sasaki (1979) have extended the weak current large–angle model, developed by Liu and Dalrymple (1978), to include lateral mixing stresses. Near the breaker line, the effects of both lateral mixing and large incident wave angle are comparable. The final formula for time–averaged bottom shear stresses can be presented as:

$$\bar{\tau}_b = \frac{1}{\pi}\rho c_f U_2 (gh)^{1/2} \begin{cases} \gamma_b\left(1 + \dfrac{h}{h_b}\sin^2\alpha_b\right), & \bar{x} \le \bar{x}_b \\[2ex] \dfrac{H}{h}\left(1 + \dfrac{h}{h_b}\sin^2\alpha_b\right), & \bar{x} > \bar{x}_b \end{cases} \tag{8.30}$$

Outside the breakers, the wave height is approximated from the long wave theory. Thus:

$$H \approx \left(\frac{h_b}{h}\right)^{1/4} H_b. \tag{8.31}$$

Note that the stress $\bar{\tau}_b$ takes the no–zero value also outside the surf zone and depends on the relative directions of the U_2 and u_b velocities.

Theoretical knowledge regarding the third term in the balance of force, mentioned above, i.e. the turbulent mixing processes, is very weak. The turbulence length scales here are based on the water depth or the surf zone width. Under the assumption that velocity fluctuations $(\tilde{u}_1, \tilde{u}_2)$ in the $\bar{x}$ and

$\bar{y}$ directions are due to wave orbital motions, then the lateral turbulent eddy stress is:

$$\bar{\tau}_L = h^{-1} T_L = -\rho \overline{\tilde{u}_1 \tilde{u}_2} = \mu_L \frac{dU_2}{d\bar{x}} = \rho \nu_L \frac{dU_2}{d\bar{x}}, \qquad (8.32)$$

in which: $\bar{\tau}_L = h^{-1} T_L$, μ_L - lateral eddy viscosity. Note that tilde symbol means velocity fluctuations on the scale of wave motion.
It is more convenient to work with the kinematic eddy viscosity, i.e.:

$$\mu_L = \rho \nu_L = \rho |\tilde{u}_1| l, \qquad (8.33)$$

where Prandtl mixing length (l) hypothesis was used and $\tilde{u}_1$ is the reference velocity. Thus, $\tilde{u}_2 = l(dU_2/dx)$ and $\bar{\tau}_L = -\rho \overline{\tilde{u}_1 l}(dU_2/dx)$.
Following the Kraus and Sasaki (1979) we assume that the characteristic velocity $\tilde{u}_1$ is proportional to the maximum orbital velocity u_{max}, while the characteristic length scale l is proportional to the distance $\bar{x}$. Therefore from eq. (8.33) we have:

$$\nu_L = \Gamma \bar{x} \, u_{max} = \frac{\pi}{2} \Gamma \bar{x} \sqrt{gh} \begin{cases} \gamma_b & \text{inside breaker line} \\ \\ \dfrac{H}{h} & \text{outside breaker line} \end{cases} \qquad (8.34)$$

The coefficient Γ is similar to the coefficient N introduce by Longuet–Higgins (1970).
Finally, the time–averaged lateral mixing force T_L takes the form:

$$T_L = h \, \tau_L = h \, \rho \, \nu_L \frac{dU_2}{d\bar{x}}. \qquad (8.35)$$

It should be noted that the application of various lateral mixing models does not produce substantial differences in the longshore velocity profiles (McDougal and Hudspeth, 1986).

Substituting the particular terms into the force balance equation (8.24) we get the longshore current profile in the nondimensional form:

$$V = \frac{U_2}{U_{2b}}, \qquad X = \frac{\bar{x}}{\bar{x}_b}, \qquad (8.36)$$

$$V = \begin{cases} \sum_{n=0}^{\infty} (A_n X + B_n X^p) X^n, & 0 < X < 1.0 \\ \\ \sum_{n=0}^{\infty} C_n X^{q-n}, & 1 < X < \infty \end{cases} \tag{8.37}$$

in which:

$$A_n = \begin{cases} \left(1 - \frac{5}{2}P\right)^{-1}, & n = 0 \\ \\ \dfrac{a_n - b^2 A_{n-1}}{1 - (n+1)(n+\frac{5}{2})P}, & n = 1, 2, \ldots \end{cases} \tag{8.38}$$

$$a_n = \begin{cases} 1.0 & n = 0 \\ \\ -0.2 \dfrac{(2n+5)(2n-3)!! \, b^{2n}}{2^n \, n!}, & n = 1, 2, \ldots \end{cases} \tag{8.39}$$

$$B_n = B_0 \begin{cases} 1.0, & n = 0 \\ \beta_n, & n = 1, 2, \ldots \end{cases} \qquad C_n = C_0 \begin{cases} 1.0, & n = 0 \\ \delta_n, & n = 1, 2 \end{cases} \tag{8.40}$$

$$b = \sin \alpha_b, \tag{8.41}$$

$$\beta_n = \frac{b^2}{(p+n)(p+n+3)P - 1.0} \beta_{n-1}, \qquad \beta_0 = 1.0, \tag{8.42}$$

$$\delta_n = \frac{b^2}{(q-n)(q-n+1/4)Q - 1.0} \delta_{n-1}, \qquad \delta_0 = 1.0. \tag{8.43}$$

The coefficients B_0 and C_0 should be calculated as follows:

$$B_0 = \frac{SS'_q - S'S_q}{S'_p S_q - S_p S'_q}, \tag{8.44}$$

$$C_0 = \frac{SS'_p - S'S_q}{S'_p S_q - S_p S'_q}, \tag{8.45}$$

where:

$$S = \sum_{n=0}^{\infty} A_n, \qquad S_p = \sum_{n=0}^{\infty} \beta_n, \qquad S_q = \sum_{n=0}^{\infty} \delta_n, \tag{8.46}$$

$$S' = \sum_{n=0}^{\infty} (n+1)A_n, \qquad S'_p = \sum_{n=0}^{\infty} (n+1)\beta_n, \qquad S'_q = \sum_{n=0}^{\infty} (q-n)\delta_n, \qquad (8.47)$$

$$P = \frac{\pi}{2} \frac{\Gamma}{c_f} \tan \beta \left(1 + \frac{3\gamma^2}{8}\right)^{-1}, \qquad Q = \frac{\pi}{2} \frac{\Gamma}{c_f} \tan \beta, \qquad (8.48)$$

$$p = -\frac{3}{4} + \sqrt{\frac{9}{16} + \frac{1}{P}}, \qquad q = -\frac{1}{8} - \sqrt{\frac{1}{64} + \frac{1}{Q}}. \qquad (8.49)$$

The reference velocity is equal to the longshore current velocity at the breaker line when the lateral mixing is neglected, i.e.:

$$U_{2b} = \frac{5}{16} \pi \gamma_b \left(1 + \frac{3}{8} \gamma_b^2\right)^{-1} \frac{\sqrt{gh_b}}{c_f} \tan \beta \sin \alpha_b. \qquad (8.50)$$

When the lateral mixing stress can be omitted, the longshore current velocity takes the form:

$$\left.
\begin{aligned}
V &= \frac{X}{(1 + X\sin^2\alpha_b)} \left[(1 - X\sin^2\alpha_b)^{1/2} - \frac{\sin^2\alpha_b}{5} X (1 - X\sin^2\alpha_b)^{-1/2} \right] \\
&\qquad \text{at} \quad 0 < X < 1 \\
V &= 0 \quad \text{at} \quad 1 < X < \infty
\end{aligned}
\right\} .(8.51)$$

It can be shown that the maximum of the longshore current velocity takes place at the ordinate $X = X_{MAX}$, which satisfies the equation:

$$\sum_{n=0}^{\infty} \left[(n+1)A_n X_{MAX}^n + (n+p)B_n X_{MAX}^{n+p-1} \right] = 0. \qquad (8.52)$$

The first approximation to the solution of eq. (8.52) is:

$$X_{MAX}^{(0)} = \left(\frac{-A_0}{pB_1} \right)^{\tilde{p}}, \qquad (8.53)$$

in which: $\tilde{p} = (p-1)^{-1}$.

Therefore, the velocity at the breaker line is obtained as:

$$V_b = \sum_{n=0}^{\infty} (A_n + B_n) = \sum_{n=0}^{\infty} C_n, \qquad (8.54)$$

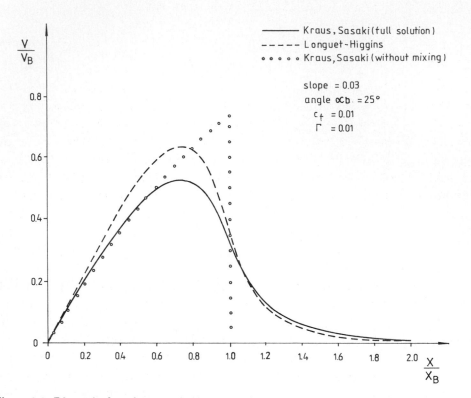

Figure 8.4: Dimensionless theoretical current profile with parameters $\Gamma = 0.0$ and 0.01 according to Kraus and Sasaki (1979), and Longuet–Higgins (1970).

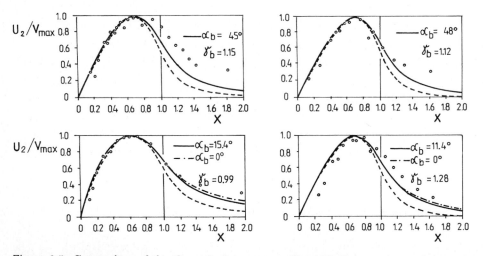

Figure 8.5: Comparison of the theoretical current profiles with the experimental data. (From Kraus and Sasaki, 1979)

while for the mean velocity in the surf zone we have:

$$\bar{V} = \sum_{n=0}^{\infty} \left(\frac{A_n}{n+2} + \frac{B_n}{n+p+1} \right). \tag{8.55}$$

In general, the longshore current velocity at the arbitrary point within the surf zone is depending upon six independent variables: γ_b, $\tan\beta$, c_f, α_b, h_b and Γ. The coefficient γ_b is in the range between 0.8 and 1.2 and the friction coefficient is of the order $(0.1 \div 3.0)\,10^{-2}$. The laboratory and field data indicate that the eddy viscosity closure coefficient Γ in eq. (8.34) is of the order $0.005 \div 0.05$. The key dimensionless parameter for the longshore current derivation is the mixing parameter P defined by (8.48). Therefore, the number of the coefficients which are needed to determine the longshore current profile can be reduced to the three parameters, i.e. P, γ_b and α_b. It should be noted that usually the parameter P is smaller than 0.5. Dimensionless profiles with mixing parameter $P = 0.0, 0.03$ ($\gamma_b = 1.0$, $\tan\beta = 0.03$, $\alpha_b = 25^\circ$, $c_f = 0.01$ and $\Gamma = 0.01$) are shown in Fig. 8.4, where the dashed line is the original theory of Longuet–Higgins (1970). Kraus and Sasaki (1979) used the velocity V_{max} and its position X_{max} to fit and compare their theory with laboratory observation as shown in Fig. 8.5 for four different conditions. Velocities are normalized by the maximum current V_{max} and distance $\bar{x}_b$. By requiring the theoretical location of the maximum velocity to match the experimental results, all curves were fitted to the data in this regard. Agreement for profile shape is very good in the region shoreward of V_{max}. The dotted line is the original model of Longuet–Higgins (1970) which gives very similar results since small angles are present.

Further extention of the long–shore current theory is discussed in the papers by Wu et al. (1985), Leontev (1988), Baum and Basco (1986). Specially, Baum and Basco have developed a numerical model for realistic, multiple bar–trough beach profiles. Each of the terms in a longshore momentum balance equation is rederived to take into account separate cross shore variations in the wave height and the water depth, as well as the wave approach angle. The model permits wave breaking and reforming; this feature in turn produces multiple peaked profiles of the current distribution as found across natural beaches.

8.4 Rip currents and coastal circulation

The analytical models of the nearshore circulation, considered as the sum of the mean and wave–induced motion and rip currents, are based on the conservation of mass and momentum equations. Usually Coriolis accelerations are neglected and the velocity field is assumed to be independent of the water depth. Thus the applicable equations are derived in the following by averaging the differential motion equations and the mass equation over depth and over time.

8.4.1   Averaging of the continuity and momentum equations

Consider the continuity equation:

$$\frac{\partial u_{t_i}}{\partial x_i} + \frac{\partial w}{\partial z} = 0, \qquad i = 1, 2, \tag{8.56}$$

in which velocity $\vec{u}_t\,(u_{t_1}, u_{t_2}, w)$ represents a sum of mean horizontal velocity U_i and wave fluctuations, i.e.:

$$u_{t_i} = U_i + u_i, \qquad w_t = w, \qquad i = 1, 2. \tag{8.57}$$

Hence, the mean velocity is defined as:

$$U_i(x, y, t) = \frac{1}{\zeta + h} \overline{\int_{-h}^{\zeta} u_{t_i}(x, y, z, t)dz}, \qquad i = 1, 2. \tag{8.58}$$

The overbar denotes an averaging over the time period T. It means that a quantity $\rho U_i(\bar{\zeta} + h)$ is the mean rate of mass flux across a vertical plane of unit width along $x_i = const$. The wave fluctuations u_i satisfy the following conditions:

$$\overline{\int_{-h}^{\zeta} u_i dz} = 0. \tag{8.59}$$

The following analysis is a modification of the procedure developed by Mei (1983). Thus, integrating the continuity equation (8.56) we get:

$$\frac{\partial}{\partial x_i} \int_{-h}^{\zeta} u_{t_i} dz + \left[-u_{t_i} \frac{\partial \zeta}{\partial x_i} + w \right]_{\zeta} - \left[u_{t_i} \frac{\partial h}{\partial x_i} + w \right]_{h} = 0. \tag{8.60}$$

On the two boundaries (free surface and sea bottom) the following kinematic conditions should apply:

$$\frac{\partial \zeta}{\partial t} + u_{t_i} \frac{\partial \zeta}{\partial x_i} = w, \qquad z = \zeta \tag{8.61}$$

$$\left[u_{t_i} \frac{\partial h}{\partial x_i} + w \right]_{-h} = 0, \qquad z = -h. \tag{8.62}$$

Because of (8.61) and (8.62), eq. (8.60) becomes:

$$\frac{\partial \zeta}{\partial t} + \frac{\partial}{\partial x_i} \int_{-h}^{\zeta} u_{t_i} dz = 0. \tag{8.63}$$

Using eq. (8.58), the time average of eq. (8.63) may be written as:

$$\frac{\partial \bar{\zeta}}{\partial t} + \frac{\partial}{\partial x_i} \left[U_i(\bar{\zeta} + h) \right] = 0, \qquad j = 1, 2. \tag{8.64}$$

As should be expected, eq. (8.64) is the same as a nonlinear equation for long waves (see eq. (4.22)).

In order to obtain the averaged momentum equation we rearrange the Navier–Stokes equation of motion as follows (Mei, 1983):

- horizontal momentum $(i, j = 1, 2)$:

$$\frac{\partial u_{t_j}}{\partial t} + \frac{\partial u_{t_i} u_{t_j}}{\partial x_i} + \frac{\partial u_{t_j} w}{\partial z} = \frac{-1}{\rho} \frac{\partial}{\partial x_i} \left(p \delta_{ij} \right), \tag{8.65}$$

- vertical momentum:

$$\frac{\partial w}{\partial t} + \frac{\partial u_{t_i} w}{\partial x_i} + \frac{\partial w^2}{\partial z} = \frac{-1}{\rho} \frac{\partial}{\partial z} \left(p + \rho g z \right). \tag{8.66}$$

Integrating the eq. (8.65) vertically and invoking the boundary conditions on the free surface and on the bottom, we get:

$$\frac{\partial}{\partial t} \int_{-h}^{\zeta} \rho u_{t_j} dz + \frac{\partial}{\partial x_i} \int_{-h}^{\zeta} \rho u_{t_i} u_{t_j} dz =$$

$$= (p)_{-h} \frac{\partial h}{\partial x_j} - \frac{\partial}{\partial x_i} \int_{-h}^{\zeta} p \delta_{ij} dz + \tau_j^{(F)} |\nabla F| - \tau_j^{(B)} |\nabla B|, \tag{8.67}$$

in which: $F(x, y, z, t) = z - \zeta(x, y, t)$ - free surface, $B(x, y, z) = z + h(x, y) = 0$ - bottom surface.

The gradients of F and B functions were defined as:

$$\nabla F = \left(-\frac{\partial \zeta}{\partial x}, -\frac{\partial \zeta}{\partial y}, 1 \right), \qquad \nabla B = \left(\frac{\partial h}{\partial x}, \frac{\partial h}{\partial y}, 1 \right) \tag{8.68}$$

and the unit outward normals:

$$\vec{n} = \frac{\nabla F}{|\nabla F|}, \qquad \vec{n} = \frac{-\nabla B}{|\nabla B|}. \tag{8.69}$$

The stresses $\tau_j^{(F)}$ and $\tau_j^{(B)}$ correspond to the surface and bottom stresses, respectively, and the stress $\tau_j^{(F)}$ denotes the total horizontal shear stress including both atmospheric pressure and shear stress.

On the free surface $(z = \zeta)$, the atmospheric unit force must be in the equilibrium with the fluid stresses, i.e.:

$$(p\delta_{ij})_\zeta \frac{\partial \zeta}{\partial x_i} = \tau_j^{(F)} |\nabla F|. \tag{8.70}$$

On the nonpermeable bottom, the stress $\tau_j^{(B)}$ satisfies the following conditions:

$$\tau_{ij} \frac{\partial h}{\partial x_i} = -\tau_j^{(B)} |\nabla B|, \qquad z = -h. \tag{8.71}$$

Physically, eq. (8.67) represents the momentum balance in a vertical column of fluid of height $(\zeta + h)$ and unit cross section. The left–hand side terms are the acceleration and the net momentum flux through the vertical sides of the column. The right–hand side terms express the summation of the pressure by the bottom to the fluid, the net surface stresses on the vertical sides, and the surface stress at the free surface and the sea bottom. After the time averaging of the eq. (8.67), we finally get:

$$\rho\left(\bar{\zeta} + h\right)\left[\frac{\partial U_j}{\partial t} + U_i \frac{\partial U_j}{\partial x_i}\right] = \bar{p}_d \frac{\partial h}{\partial x_j} - \rho g(\bar{\zeta} + h)\frac{\partial \bar{\zeta}}{\partial x_j} +$$

$$- \frac{\partial S_{ij}}{\partial x_i} + \bar{\tau}_j^{(F)} |\nabla F| - \bar{\tau}_j^{(B)} |\nabla B| \tag{8.72}$$

where: $\bar{p}_d$ - mean dynamic pressure at the bottom, i.e. $\bar{p}_d = (\bar{p})_{-h} - \rho g\left(\bar{\zeta} + h\right)$ and:

$$S_{ij} = \overline{\int_{-h}^{\zeta} (p\,\delta_{ij} + \rho u_i\,u_j)\,dz} - \frac{1}{2}\rho g(\bar{\zeta} + h)^2 \delta_{ij} \tag{8.73}$$

is the (i, j) component of the stress tensor representing the excess momentum fluxes.

The term $\bar{p}_d\,(\partial h/\partial x_j)$ in eq. (8.72) is $O|\nabla h|$ times smaller than the largest remaining terms and may be ignored.

Therefore eq. (8.72) becomes:

$$\frac{\partial U_j}{\partial t} + U_i \frac{\partial U_j}{\partial x_i} = -g \frac{\partial \bar{\zeta}}{\partial x_j} - \frac{1}{\rho(\bar{\zeta} + h)} \frac{\partial S_{ij}}{\partial x_i} +$$

$$+ \frac{\bar{\tau}_j^{(F)}}{\rho(\bar{\zeta} + h)} - \frac{\bar{\tau}_j^{(B)}}{\rho(\bar{\zeta} + h)}, \tag{8.74}$$

when the surface and bottom slopes are regarded as small. Hence:

$$\mid \nabla F \mid = 1 + O(\nabla \zeta)^2 \qquad \text{and} \qquad \bar{\tau}_j^{(F)} \mid \nabla F \mid = \bar{\tau}_j^{(F)} \left[1 + O(\nabla \zeta)^2 \right], \tag{8.75}$$

and:

$$\mid \nabla B \mid = 1 + O(\nabla h)^2 \qquad \text{and} \qquad \bar{\tau}_j^{(B)} \mid \nabla B \mid = \bar{\tau}_j^{(B)} \left[1 + O(\nabla h)^2 \right]. \tag{8.76}$$

Eq. (8.74) will be used in the next Section to model the nearshore circulation.

8.4.2 Modelling of the coastal circulation

The longshore current, discussed in Section 8.3, can form together with rip currents a two–dimensional coastal current system within and beyond the surf zone, which is known as the *nearshore circulation*. The mechanisms proposed to explain rip currents are of two basic categories:

  - wave - wave and wave - current interactions, and
  - bottom topography variation.

The analytic solutions of the first category are developed to predict rip current spacing. The time–averaged horizontal motion equations, and continuity, as well as the energy equation are usually expressed in terms of an equilibrium state plus a small perturbation for the four variables: stream function Ψ, mean sea level $\bar{\zeta}$, water depth h and energy E (Basco, 1982). The perturbation equations are then subjected to small disturbances in each of the perturbation variables. As the result, the non–dimensional rip current spacing λx_b ($\lambda = 2\pi/L_r$ - longshore wave number for the circulation, L_r - rip current spacing), x_b - surf zone width) can be approximated as (Dalrymple, 1978):

$$\lambda x_b \approx \frac{1}{A_D} + 2.8, \tag{8.77}$$

in which: $A_D = (\gamma_b \tan \beta)/(64 c_f)$.

Note that this result yields an increase in L_r with wave height as γ_b is directly proportional to H_b.

The majority of rip currents observed on coastal lines suggest however, that the longshore bottom topography variations provide the basic mechanism to create the nearshore circulation. Moreover, the surface stresses $\bar{\tau}_j^{(F)}$ can be neglected. If the bottom variability is periodic, then also periodic nearshore circulation cells may form. Let us consider the steady mean flow only and the current velocity $U(U_1, U_2)$ is assumed to be independent of the water depth. Therefore, the eq. (8.64) and (8.74) apply, i.e.:

$$g \frac{\partial \bar{\zeta}}{\partial x} = -\frac{1}{\rho(\bar{\zeta} + h)} \left(\frac{\partial S_{xx}}{\partial x} + \frac{\partial S_{xy}}{\partial y} \right) - \frac{\bar{\tau}_x^{(B)}}{\rho(\bar{\zeta} + h)}, \tag{8.78}$$

$$g \frac{\partial \bar{\zeta}}{\partial y} = -\frac{1}{\rho(\bar{\zeta} + h)} \left(\frac{\partial S_{xy}}{\partial y} + \frac{\partial S_{yx}}{\partial x} \right) - \frac{\bar{\tau}_y^{(B)}}{\rho(\bar{\zeta} + h)}, \tag{8.79}$$

$$\frac{\partial}{\partial x} \left[U_1 \left(\bar{\zeta} + h \right) \right] + \frac{\partial}{\partial y} \left[U_2 \left(\bar{\zeta} + h \right) \right] = 0. \tag{8.80}$$

In eqs. (8.78) - (8.79) the nonlinear term was omitted under the assumption that a mean current is of second order in the mean surface slope. We rewrite the components of the radiation stress tensor S (eq. (4.23)) as follows:

$$S_{xx} = \frac{1}{8} \rho g H^2 \left[\left(2m - \frac{1}{2} \right) \cos^2 \Theta + \left(m - \frac{1}{2} \right) \sin^2 \Theta \right], \tag{8.81}$$

$$S_{yy} = \frac{1}{8} \rho g H^2 \left[\left(2m - \frac{1}{2} \right) \sin^2 \Theta + \left(m - \frac{1}{2} \right) \cos^2 \Theta \right], \tag{8.82}$$

$$S_{xy} = S_{yx} = \frac{1}{16} \rho g H^2 m \sin 2\Theta. \tag{8.83}$$

When the shallow water approximation ($m \rightarrow 1.0$) is used, eqs. (8.81) - (8.83) yield:

$$S_{xx} = \frac{1}{16} \rho g H^2 \left(3\cos^2\Theta + \sin^2\Theta\right) \Big\}$$

$$S_{xy} = \frac{1}{16} \rho g H^2 \left(3\sin^2\Theta + \cos^2\Theta\right) \Big\} \tag{8.84}$$

$$S_{xy} = S_{yx} = \frac{1}{16} \rho g H^2 \sin 2\Theta \Big\}$$

Additionally, for time–averaged bottom shear stress, $\bar{\tau}^{(B)}$, we assume:

$$\bar{\tau}_x^{(B)} = c_f \, \rho \, U_1 \, \bar{u}_b, \tag{8.85}$$

and

$$\bar{\tau}_y^{(B)} = c_f \, \rho \, U_2 \, \bar{v}_b, \tag{8.86}$$

in which: $\bar{u}_b$, $\bar{v}_b$ - components of the mean bottom velocity induced by waves. If the linear wave theory is applied, eqs. (8.85) and (8.86) yield:

$$\bar{\tau}_x^{(B)} = 2c_f \frac{H U_1}{(\zeta + h)T \sinh(kh)} \Big\}$$

$$\bar{\tau}_y^{(B)} = 2c_f \frac{H U_2}{(\zeta + h)T \sinh(kh)} \Big\} \tag{8.87}$$

Consider the mass transport stream function, Ψ, in the form:

$$\frac{\partial \Psi}{\partial x} = U_2 \, h, \qquad \frac{\partial \Psi}{\partial y} = -U_1 \, h. \tag{8.88}$$

Differentiating eqs. (8.78) in y and eq. (8.79) in x, and adding these equations, after some manipulations we get (Noda, 1974):

$$\frac{\partial^2 \Psi}{\partial x^2} + \frac{\partial^2 \Psi}{\partial y^2} + \frac{1}{F} \frac{\partial F}{\partial y} \frac{\partial \Psi}{\partial y} + \frac{1}{F} \frac{\partial F}{\partial x} \frac{\partial \Psi}{\partial x} = \frac{g}{F} \left\{ \frac{\partial}{\partial y} \left[\frac{1}{h} \left(\frac{\partial \tilde{S}_{xx}}{\partial x} + \frac{\partial \tilde{S}_{xy}}{\partial y} \right) \right] + \right.$$

$$\left. - \frac{\partial}{\partial x} \left[\frac{1}{h} \left(\frac{\partial \tilde{S}_{yy}}{\partial y} + \frac{\partial \tilde{S}_{xy}}{\partial x} \right) \right] \right\}, \tag{8.89}$$

where:

$$F = 2 c_f \frac{H}{h^2 T \sinh(kh)}, \tag{8.90}$$

$$\tilde{S}_{xx} = H^2 \left[\frac{3}{16} \cos^2 \Theta + \frac{1}{16} \sin^2 \Theta \right]$$

$$\tilde{S}_{yy} = H^2 \left[\frac{3}{16} \sin^2 \Theta + \frac{1}{16} \cos^2 \Theta \right] \Bigg\} . \qquad (8.91)$$

$$\tilde{S}_{xy} = \tilde{S}_{yx} = \frac{1}{16} H^2 \sin 2\Theta$$

Wave heights must be specified a priori throughout the field of solution because the local wave energy appears in the radiation stress terms. This requires use of wave shoaling, refraction and reflection theory outside the breakers, and some for energy dissipation in the surfzone (see Chapter 3).

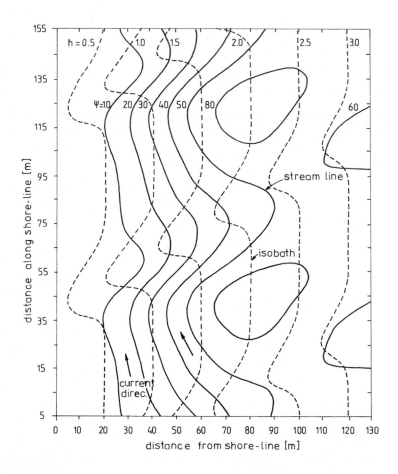

Figure 8.6: The pattern of mass transport stream function for given bathymetry

Moreover, for steady flows, the convective accelerations and lateral turbulent mixing stresses were neglected.

In order to solve eq. (8.89) the finite–difference methods were employed, however, no accuracy analysis of these methods is presented. In Figure 8.6 the pattern of the mass transport stream function Ψ for the given bathymetry, calculated according this equation is illustrated . The shortcomings of this model developed by Noda (1974) have been addressed in a series of developments by Ebersole and Dalrymple (1980), Mei and Liu (1977), and Sasaki (1984). In the paper by Wu and Liu (1984) a perturbation method was developed to investigate the effects of nonlinear inertial forces on breaking–induced nearshore currents for the bottom topography varying periodically in the shoreline direction. The lateral turbulent mixing was ignored. When the nonlinear terms are included, an advective shift as well as amplitude reduction in the current pattern is observed. Numerical experiments (Wu et al., 1985) indicate that the Γ coefficient for the nonlinear model to fit the data is larger than that for the linear model, and the c_f value on the contrary is smaller for the nonlinear model.

All numerical models presently available have serious limitations. In earlier versions, the important terms in the momentum balance equations were omitted (e.g. convective accelerations, lateral mixing etc.). On the other hand, the later models simply lack the numerical accuracy due to excessive truncation errors in the algorithm employed (Basco, 1982). The assumption of the independence of velocity field on the water depth yields the only approximate result. However, the flow pattern in the coastal zone is three–dimensional and the vertical profile of the current should be considered. The cross–shore net flow pattern in the coastal zone is a balance between the mass transport directed inshore in breaking waves and the seawards oriented return flow, called the *undertow*. The field experiments on very gentle sand slopes (Leontev and Speransky, 1980) indicate that the characteristic current velocity is of order 0.1 m/s. The highest velocities (~ 0.3 m/s) are observed in the surf zone. Usually in the top layer, the flow is shorewards while in the bottom layer the compensating flow is directed offshore. The thickness of the inshore flow is slightly increasing in the breaking area. The increasing of this flow may cause the deficit of flow, seawards the breakers. This deficit is probably related to the observed rise of the compensating flow (Fig. 8.7). The direction of the flow in the bottom layer depends on the local bottom slope. For the slope of order 0.04 $\div$ 0.06, the slow drift shorewards is observed. In other case the compensating flow is arising. However, the above description is only qualitative and it can be quite different in particular cases (e.g. due to rip current).

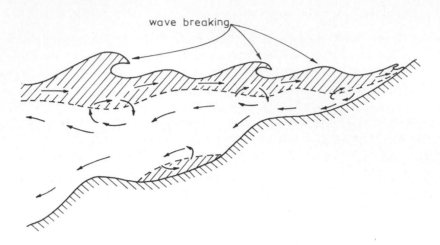

Figure 8.7: Vertical structure of flow.

In order to clarify the mechanisms generating the current pattern we consider the two–dimensional regular progressive waves propagating to the shore. Thus, eq. (8.74) becomes:

$$\frac{\partial S_{xx}}{\partial x} = -\rho g \left(h + \bar{\zeta} \right) \frac{\partial \bar{\zeta}}{\partial x} - \bar{\tau}_x^{(B)}. \tag{8.92}$$

The force balance expressed by (8.92) implies that in the surf zone the decrease of radiation stress is partly offset by an increase $\bar{\zeta}$. This set–up generates a seaward oriented pressure force on the fluid. According to Svendsen (1984), the undertow is a seaward oriented mean current created in a surf zone, essentially below wave trough, as compensation for the mass flux created between wave trough and crest by the breakers. The undertow is associated with the large shear stresses, directed against the flow, i.e. shorewards. Thus, the driving mechanism for the undertow can be expressed as (Svendsen, 1984):

$$\frac{\partial \overline{u_t^2}}{\partial x} = -g \frac{\partial \bar{\zeta}}{\partial x} + \frac{1}{\rho} \frac{\partial \overline{\tau_{zx}}}{\partial z}, \tag{8.93}$$

in which: $u_t = U_1(z) + u(z,t)$, U_1 - mean (undertow) velocity, u - oscillatory part of the flow.
For the horizontal turbulent shear stresses, the simple eddy viscosity model is assumed, i.e.:

$$\tau_{zx} = \rho \, \nu_t \, \frac{\partial U_1}{\partial z}. \tag{8.94}$$

Therefore, eq. (8.93) may be rewritten as:

$$\frac{\partial}{\partial z} \left[\nu_t \, \frac{\partial U_1}{\partial z} \right] = \frac{\partial}{\partial x} \left[\rho \, \overline{u^2} + \rho \, g \, \bar{\zeta} \right]. \tag{8.95}$$

A general solution to eq. (8.95) can be obtained under the boundary conditions expressing the depth integrating continuity, no slip condition at the bottom and the conditions within the boundary layer. As a result we obtain the parabolic profile for the Eulerian mean velocity below wave trough what is in agreement with the experimental observations by Stive and Wind (1982) and Hansen and Svendsen (1984).

8.5 Mass transport in shallow water

Usually the wave motion is assumed to be periodic and the wave profile is assumed to be that of a steady state. However, for solving the nonlinear problem, two additional conditions are necessary. These are related to the rotationality and the mass transport. The assumption of zero mass transport yields the closed orbit theories (e.g. Gerstner deep water solution or power series solution of Boussinesq for shallow water) and the rotational motion. The wave motion can also be assumed to be irrotational, in which case a mass transport distribution is found as a result of nonlinearity. It should be noted that the linear solution for the irrotational motion predicts a closed orbit (see eq. (2.65)) and no mass transport. For a given mass transport distribution which is a function of the vertical coordinate, two conditions must be satisfied:

1. The net average flow across any section must be zero in order to satisfy continuity, i.e.:

$$\int_{-h}^{0} \bar{u}_1(z) dz = 0. \tag{8.96}$$

2. Due to presence of viscosity:

$$\bar{u}_1(z) = 0 \qquad \text{at} \qquad z = -h, \tag{8.97}$$

in which $\bar{u}_1(z)$ is the mean drift velocity (or mass transport velocity).

The method by which eq. (8.96) is satisfied cannot be arbitrarily chosen; it must reflect a solution of the equations of motion with proper boundary conditions. According to Longuet–Higgins (1953), it is impossible to solve the Navier–Stokes equation when the effects of convective acceleration and those of viscosity are retained. For the two–dimensional wave problem this equation becomes:

$$\left(\frac{\partial}{\partial t} + u\frac{\partial}{\partial x} + w\frac{\partial}{\partial z} - \nu\nabla^2\right)\nabla^2\psi = 0, \qquad (8.98)$$

in which: ψ - stream function, i.e.:

$$(u, w) = \left(\frac{\partial\psi}{\partial z}, \ -\frac{\partial\psi}{\partial x}\right), \qquad (8.99)$$

and ν - kinematic coefficient of viscosity.

The second and third terms in eqs. (8.98) represent the rate of change of the vorticity (with the accuracy to sign) at a fixed point due to convection. The last term represents the rate of change of the vorticity due to viscous diffusion and it is similar to a term in the heat conduction equation. If we omit the convective acceleration and retaining viscosity, eq. (8.98) yields:

$$\left[\frac{\partial}{\partial t} - \nu\nabla^2\right]\nabla^2\psi = 0. \qquad (8.100)$$

The solution to eq. (8.100) (called the *conduction solution*) is valid only when $A/\delta \ll 1.0$ (A - wave amplitude, δ - thickness of the boundary layer). For waves of resonable period and height this parameter is certainly greater than one. However, the solution will nevertheless be assumed valid within the bottom boundary layer in which the Lagrangian mass transport velocity $\bar{u}_L$ becomes:

- *purely progressive wave*

$$\bar{u}_L = \frac{kwA^2}{4\sinh^2(kh)}\left[-8e^{-\xi}\cos\xi + 3\,e^{-2\xi} + 5\right], \qquad (8.101)$$

in which: $\xi = z + h/\delta$.

- *standing waves*

$$\bar{u}_L = \frac{kwA^2}{2\sinh^2 kh}\ \sin(2kx)\left[8\,e^{-\xi}\sin\xi + 3\,e^{-2\xi} - 3\right]. \qquad (8.102)$$

In Fig. 8.8 the mass transport profiles within Stokes boundary layer near the bottom, for the progressive and standing waves, are presented ($T = 10\,s$; $A = 1.0\,m$; $h = 5\,m$ and $2kx = \pi/2$).

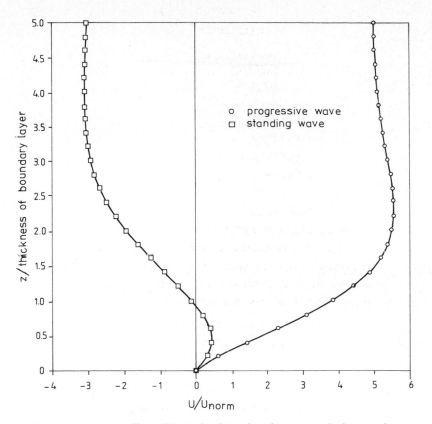

Figure 8.8: Mass transport profile within Stokes boundary layer near the bottom for progressive and standing waves.

The transport velocity is normalized against $k\omega A^2/4\sinh^2(kh) \approx 0.06\,m/s$. For progressive wave, the mass transport velocity $\bar{u}_L(z)$ is always positive, and when ξ tends to infinity, it tends to the value $5/4\,(k\omega A^2/\sinh^2(kh))$. For the standing waves, as ξ tends to infinity, the normalized velocity $\bar{u}_L$ tends to -3.

For the interior of fluid where viscous effects are negligible, the *conduction solution* is given by (Longuet–Higgins, 1953):

$$\bar{u}_L = \omega\,k\,A^2\,F\left(\frac{z}{h}\right),\qquad (8.103)$$

in which:

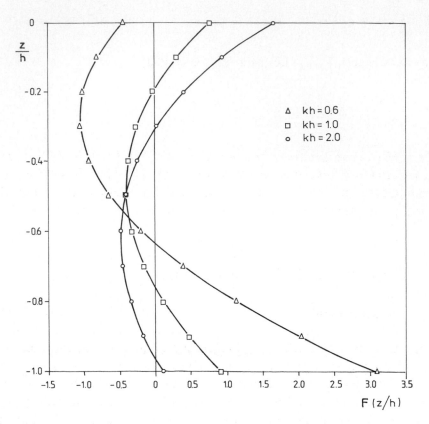

Figure 8.9: The mass transport in the interior of fluid. (From Longuet–Higgins, 1953)

$$F\left(\frac{z}{h}\right) = \frac{1}{4\sinh^2 kh}\,[3 + 2\cosh 2k(z+h)] +$$

$$+ \ kh\,\sinh(2kh)\left[1 + 4\left(\frac{z}{h}\right) + 3\left(\frac{z}{h}\right)^2\right] +$$

$$+ \ 3\left[\frac{\sinh 2kh}{2kh} + \frac{3}{2}\right]\left[\left(\frac{z}{h}\right)^2 - 1.0\right]. \tag{8.104}$$

The solution (8.103) satisfies the condition of the zero net mass flux through-out the depth. The function $F\,(z/h)$ for $kh = 0.6, 1.0$ and 2.0 is shown in Fig. 8.9.

Omitting the viscous terms but retaining the convective accelerations in eq. (8.98), we get:

$$\left[\frac{\partial}{\partial t} + u \frac{\partial}{\partial x} + w \frac{\partial}{\partial z}\right] \nabla^2 \Psi = 0. \tag{8.105}$$

In this case, a solution is (Longuet–Higgins, 1953):

$$\bar{u}_L = \frac{kA^2}{4 \sinh^2(kh)} H(z), \tag{8.106}$$

in which $H(z)$ is an arbitrary function of z satisfying the specified conditions at $z = 0$ and $z = -h$. The solution describes an arbitrary horizontal flow depending only on z, which can, however be defined further if the conditions at $x = \pm\infty$ are known (e.g. at the wave maker and the wave absorber).

8.6 Tidal currents in coastal zone

8.6.1 Oceanic tides

The rhythmic rise and fall of water, known as the tide, is a characteristic feature of many coastal areas. It is evident that these vertical displacements of the surface are accompanied by horizontal displacements of the water masses, called *tidal currents*. The most commonly encountered constituent of tide denoted by M_2, is that interval between two successive high waters or between two successive low waters is 12 h and 25 min. Consequently, the high and low water is retarded from one day to the next by 50 min because the moon is also retarded daily by 50 min on its passage through the meridian. The second important tide constituent, denoted by S_2, is resulting from the sun attraction force. This constituent has a period of 12 h 00 min, alternately reinforcing and opposing the lunar semi–durnal constituent. It is clear that the height of the high and low water depends upon the relative position of the sun, the moon and the earth. When the earth, the moon and the sun are aligned, the effects of the moon and the sun are in phase, causing the tides of maximum range, known as *spring tides*. This happens at new and full moon. The tides of the minimum range, *neap tides*, occur near the first and third quarters of the moon. The complete list of all components is very long and comprises about 100 long periodic, 160 diurnal, 115 semi–diurnal and 14 one–third diurnal terms (Defant, 1961). In practice, only more important harmonic constituents are taken into account. They are shown in Table 8.1. The detail analysis of the oceanic tides can be found in many text books (Defant, 1961; Godin, 1972) and it will be not repeated here.

Table 8.1: The basic harmonic constituents of tides.

Class	Constituent	Symbol	Period (hours)
Semidiurnal	Principal lunar	M_2	12.25
	Principal solar	S_2	12.00
	Larger lunar elliptic	N_2	12.66
	Luni–solar	K_2	11.97
Diurnal	Luni–solar	K_1	23.93
	Principal lunar	O_1	26.87
	Principal solar	P_1	24.07

8.6.2   Shallow water tides

Tidal waves belong to the class of long waves with wave length of hundreds of kilometers. When the tides, generated by external forces on terrestrial waters, penetrate into the shelf, their propagation is influenced by the shallow water of the coastal areas. The tide trough is retarded more than its crest as a result of the difference in depth at high and low water, which effects the frictional forces. The shallow water constituents are generated by effects which are described by the non–linear terms in the equation of motion. The most appropriate equations system for the tidal waves propagation in the shallow water is the system of depth–integrated equations developed in Section 8.4. After introduction of the Coriolis forces, eq. (8.74) becomes:

$$\frac{\partial U_1}{\partial t} + U_1\frac{\partial U_1}{\partial x} + U_2\frac{\partial U_1}{\partial y} - fU_2 = -g\frac{\partial \bar{\zeta}}{\partial x} - \frac{\tau_x^{(B)}}{\rho(\bar{\zeta}+h)}, \tag{8.107}$$

$$\frac{\partial U_2}{\partial t} + U_1\frac{\partial U_2}{\partial x} + U_2\frac{\partial U_2}{\partial y} + fU_1 = -g\frac{\partial \bar{\zeta}}{\partial y} - \frac{\tau_y^{(B)}}{\rho(\bar{\zeta}+h)}. \tag{8.108}$$

in which: $f = 2\Omega\sin\varphi$.

In eqs. (8.107) and (8.108), the radiation stress tensor expressing the current–wave interaction and surface stresses $\bar{\tau}^{(F)}$ were omitted. This interaction will be discussed in the next Section. Similarly, the continuity equation takes the form:

$$\frac{\partial \bar{\zeta}}{\partial t} + \frac{\partial}{\partial x}\left[(\bar{\zeta}+h)U_1\right] + \frac{\partial}{\partial y}\left[(\bar{\zeta}+h)U_2\right] = 0. \tag{8.109}$$

The resultant bottom stress $\tau^{(B)}$ may be related to the bottom current U_b by the quadratic law, i.e. $\tau^{(B)} = c_f \rho U_b^2$. A typical value of the friction coefficient would be $c_f = 2 \cdot 10^{-3}$. For the bottom current U_b of $0.5\,m\,s^{-1}$, this would give a bottom stress $\tau^{(B)}$ of approximately $0.5\,Nm^{-2}$.

The equations (8.107) - (8.109) are the basis for the study of tides in shallow seas. They are nonlinear and the numerical techniques are available to solve them (Dronkers, 1964). To gain some insight into the tidal waves propagation in the shallow waters, we make certain assumptions. If the acceleration terms are small, the terms like $U_i\,\partial U_j/\partial x_i$ can be neglected. Moreover, if the tidal elevation $\bar{\zeta}$ is small, compared with the water depth h, then the term $(h + \bar{\zeta})$ can be replaced by h.

It would be useful to consider first the tidal waves as free waves. With these linearising assumption, eqs. (8.107) - (8.109) become:

$$\left.\begin{array}{rcl}
\dfrac{\partial U_1}{\partial t} - fU_2 & = & -g\dfrac{\partial \bar{\zeta}}{\partial x} \\[3mm]
\dfrac{\partial U_2}{\partial t} + fU_1 & = & -g\dfrac{\partial \bar{\zeta}}{\partial y} \\[3mm]
h\left(\dfrac{\partial U_1}{\partial x} + \dfrac{\partial U_2}{\partial y}\right) & = & -\dfrac{\partial \bar{\zeta}}{\partial t}
\end{array}\right\}. \tag{8.110}$$

The terms fU_1 and fU_2 in eq. (8.110) represent the Coriolis effect. This is due to the rotation of the earth and can be regarded as a force acting perpendicular to the direction of motion. The direction of the Coriolis force is to the right of the current velocity in the northern hemisphere and to the left in the southern hemisphere. In order to find the current velocities and the surface elevation, we consider a progressive wave travelling in a wide channel with straight banks and a constant depth. If we assume that the water particles move in the longitudinal direction only ($U_2 = 0$), we get:

$$\left.\begin{array}{rcl}
\dfrac{\partial U_1}{\partial t} & = & -g\dfrac{\partial \bar{\zeta}}{\partial x} \\[3mm]
fU_1 & = & -g\dfrac{\partial \bar{\zeta}}{\partial y} \\[3mm]
h\dfrac{\partial U_1}{\partial x} & = & -\dfrac{\partial \bar{\zeta}}{\partial t}
\end{array}\right\}, \tag{8.111}$$

Here the x-axis is in the longitudinal direction of the channel and the y-axis is perpendicular to the bank with $y = 0$ at the right bank.

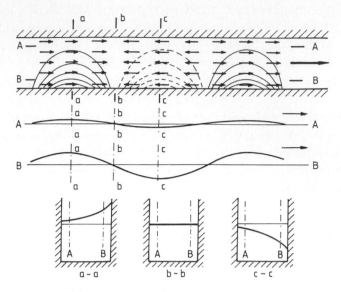

Figure 8.10: Kelvin wave in channel in the northern hemisphere.

Eliminating U_1 from eqs. (8.111) leads:

$$\frac{\partial^2 \bar{\zeta}}{\partial t^2} = gh \frac{\partial^2 \bar{\zeta}}{\partial x^2}, \tag{8.112}$$

and by eliminating ζ from eqs. (8.111) a similar equation is obtained for U_1:

$$\frac{\partial^2 U_1}{\partial t^2} = gh \frac{\partial^2 U_1}{\partial x^2}. \tag{8.113}$$

A simple harmonic form of the progressive wave can be expressed by:

$$\left.\begin{aligned}
\bar{\zeta} &= A e^{-my} \cos(kx - \omega t) \\
U_1 &= U e^{-my} \cos(kx - \omega t)
\end{aligned}\right\}, \tag{8.114}$$

where: $U = \sqrt{g/h}\, A$ and $m = f/\sqrt{gh}$.

Equations (8.114) represent a Kelvin wave travelling in the x direction. Due to Coriolis effect the tidal amplitude decreases exponentially with increasing distance from the bank. A Kelvin wave in the longitudinal channel in the northern hemisphere is illustrated in Fig. 8.10. In order to get an idea of dimensions and speed of travel of Kelvin waves, we consider the channel of water depth $h = 50\,m$, situated in latitude $30^\circ N$. For the M_2 tide, $T =$

$12h\ 25\ min$, the wave length $L = 990\ km$. In latitude $30°N$, the wave number $m = 0.33 \cdot 10^{-5} m^{-1}$. Therefore, the amplitude would decrease by a factor $e(= 2.78)$ in a distance $303\ km$, measured to the left along the wave front. Taking $A = 1\ m$, eq. (8.114) gives $U \approx 0.45\ m/s$. Usually Kelvin waves are highly modified due to irregular bed topography and the effect of bottom friction.

Other wave–like simple solution of equations (8.110) is Sverdrup wave in which the crests are horizontal, i.e. $\partial\zeta/\partial y = 0$ and $\partial U_2/\partial y = 0$ in (8.110). As the result we obtain:

$$\left.\begin{aligned}
\zeta &= A\cos(kx - \omega t) \\[2mm]
U_1 &= \sqrt{\frac{g}{h}}\sqrt{\frac{1}{1 - s^2}}\cos(kx - \omega t) \\[2mm]
U_2 &= -\sqrt{\frac{g}{h}}\sqrt{\frac{s^2}{1 - s^2}}\sin(kx - \omega t)
\end{aligned}\right\}, \tag{8.115}$$

in which:

$$s = \frac{f}{\omega}, \qquad C^2 = gh\left[1 - \left(\frac{f}{\omega}\right)^2\right]^{-1}. \tag{8.116}$$

Eq. (8.116) indicates that the velocity of Sverdrup waves is greater than that of Kelvin waves. The velocity components U_1 and U_2 are $90°$ out of phase and the end–points of the resulting velocity vector lie on an ellipse. The direction of the maximum velocity of the current coincides with the direction of progress of the waves.

The effect of friction on tidal waves in the shallow water is usually important. For the special case of free, one–dimensional wave with the small tidal elevations, eqs. (8.107), (8.108) and (8.109) can be rewritten in the form:

$$\left.\begin{aligned}
\frac{\partial U_1}{\partial t} &= -g\frac{\partial\bar\zeta}{\partial x} - \frac{\tau_x^{(B)}}{gh} \\[2mm]
\frac{\partial\bar\zeta}{\partial t} &+ h\frac{\partial U_1}{\partial x} = 0
\end{aligned}\right\}. \tag{8.117}$$

If the tidal constituent of velocity is assumed to be harmonic in time, the expression for $\tau_x^{(B)}$ can be linearized, i.e.:

$$\tau_x^{(B)} = K\rho U_1, \tag{8.118}$$

in which: U_1 - bottom velocity, and:

$$K = \frac{8}{3\pi} c_f |U_1|.$$

(8.119)

The solution for U_1 is (Bowden, 1983):

$$U_1 = |U_1| e^{-\mu x} \cos(kx - \omega t - \varphi),$$

(8.120)

where:

$$|U_1| = \frac{\omega A}{h (c_f^2 + \mu^2)^{1/2}}, \qquad \mu = \frac{4}{3\pi} \frac{c_f |U_1|}{h \sqrt{gh}}, \qquad \tan \varphi = \frac{\mu}{c_f}.$$

(8.121)

Eq. (8.120) represents the damped Kelvin wave travelling in the x - direction. Thus:

$$\bar{\zeta} = A e^{-\mu x} \cos(kx - \omega t).$$

(8.122)

The response of the more complicated areas, such as partially enclosed seas and bays to the driving force exerted by the ocean tide can be estimated by the numerical modelling. The sea area under consideration is represented by a network of points, at which the elevations of sea surface and current components are to be calculated, and appropriate boundary conditions are applied. A number of such modelling techniques are described by Ramming and Kowalik (1980).

8.7   Interactions of waves and currents

In many real problems the fluid upon which waves appear is flowing and the waves are modified by current. The transformation of sea waves by currents is a significant physical process in many coastal and offshore areas, for instance near river mouths and tidal inlets, in the surf zone along beaches under storm conditions, and where wind waves meet major ocean currents. The inclusion of the current–wave interaction is of great importance for phenomena such as sediment transport, ship navigation, and forces on sea structures. Physically, the interaction between short waves and currents is a problem of wave propagation in an inhomogeneous, non–isotropic, dispersive and dissipative moving medium. During the waves propagation in such medium, the kinematics and dynamics of the water particles are changed substantially in comparison with the absence of current. In following Sections we will consider both effects.

8.7.1   Waves on uniform currents

We shall consider steady progressive waves (in the linear approximation) propagating on the constant water depth h at the angle Θ against the positive x-axis. We assume that the uniform current $\vec{U}(U_1, 0, 0)$ is parallel to this axis. The variation of Lagrangian $\bar{\mathcal{L}}$ (1.55) in δE yields the dispersion relation in the form:

$$\left(\omega - \vec{k} \cdot \vec{U}\right)^{1/2} = gk \, \tanh(kh), \tag{8.123}$$

or:

$$\omega - k \left|\vec{U}\right| \cos \Theta = \pm \sigma, \tag{8.124}$$

in which:

$$k = \left|\vec{k}\right|, \qquad \sigma = [gk \, \tanh(kh)]^{1/2}. \tag{8.125}$$

The frequency ω is the observed frequency in the constant frame of reference, while σ is the intrinsic frequency or the frequency in the frame of reference moving with uniform velocity $\vec{U}$. Let:

$$a = \omega \sqrt{\frac{h}{g}}, \qquad b = \frac{\left|\vec{U}\right| \cos \Theta}{\sqrt{\omega h}}, \qquad \sigma^* = (kh \, \tanh kh)^{1/2}. \tag{8.126}$$

Thus, eq. (8.124) may be rewritten as:

$$a - b(kh) = \pm \, \sigma^*. \tag{8.127}$$

The easiest way to solve the dispersion relation (8.127) for $\vec{k}$ is to consider of the intersection of plane (Peregrine, 1976):

$$\chi = a - b(\Theta)kh, \tag{8.128}$$

with the surface of revolution:

$$\chi = \pm \, \sigma^* \tag{8.129}$$

in (kh, χ) space; the rotation is taken against the χ-axis. If $\vec{U} = 0$ or $\Theta = 90^\circ$, the $b = 0$ and $\omega = \sigma$. Therefore, if vector $\vec{k}$ is perpendicular to $\vec{U}$ then the current does not affect the solution. For any other direction Θ, a diametral

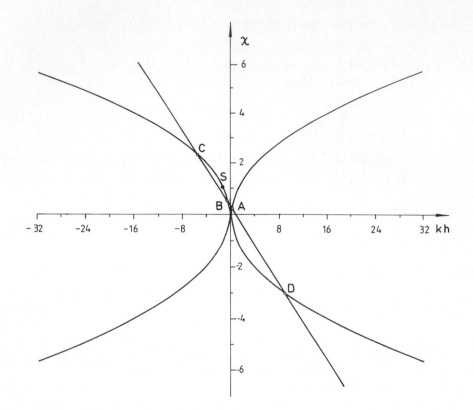

Figure 8.11: Solution of dispersion relation for wave propagating on the current.

section of (8.129) yields a curve as shown in Fig. 8.11. In that diametral plane, four solution points A, B, C and D can be distinguished. The point A in which $\chi = \sigma_A > 0$ and $(kh)_A > 0$, corresponds to waves with a component of $\vec{k}$ in the direction of current. The observed frequency ω is greater than the intrinsic frequency σ_A. Substituting $\sigma = \sigma_A$ into (8.124) and dividing by the wave number k_A we get:

$$C_A^{(a)}(k_A) = C_A(k_A) + \left|\vec{U}\right| \qquad \text{at} \qquad \Theta = 0^o, \qquad\qquad (8.130)$$

in which: C_A - phase frequency corresponding to the frequency σ_A, i.e. $C_A = \sigma_A/k_A$, $C_A^{(a)}$ - absolute phase velocity observed in the steady reference frame.

In the points B and C we have $\sigma_A > 0$ and $kh < 0$. In particular, the point B corresponds to wave propagating against the current, i.e.:

$$C_B^{(a)}(k_B) = C_B(k_B) - \left|\vec{U}\right| \qquad \text{at} \qquad \Theta = 180^o. \qquad\qquad (8.131)$$

The relations (8.130) and (8.131) clearly exhibit the Doppler effect.

In order to explain the physical behaviour of waves corresponding to the point C, we select first the point S at the curve of revolution. In this point, the line (8.128) is tangential to the curve (8.129). Thus:

$$C_g(k_S) + \left|\vec{U}\right| \cos \Theta = 0. \tag{8.132}$$

Under the condition (8.132), the wave energy is either at rest or moving perpendicular to the current. In the point B, lying to the right of the point S, we have:

$$\left|\vec{U}\right| < |C_g(k_B)| < |C_B(k_B)|, \tag{8.133}$$

while at the point C is:

$$|C_g(k_B)| < \left|\vec{U}\right| < |C_B(k_B)|. \tag{8.134}$$

The wave is now propagating against the current in the sense that its crest moves upstream but its energy is being swept downstream.

The point D corresponds to waves with $\sigma_D < 0$. The direction of wave propagation is upstream but the wave is being swept downstream faster than its phase velocity, i.e.:

$$\left|\vec{U}\right| > |C_D(k_D)| > |C_g(k_D)|. \tag{8.135}$$

It should be added that the difference between the dispersion relation for waves on current and for waves on still water can be an appreciable source of error if measurements of pressure at the bottom are used to deduce surface wave amplitudes (Peregrine, 1976).

8.7.2 Waves on slowly varying currents

Usually the current systems in the coastal zone are non–uniform and non–steady. The variations of a current can change all the parameters describing a wave train. If the waves propagate onto a faster or slower flow, the frequency will remain constant, but the wave length will either increase or decrease. Moreover, for the non–zero angle between the wave and current vectors, a change in current produces the refraction of the waves and the energy transfer between the current and the waves.

In a study of waves on a current with large–scale variation it is natural to start by specifying the ray conservation principle. Under the assumption of slowly varying current (eq. (1.13)), from eqs. (1.18) and (8.123) we get:

$$\frac{\partial \vec{k}}{\partial t} + \nabla_h \left[\sigma(k, h) + \vec{k} \cdot \vec{U} \right] = 0. \tag{8.136}$$

If the direction of the phase velocity is taken in the direction of the local wave–number, specified by the unit vector $\vec{l} = \vec{k}/\left|\vec{k}\right|$, eq. (8.136) yields:

$$\frac{\partial \vec{k}}{\partial t} + \left(C_g + \vec{l} \cdot \vec{U} \right) \nabla k + \left[\frac{\partial \sigma}{\partial h} \nabla h + k \nabla \left(\vec{l} \cdot \vec{U} \right) \right] = 0. \tag{8.137}$$

For a steady wave train, the variations of the wave–number along the wave ray and the variations of the rays direction Θ can be rewritten in the form (Phillips, 1977):

$$\frac{\partial k}{\partial s} = \left(C_g + \vec{l} \cdot \vec{U} \right)^{-1} \left[\frac{\partial \sigma}{\partial h} \frac{\partial h}{\partial s} + k \frac{\partial}{\partial s} \left(\vec{l} \cdot \vec{U} \right) \right], \tag{8.138}$$

$$\frac{\partial \Theta}{\partial s} = -\left(C_g + \vec{l} \cdot \vec{U} \right)^{-1} \left[\frac{1}{k} \frac{\partial \sigma}{\partial h} \frac{\partial h}{\partial n} + \frac{\partial}{\partial n} \left(\vec{l} \cdot \vec{U} \right) \right], \tag{8.139}$$

since $\nabla \times \vec{k} = 0$.

As the ray pattern is known, the consequent amplitude change can be found by assuming conservation of the wave action (see Section 1.4). Adopting the definitions given in Fig. 8.12 we get (Jonsson et al., 1970):

$$\frac{\partial}{\partial x} \left[\frac{E}{\sigma} (U \cos \delta + C_g \cos \alpha) \right] + \frac{\partial}{\partial y} \left[\frac{E}{\sigma} (U \sin \delta + C_g \sin \alpha) \right] = 0. \tag{8.140}$$

The absolute group velocity $\vec{C}_g^{(a)} = \vec{C}_g + \vec{U}$ defines the wave ray direction and:

$$\tan \mu = \frac{U \sin \delta + C_g \sin \alpha}{U \cos \delta + C_g \cos \alpha}. \tag{8.141}$$

The wave ray is parallel to the wave orthogonal only when $U = 0$ and $\alpha = \delta$.

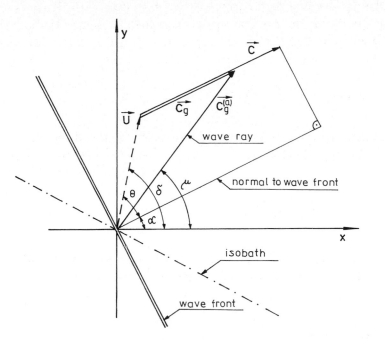

Figure 8.12: Velocities and angles. Definition sketch.

8.7.3 Waves on current varying in horizontal plane

In a wave train propagating on a current in which the surface velocity varies, the excess momentum flux produces an interchange of energy between waves and current. For the simplicity we assume that $\vec{U} = U(U_1(x), 0, 0)$. Such current describes the flow in channels and rivers where the velocity is changing in response to the depth $h(x)$. Since the current is steady, the dispersion relation becomes:

$$\sigma^2 = (\omega - k \cdot U_1)^2 = gk \tanh kh. \tag{8.142}$$

For a given U_1 and h values, eq. (8.142) is only needed to determine k. Let us consider the simpler case of the deep water. Thus:

$$\omega = k(C + U_1) = const = K. \tag{8.143}$$

We take the constant K equal to ω_0 which is the wave frequency at a location where $U_1 = 0$, i.e. $\omega_0 = (gk_0)^{1/2}$, $C_0 = (g/k_0)^{1/2}$. Eq. (8.143) can be rewritten in the form:

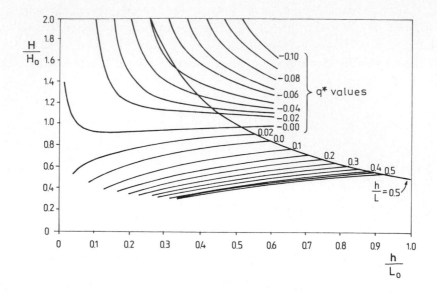

Figure 8.13: Influence of the current on the wave height transformation.(From Jonsson et al., 1970)

$$\left(\frac{C}{C_0}\right)^2 - \left(\frac{C}{C_0}\right) - \frac{U_1}{C_0} = 0, \tag{8.144}$$

having a solution:

$$\frac{C}{C_0} = \frac{1}{2} + \frac{1}{2}\left(1 + \frac{4U_1}{C_0}\right)^{1/2}. \tag{8.145}$$

The solution (8.145) indicates that the opposite current velocity $U_1 = (-1/4)C_0 = (-1/2)C = -C_g$ corresponds to the kinematic limit. Thus, if the current velocity is greater and opposite to the local group velocity of the waves, the energy can no longer be propagated against the stream.

In order to describe the wave height variation we adopt the two–dimensional current–wave system. We assume that the absolute wave period, water discharge, the bottom topography, and the deep–water wave height (the current velocity is zero) are equal to $T_a, q, h(x), H_0$, respectively. The wave action conservation principle (8.140) yields the variation of the wave height H in the function (h/L_0) as shown in Fig. 8.13. The non–dimensional discharge q^* is defined as $q^* = q/C_0L_0$, in which $C_0 = (g/2\pi)T_a$ and $L_0 = C_0T_a$. The negative values of the discharge correspond to the waves moving against the currents, while for the positive values of q, the directions of waves and currents are parallel.

Consider now the steady current in which the velocity is varying with x, i.e. $\vec{U} = (0, U_2(x),))$. For simplicity the linear shallow water theory will be used. We define $u(x, y, t)$, $v(x, y, t)$ as the velocities induced by waves and $h = h(x)$ as the water depth. Assuming small amplitudes, the linearized equations of motion and continuity are given by (see also Section 4.4):

$$
\left.
\begin{aligned}
\frac{\partial u}{\partial t} + U_2(x) \frac{\partial u}{\partial y} &= -g \frac{\partial \zeta}{\partial x} \\[1em]
\frac{\partial v}{\partial t} + U_2(x) \frac{\partial v}{\partial y} + u \frac{\partial U_2(x)}{\partial x} &= -g \frac{\partial \zeta}{\partial y} \\[1em]
\frac{\partial \zeta}{\partial t} + U_2(x) \frac{\partial \zeta}{\partial y} + \frac{\partial (hu)}{\partial x} + h \frac{\partial v}{\partial y} &= 0
\end{aligned}
\right\}.
\tag{8.146}
$$

For uniformly periodic motion in time and in y-direction we assume:

$$
(u, v, \zeta) = [u(x), v(x), \zeta(x)] \, e^{i(my - \omega t)},
\tag{8.147}
$$

in which: m - wave number in the $+y$ direction.
Substituting (8.147) into (8.146) yields (Mei and Lo, 1984; Kirby, 1986):

$$
\frac{\partial^2 \zeta}{\partial x^2} + \left[\frac{1}{h} \frac{dh}{dx} + \frac{2m}{\sigma} \frac{dU_2}{dx} \right] \frac{d\zeta}{dx} + \left(\frac{\sigma^2}{gh} - m^2 \right) \zeta = 0.
\tag{8.148}
$$

In order to simplify the further calculations we assume that the region under consideration can be divided into subregions of constant depth h and current U_2, separated by vertical vortex sheets. Therefore, the wave motion in each region may be treated as irrotational and the potential theory can be used. In any (i) region we have:

$$
\frac{\partial^2 \zeta_i}{\partial x^2} + \left(\frac{\sigma^2}{gh} - m^2 \right) \zeta_i = 0.
\tag{8.149}
$$

The solution of eq. (8.149) is:

$$
\zeta_i = a \, \exp(i l_i \, x) + b \, \exp(-i l_i \, x),
\tag{8.150}
$$

in which:

$$
l_i = \left(\frac{\sigma_i^2}{g \, h_i} - m^2 \right)^{1/2}.
\tag{8.151}
$$

with σ_i - intrinsic frequency relative to the current.

Using the result of Kirby (1986) we consider only the scattering of an freely propagating incident wave from region 1 ($x < 0$) at a discontinuity of depth and current at $x = 0$. Thus:

$$\left. \begin{array}{ll} \zeta_i & = \exp(i\,l_1\,x), \qquad \zeta_R = K_R \exp(-i\,l_1\,x) \\ \zeta_T & = K_T \exp(i\,l_2\,x) \end{array} \right\}, \tag{8.152}$$

in which:

$$l_1 = \left[\frac{\omega^2}{gh_1} - m^2 \right]^{1/2}, \qquad l_2 = \left[\frac{(\omega - mU_2)^2}{gh_2} - m^2 \right]^{1/2}; \tag{8.153}$$

ζ_i, ζ_R and ζ_T are the incident, reflected and transmitted waves, respectively. In the region 2 ($x > 0$), the water depth and current velocity are assumed to be equal to h_2 and U_2. At the discontinuity ($x = 0$) we assume (Kirby, 1986):

$$\frac{h_1}{\sigma_1^2} \frac{\partial \zeta_1}{\partial x} = \frac{h_2}{\sigma_2^2} \frac{\partial \zeta_2}{\partial x} \qquad \text{at} \qquad x = 0, \tag{8.154}$$

or:

$$\frac{h_1 \, l_1}{\sigma_1^2} (1 - K_R) = \frac{h_2 \, l_2}{\sigma_2^2} K_T. \tag{8.155}$$

Furthermore, we require the continuity of ζ. Thus:

$$1 + K_R = K_T. \tag{8.156}$$

Solving for K_R and K_T gives:

$$K_R = \left(1 - \frac{h_2}{h_1} \frac{\sigma_1^2}{\sigma_2^2} \frac{l_2}{l_1} \right) \left(1 + \frac{h_2}{h_1} \frac{\sigma_1^2}{\sigma_2^2} \frac{l_2}{l_1} \right)^{-1}, \tag{8.157}$$

$$K_T = 2 \left(1 + \frac{h_2}{h_1} \frac{\sigma_1^2}{\sigma_2^2} \frac{l_2}{l_1} \right)^{-1}. \tag{8.158}$$

The matching conditions method, described above, has been used by Mei and Lo (1984) to study the scattering of incident waves on the *top-hat* current defined by:

$$U_2 = \begin{cases} 0, & |x| > a \\ U_2, & |x| \le a \end{cases} \tag{8.159}$$

Kirby et al. (1987) have used the similar approach to study the interaction of the surface waves and the current flowing along a submerged trench. In Section 3.6 it was shown that a strong resonance appears in the neighbourhood of $2k/\lambda = 1$ (λ - bar wave number), leading to greatly enhanced reflection from the long–crested underwater bar field. The presence of an ambient current field has an effect on the reflective characteristics of the bars. Moreover, the current produces a significant shift of the resonant frequencies and intensifies the reflection of waves due to the additional effect of the perturbed current field (Kirby, 1988).

Let us extent the analysis given above on the case of random waves propagating from a spatially homogeneous region into an inhomogeneous region with a nonuniform depth profile h(x) and traversing a steady nonuniform current field, $\vec{U} = (0, U_2(x), 0)$. We can interrelate the wave number spectral density $F(\vec{k}; \vec{x}, t)$ and the directional spectral density $S_1(\omega, \Theta; \vec{x}, t)$ by:

$$F(\vec{k}; \vec{x}, t) = \frac{\omega}{k} \frac{\partial \Omega}{\partial k} S_1(\omega, \Theta; \vec{x}, t), \tag{8.160}$$

where:

$$\omega = \Omega(\vec{k}, f) = \vec{U} \cdot \vec{k} + \sigma. \tag{8.161}$$

Under the conservative conditions, the energy balance reduces to (see also eq. (1.74)):

$$\left[\frac{\partial}{\partial t} + \nabla_h \left(\vec{U} + \vec{C}_g \right) \right] \left(\frac{F}{\sigma} \right) = 0. \tag{8.162}$$

Thus:

$$\frac{F}{\sigma} = \frac{\omega}{k} \frac{\partial \Omega}{\partial k} \frac{S_1}{\sigma} = const, \tag{8.163}$$

or:

$$\frac{F}{\omega - \vec{U} \cdot \vec{k}} = \frac{\omega}{k} \frac{\partial \Omega}{\partial k} \frac{S_1(\omega, \Theta)}{\omega - \vec{U} \cdot \vec{k}} = const. \tag{8.164}$$

To specify the space–dependent quantities, we designate the deep water values that are spatially homogeneous in the absence of currents by the subscript (∞). Hence, the simplest form of the constant in (8.164) is $(F/\omega)_\infty$ or $(C_g S_1/k)_\infty$. In terms of these quantities, eq. (8.164) yields:

$$F(\vec{k}) = \left(1 - \frac{\vec{U} \cdot \vec{k}}{\omega}\right) F_\infty(\vec{k}_\infty) \qquad (8.165)$$

and

$$S_1(\omega, \Theta) = \frac{k(C_g)_\infty \left(1 - \omega^{-1}\vec{U} \cdot \vec{k}\right)}{k_\infty (\partial \Omega/\partial k)} S_{1_\infty}(\omega, \Theta_\infty). \qquad (8.166)$$

Physically, the functions $F(\vec{k})$ and $S_1(\omega, \Theta)$ should be greater than zero. Therefore, the condition:

$$\left(1 - \frac{\vec{U} \cdot \vec{k}}{\omega}\right) \geq 0, \qquad (8.167)$$

must be satisfied. Moreover, the Snell's law gives:

$$|\sin \Theta| \leq 1.0, \qquad (8.168)$$

where the angle Θ, between wave orthogonal and normal to the bottom contour, is taken to be positive clockwise.
Using eqs. (8.161), (8.167) and (8.168), we get:

$$\left(1 - \frac{\vec{U} \cdot \vec{k}}{\omega}\right) \geq [\tanh(kh) \, |\sin \Theta_\infty| \,]^{1/2}. \qquad (8.169)$$

When (8.169) becomes an equality, $|\sin \Theta| = 1$, we assume that the associated spectral component is totally reflected. As the denominator of eq. (8.166) represents the component of the transport velocity along the orthogonal, the following condition must be satisfied:

$$\frac{\partial \Omega}{\partial k} = \vec{C}_g + \vec{U} \cdot \frac{\vec{k}}{k} > 0. \qquad (8.170)$$

Therefore, local group velocity $\vec{C}_g$ must be opposite in direction and larger in magnitude relative to the horizontal current component, $\vec{U} \cdot \vec{k}/k$, in the direction of wave propagation. When $\vec{C}_g = -\vec{U} \cdot \vec{k}/k$, the associated spectral component can no longer propagate against the current in that direction and the spectral magnitude, S_1, becomes infinity. This fact suggests that energy of these spectral components will be dissipated by wave breaking. In the same time, the spectral quantity F over $\vec{k}$ - space remains always bounded. In general, the transformation of the spectral density F is entirely due to the current interaction in contrast with the transformation of the spectral density S_1 that involves the combined current–depth effects.

Following Tayfun et al. (1976), we summarize the main results for the current $\vec{U} = (0, U_2(x), 0)$ as:

dispersion relation:

$$\omega^2 \left[1 - \frac{U_2(x)}{C_\infty} \sin \Theta_\infty \right]^2 = gk \tanh kh, \tag{8.171}$$

phase speed relative to the current:

$$C = C_\infty \left[1 - \frac{U_2(x)}{C_\infty} \sin \Theta_\infty \right]^{-1} \tanh kh, \tag{8.172}$$

angle and wave numbers:

$$\sin \Theta = \frac{k_\infty}{k} \sin \Theta_\infty = \tanh kh \left[1 - \frac{U_2(x)}{C_\infty} \sin \Theta_\infty \right]^{-2} \sin \Theta_\infty, \tag{8.173}$$

$$k = k_\infty \left[1 - \frac{U_2(x)}{C_\infty} \sin \Theta_\infty \right]^2 (\tanh kh)^{-1}, \tag{8.174}$$

absolute group velocity:

$$\frac{\partial \Omega}{\partial k} = \frac{\omega}{k} \left[m + (1 - m) \frac{U_2(x)}{C_\infty} \sin \Theta_\infty \right], \tag{8.175}$$

spectral density:

$$S_1(\omega, \Theta) = \frac{k(C_g)_\infty \left(1 - \omega^{-1} \vec{U}_2 \cdot \vec{k} \right)}{k_\infty (\partial \Omega / \partial k)} S_{1_\infty}(\omega, \Theta_\infty). \tag{8.176}$$

8.7.4   Current velocity varying with depth

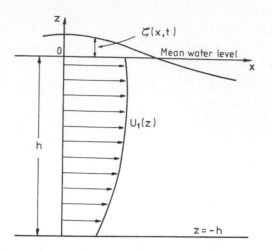

Figure 8.14: Definition of the coordinate system

As we have shown in previous Sections, the time scale of temporal variations in currents is mostly much longer than the period of gravity waves, so that a steady current is a reasonable assumption. Similarly, the spatial scale of horizontal variations in currents is frequently much larger than the wave length of wind waves. However, the same cannot be said for current velocity variations in the vertical direction. The main difficulty in attempting an analytical treatment is that the governing equation cannot be solved exactly for general wave numbers and frequencies unless the vorticity is constant.

The simplest representation of the vertical profile of Eulerian horizontal velocity $U_1(z)$ is slip at the bed and a uniform distribution in the vertical. The motion is now irrotational and the steady irrotational wave theory can be used. Any other $U_1(z)$ profile involves a vertical distribution of vorticity and rotational flow. If the current varies linearly with depth (the vorticity is constant), the solution was obtained by Jonsson et al. (1978).

Let us discuss now the interaction between a regular linear wave train and a steady shear current with arbitrary profile in two dimensional plane (Fig. 8.14). The surface elevation $\zeta(x,t)$ is sinusoidal, i.e.:

$$\zeta(x,t) = a \, \cos(kx - \omega t) \tag{8.177}$$

and the total velocity field $(u_T(x,z,t), w(x,z,t))$ can be represented as:

$$\left. \begin{array}{rcl} u_T(x,z,t) & = & U_1(z) + u(z)\cos(kx - \omega t) \\[2mm] w_T(x,z,t) & = & w(z)\sin(kx - \omega t) \end{array} \right\}.$$ (8.178)

The velocity $U_1(z)$ corresponds to the current when no waves are present.

The hydrodynamic stability theory provides a convenient equation for wave–current interaction in terms of a vertical velocity w (Peregrine, 1976):

$$\frac{d^2 w}{dz^2} - \left[k^2 - \frac{k}{\omega - kU_1} \frac{d^2 U_1}{dz_1} \right] w = 0.$$ (8.179)

The appropriate boundary conditions to be satisfied by $w(z)$ are:

$$w(z) = 0 \quad \text{on} \quad z = -h,$$ (8.180)

$$(\omega - kU_1)^2 \frac{dw}{dz} + k\,(\omega - kU_1)\,w\,\frac{dU_1}{dz} - gk^2 w = 0 \quad \text{on} \quad z = 0,$$ (8.181)

$$w(z) = a(\omega - kU_1) \quad \text{on} \quad z = 0.$$ (8.182)

The first condition (8.181) expresses the dispersion relation and the second (8.182) is simply the kinematic free surface condition. The system (8.179) - (8.182) for k and $w(z)$ should be solved under the assumption that ω, a, k and $U_1(z)$ are known. Particularly, the relation between wavelike horizontal velocity $u(z)$ and velocity $w(z)$ can be obtained from the continuity equation, i.e.:

$$u(z) = \frac{1}{k} \frac{dw}{dz}.$$ (8.183)

However, the system of equations (8.179) - (8.182) cannot be solved analytically for arbitrary wave number k and frequency ω unless the current is depth independent or it varies linearly with depth. For general case, it is necessary to resort to numerical solution of equations (8.179) - (8.182). The comparison between the predictions of a numerical model and experimentally measured values of the wave length and velocity profiles has shown a good agreement (Thomas, 1981). In the experiments, an adverse current containing an arbitrary distribution of vorticity was generated. In Figure 8.15 the comparison of the predicted and measured values of the maximum horizontal wavelike component $u(z)$ was shown.

A little work exists for rotational currents and the theory cannot be used to include arbitrary distributions of vorticity and finite–amplitude waves. For

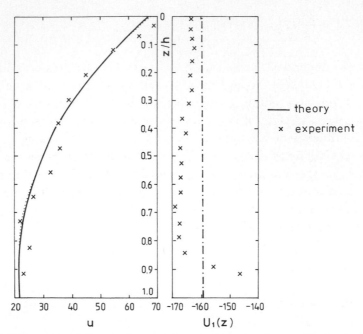

Figure 8.15: Comparison between theory and experiment for the amplitude of wavelike component $u(z)$ under a regular wave interacting with horizontal current $\bar{U}_1 = 0.1598\ m/s$. (From Thomas, 1981)

example in Chapter 2, the influence of the uniform current on the wave length of the 5th order Stokes waves was presented. A more general profile $U_1(z)$ involves a vertical distribution of vorticity and rotational flow. Therefore, let us consider a current velocity as:

$$U_1(z) = \bar{U}_1 + \Omega_0 \left(z + \frac{1}{2} h \right), \tag{8.184}$$

which permits some measure of the influence of more realistic current velocity profiles. In eq. (8.184), the vorticity Ω_0 remains constant throughout the fluid domain. The corresponding boundary value problem in terms of the stream function ψ becomes (Kishida, 1986):

$$\frac{\partial^2 \psi}{\partial x^2} + \frac{\partial^2 \psi}{\partial z^2} = \Omega_0, \tag{8.185}$$

$$\psi = 0 \qquad \text{at} \qquad z = 0, \tag{8.186}$$

$$\frac{1}{2} \left(\frac{\partial \psi}{\partial x} \right)^2 + \frac{1}{2} \left(\frac{\partial \psi}{\partial z} \right)^2 - \Omega_0\, Q + g\zeta = R \qquad \text{at} \qquad z = \zeta, \tag{8.187}$$

$$\psi(x, z) = \psi\left(x + \frac{2\pi}{k}, z\right).$$ (8.188)

Expanding the dependent variables - ψ, ζ, Q and R - into power series of small
parameter $\epsilon = (1/2)kH$, the successive problems of order ϵ^i can be obtained.
Higher order theory predicts a reduction in the wave number for an assist-
ing current and increasing of it for an opposing current, which is well known
result. The influence of vorticity is demonstrated by decreasing of the wave
number as the vorticity increases. Vorticity at realistic values is generally
unimportant for the kinematic and dynamic parameters (wave number, surface profile, horizontal velocity and acceleration, dynamic pressure and energy
etc.). However, the influence of a uniform coflowing current on higher order
waves is important and cannot be neglected.

8.8   References

Basco, D.R., 1982. Surf zone currents. 1. State of knowledge. CERC,
Misc. Rep., 82-7, 243 pp.

Baum, S.K. and Basco, D.R., 1986. A numerical investigation of the
longshore current profile for multiple bar trough beaches. *Proc. 20th
Coastal Eng. Conf.*, 1: 971–985.

Bowden, K.F., 1983. *Physical oceanography of coastal waters.* Ellis Horwood Limited, 302 pp.

Bowen, A.J., 1969. The generation of longshore currents on a plain beach.
Jour. Mar. Res., 27: 206–215.

Dalrymple, R.A., 1978. Rip currents and their causes. *Proc. 16th Coastal
Eng. Conf.*, 2: 1414–1427.

Defant, G., 1961. *Physical oceanography.* II. Pergamon Press, 598 pp.

Dronkers, J.J., 1964. *Tidal computations in rivers and coastal waters.*
North–Holland Publ. Company, Amsterdam, 518 pp.

Druet, Cz., 1978. *Hydrodynamika morskich budowli i akwenow portowych.*
Wydawnictwo Morskie, Gdansk, 390 pp. (in Polish).

Ebersole, B.A. and Dalrymple, R.A., 1980. Numerical modelling of
nearshore circulation. *Proc. 17th Coastal Eng. Conf.*, 1: 2710–2725.

Felzenbaum, A.I., 1976. Obliczanie pradow w strefie przybrzeznej. *Studia
i Mater. Oceanol.*, KBM, 16: 65–95, (in Polish).

Godin, G., 1972. *The analysis of tides.* Liverpool University Press, 264 pp.

Hansen, J.B. and Svendsen, I.A., 1984. A theoretical and experimental study of undertow. *Proc. 19th Coastal Eng. Conf.,* 2246–2262.

James, I.D., 1974. A non–linear theory of longshore currents. *Estuarine and Coastal Mar. Sci.,* 2: 235–249.

Jonsson, I.G., Skougaard, C. and Wang, J.D., 1970. Interaction between waves and currents. *Proc. 12th Coastal Eng. Conf.,* 1: 489–507.

Jonsson, I.G., Brink–Kjaer, O. and Thomas, G.P., 1978. Wave action and set–down for waves on a shear current. *Jour. Fluid Mech.,* 87: 401–416.

Kirby, J.T., 1986. Comments on "the effects of a jet–like current on gravity waves in shallow water". *Jour. Phys. Oceanogr.,* 16: 395–397.

Kirby, J.T., 1988. Currents effects on resonant reflection on surface water waves by sand bars. *Jour. Fluid Mech.,* 186: 501–520.

Kirby, J.T., Dalrymple, R.A. and Seo, S.N., 1987. Propagation of obliquely incident water waves over a trench. Part 2. Currents flowing along the trench. *Jour. Fluid Mech.,* 176: 95–116.

Kishida, N., 1986. Nonlinear wave–current interaction. Report UCB/HEL - 86/05, Dep. Civil Eng., Univ. Berkeley.

Kraus, N.C. and Sasaki, T.O., 1979. Effects of wave angle and lateral mixing on the longshore current. *Coastal Eng. in Japan,* 22: 59–74.

Leontev, I.O., 1988. Randomly breaking waves and surf–zone dynamics. *Coastal Eng.,* 12: 83–103.

Leontev, I.O. and Speransky, N.S., 1980. Issledovaniye kompensatsennogo protivotecheniya v beregovoy zonie. *Vodnye Resursy,* 3: 122–131, (in Russian).

Liu, P.L.F. and Dalrymple, R.A., 1978. Bottom friction stresses and longshore currents due to waves with large angles of incidence. *Jour. Marine Res.,* 36: 357–375.

Longuet–Higgins, M.S., 1953. Mass transport in waters. *Phil. Trans.Roy. Soc. London,* A245: 535–581.

Longuet–Higgins, M.S., 1970. Longshore currents generated by obliquely incident sea waves. *Jour. Geoph. Res.,* 75: 6778–6789.

Longuet–Higgins, M.S. and Stewart, R.W. 1962. Radiation stress and mass transport in gravity waves with application to "surf beats". *Jour. Fluid Mech.*, 13: 481–504.

McDougal, W.G. and Hudspeth, R.T., 1986. Influence of lateral mixing on longshore currents. *Ocean Eng.*, 13: 419–433.

Mei, C.C., 1983. *The applied dynamics of ocean surface waves.* A Wiley–Inter–Science Publication, New York, 734 pp.

Mei, C.C. and Liu, P.L.F., 1977. Effects of topography on the circulation in and near the surf zone; linearized theory. *Jour. Estuarine and Coastal Mar. Sci.*, 5: 25–37.

Mei, C.C. and Lo, E., 1984. The effects of a jet–like current on gravity waves in shallow water. *Jour. Phys. Oceanogr.*, 14: 471–477.

Noda, E.K., 1974. Wave–induced nearshore circulation. *Jour. Geoph. Res.*, 79: 4097–4106.

Peregrine, D.H., 1976. Interaction of water waves and currents. *Adv. Appl. Mech.*, 16: 9–117.

Phillips, O.M., 1977. *The dynamics of the upper ocean.* Cambridge Univ. Press, 336 pp.

Ramming, H.G. and Kowalik, Z., 1980. *Numerical modelling of marine hydrodynamics.* Elsevier Scientific Publishing Company, Amsterdam, 368 pp.

Sasaki, M., 1984. Rip currents of free jet type. *Coastal Eng. in Japan*, 27: 139–150.

Shadrin, I.F., 1972. *Techeniya beregovoy zony bezprilivnogo morya.* Izd. Nauka, Moskva, 125 pp. (in Russian).

Stive, M.J.F. and Wind, H.G., 1982. A study of radiation stress and set up in the nearshore region. *Coastal Eng.*, 6: 1–25.

Svendsen, Ib.A., 1984. Mass flux and undertow in the surf zone. *Coastal Eng.*, 8: 347–365.

Tayfun, M.A., Dalrymple, R.A., Yang, C.Y., 1976. Random wave–current interaction in water of varying depth. *Ocean Eng.*, 3:403–420.

Thomas, G.P., 1981. Wave–current interactions: an experimental and numerical study. Part 1. Linear waves. *Jour. Fluid Mech.*, 110: 457–474.

Thornton, E.B., 1970. Variation of longshore current across the surf zone. *Proc. 12th Coastal Eng. Conf.,* 1: 291–308.

Wu, C.S. and Liu, P.L.F., 1984. Effects of nonlinear inertial forces on nearshore currents. *Coastal Eng.,* 8: 15–32.

Wu, C.S., Thornton, E. and Guza, R.T., 1985. Waves and longshore currents: Comparison of a numerical model with field data. *Jour. Geoph. Res.,* C90: 4951–4958.

Chapter 9

VARIATION OF SEA LEVEL

9.1　General remarks

The influence of water depth on surface processes is generally negligibly small for the deep oceans. When waves enter shallow water over a shoaling bottom, the characteristic vertical scale of wave motion is comparible with the actual water depth. Therefore, the precise evaluation of the local water depth plays an important role in selecting the most adequate method of computation for many hydrodynamic processes in the coastal zone.

The actual water depth in the arbitrary point of the coastal zone is a summation of many constituents. Among them, the most important are:

- geometric water depth,
- influence of the wave motion,
- tidal oscillation,
- seiching,
- storm surges.

Let us go into a more detailed consideration of these mechanisms.

9.2　Variation of mean water level due to waves

Waves approaching a sloping coastline undergo modifications in their main parameters (wave height H, wave length L, crest angle α, stillwater depth h) resulting from shoaling, refraction, diffraction, reflection, and breaking processes. Therefore, spacial changes in the radiation stress components must result and a new time–averaged equilibrium will be established for the time–averaged water level. Longuet–Higgins and Stewart (1964) by considering the conservation of momentum flux in incident wave train, derived theoretical expression for the changes in sea level:

$$\frac{dS_{xx}}{dx} + \rho g \left(h + \bar{\zeta} \right) \frac{d\bar{\zeta}}{dx} = 0, \tag{9.1}$$

If S_{xx} is known, eq. (9.1) can be integrated to yield a mean–water–level set–down outside the breaker line and a mean–water–level set–up in the surf zone. Inside the break point wave energy decreases shoreward, leading to a decrease in the radiation stress. For spilling type breakers on dissipative beaches, the assumption commonly employed is that the breaker index γ_b (ratio of breaking wave height to mean depth at breaking):

$$\gamma_b = \frac{H_b}{(h + \bar{\zeta})} \approx \left(\frac{H}{h}\right)_b, \tag{9.2}$$

remains a fixed ratio throughout the entire surf zone. Under the assumption that linear wave theory is applicable to compute radiation stress component, $S_{xx} = (3/2)E$ (see eq. (8.21)), we get:

$$S_{xx} = \frac{3}{16}\rho g \gamma_b^2 (h + \bar{\zeta})^2. \tag{9.3}$$

Using (9.3) in momentum balance equation (9.1) yields:

$$\left(1 + \frac{3\gamma_b^2}{8}\right)\frac{d\bar{\zeta}}{dx} = \frac{-3\gamma_b^2}{8}\frac{dh}{dx}, \tag{9.4}$$

or

$$\frac{d\bar{\zeta}}{dx} = -\left(1 + \frac{8}{3\gamma_b^2}\right)^{-1}\tan\beta. \tag{9.5}$$

Therefore, the mean water surface slope is proportional to the beach slope (Fig. 8.3). The total rise of the mean water level in the surf zone can be calculated by integrating (9.4); thus:

$$\bar{\zeta} = -\frac{3}{8}\gamma_b^2\left(1 + \frac{3}{8}\gamma_b^2\right)^{-1}h(x) + C. \tag{9.6}$$

Evaluating the constant at $x = x_b$, the breaker line, where $\bar{\zeta} = \bar{\zeta}_b$, gives:

$$\bar{\zeta}(x) = \bar{\zeta}_b + \frac{3}{8}\gamma_b^2\left(1 + \frac{3}{8}\gamma_b^2\right)^{-1}[h_b - h(x)]. \tag{9.7}$$

Using shallow water theory and eq. (8.23), $\bar{\zeta}_b$ becomes:

$$\bar{\zeta}_b = -\frac{1}{16}\gamma_b H_b. \tag{9.8}$$

For the maximum set–up at the shoreline, eq. (9.7) shows that:

$$\left(1 + \frac{3}{8}\gamma_b^2\right)^{-1}\bar{\zeta}_{max} = \bar{\zeta}_b + \frac{3}{8}\gamma_b^2\left(1 + \frac{3}{8}\gamma_b^2\right)^{-1}h_b, \tag{9.9}$$

or

$$\bar{\zeta}_{max} \approx \bar{\zeta}_b + \frac{3}{8}\gamma_b^2\, h_b. \tag{9.10}$$

Substituting (9.8) into (9.10) we get:

$$\bar{\zeta}_{max} \approx \frac{5}{16}\gamma_b\, H_b. \tag{9.11}$$

It means that the mean water level at the shoreline is about 25% of breaker wave height due to wave set–up.

The more general case is when waves approach the beach at an angle. The magnitude of theoretical set–down and set–up is shown to depend on wave angle. Outside the breaker line, the wave height is determined by assuming a constant energy flux and Snell's law (3.7). Including that fact into eq. (8.23) and expressing wave height H and wave angle θ in terms of deepwater values, we obtain the extention of eq. (8.23) in the form:

$$\bar{\zeta} = -\frac{H_0^2}{16}\frac{k}{m}\frac{\coth^2(kh)}{\sinh(2kh)}\frac{\cos\theta_0}{\cos\theta}, \tag{9.12}$$

Similarly, the solution for spilling breakers with γ_b constant across the surf zone becomes:

$$\frac{d\bar{\zeta}}{dx} = \left(1 + \frac{8}{3\gamma_b^2\cos^2\theta}\right)^{-1}\frac{dh}{dx}. \tag{9.13}$$

The set–up is now no longer a constant proportion to the beach slope as with normal wave incidence.

All formulas given above were based on the radiation stresses computed from linear wave theory for regular waves. In the irregular wave surf zone there is no single breaker line, since at each location only a percentage of waves passing have broken. As the wave breaking is highly nonlinear and occurs to individual waves in physical space and not to individual spectral components, the model based on a wave–by–wave height theoretical and empirical probability distribution for individual waves in the space time domain is usually used (Battjes, 1974). At each depth, the limiting wave height H_b

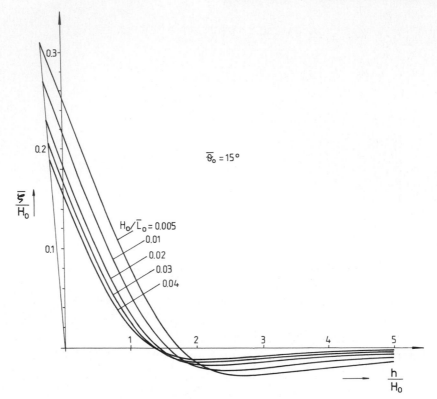

Figure 9.1: Set–down and set–up curves for a deepwater Pierson–Moskowitz spectrum, and for various wave steepness. (From Battjes, 1974)

can be defined as a wave height, which cannot be exceeded by the individual waves of the random wave field and that those wave heights which in the absence of breaking would exceed H_b are reduced by breaking to the value H_b. The energy variation in the surf zone thus results from clipping a fictitious wave height distribution which is present if breaking did not occur. The distribution for the fictitious wave heights H_f is assumed to follow the Rayleigh probability distribution:

$$
P(H) = prob\,\{H \leq H_0\} = \begin{cases} 0, & H < 0 \\ 1 - \exp\left(-\dfrac{H_0^2}{\bar{H}_f^2}\right), & 0 \leq H \leq H_m \\ 1, & H \geq H_m \end{cases} \tag{9.14}
$$

in which: H - stochastic wave height, $\quad H_0$ - wave height of interest, $\quad \bar{H}_f^2$ - mean square value of fictitious wave height, $\quad H_m$ - maximum possible wave height in the surf zone, i.e., H_b.
Additionally it is assumed that the effects of variability of wave period and

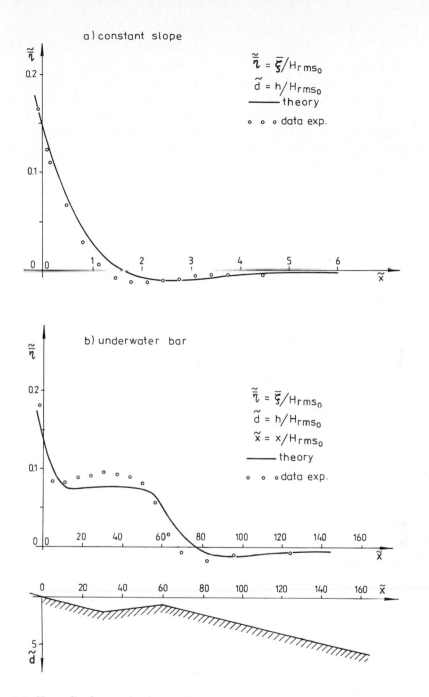

Figure 9.2: Normalized mean level variation for two types of beach profile. (From Battjes and Stive, 1985)

wave direction on breaker heights are negligible. The radiation stress is assumed to reduce by breaking in the same proportion as the total energy (Basco, 1982).

Substituting the radiation stress into momentum balance equation (9.1) we get set–down and set–up curves for the given deep–water spectrum. In Fig. 9.1 the results of calculations for a plane beach and various wave steepness are presented (Battjes, 1974). The horizontal axis is stillwater depth, h, normalized by deepwater wave height H_0. The deepwater spectrum is assumed to be a Pierson–Moskowitz spectrum and the mean wave direction is $\bar{\theta}_0 = 15^o$. It should be added that wave refraction reduces wave set–up, as expected and the mean water surface profile is almost linear near the maximum set–up line.

The mean water level variation in the surf zone can be also estimated using the Battjes and Janssen model described in Chapter 7. The results for a plane beach and an idealized bar profile are given in Fig. 9.2, according to Battjes and Stive (1985). The agreement between theory and experiment is remarkably good in view of the complexity of the physical processes involved.

9.3   Variation of mean water level due to tides

Water level variations at the coast are also influenced by tidal waves. Specially, the response of a partially enclosed body of water, such as a gulf, to the driving exerted by the ocean tide can be quite substantial. However, let us first discuss the tidal variation at the open coast line. In Chapter 8 it was shown that the surface of the undamped Kelvin wave in channel of constant water depth takes the form:

$$\bar{\zeta} = A \exp\left(\frac{-fy}{\sqrt{gh}}\right) \cos(kx - \omega t), \tag{9.15}$$

where: A - wave amplitude, f - Coriolis coefficient.
As the wave amplitude is unrealisticly increasing to infinity in a direction normal to the direction of propagation, eq. (9.15) is modelling rather properly a Kelvin wave travelling parallel to a coast, with the coast on its right–hand side in the northern hemisphere.

As opposed to Kelvin waves, the crests of the Sverdrup waves are horizontal and surface ordinates simply become:

$$\bar{\zeta} = A \cos(kx - \omega t). \tag{9.16}$$

If the restriction $U_2 = 0$ in eq. (8.110) is not imposed, other wave–like solutions are possible. One type is the Poincare wave, in which the amplitude varies sinusoidally in the transverse direction. Thus (Defant, 1961; Druet and Kowalik, 1970):

$$\zeta(x, y, t) = A \cos(ny) \cos(kx - \omega t), \tag{9.17}$$

$$U_1 = \frac{g A \omega f}{\omega^2 - f^2} \left[\frac{k}{f} \cos(ny) - \frac{n}{\omega} \sin(ny) \right] \cos(kx - \omega t), \tag{9.18}$$

$$U_2 = \frac{g A \omega f}{\omega^2 - f^2} \left[\frac{k}{\omega} \cos(ny) - \frac{n}{f} \sin(ny) \right] \sin(kx - \omega t). \tag{9.19}$$

Near a coast, due to sea bed influence, part of the energy of tidal waves approaching at an angle tends to become trapped in the coastal region. On the other hand, the reflected wave moving away from the coast will also be refracted back towards it and a caustic line may exist beyond which the energy cannot escape.

A standard harmonic representation of the shallow water tidal records, using ordinary tidal constituents, fails to reproduce the records adequately. A careful inspection of such records shows that they contain a very large number of harmonics. This is mainly due to non–linear effects in tidal wave motion. In the particular instance when two harmonics are M_2 and S_2, the friction term will cause the appearance of the new harmonics M_6, S_6, $2MS_6$, $2SM_6$, $2MS_2$, and $2SM_2$ (Godin, 1972). The problem is going, to be even more complicated when the response of a partially enclosed body of water to the ocean tide is considered. The observed oscillation of the free surface is influenced strongly by the natural modes of oscillation of water body and the resonance with the tidal periods can appear. In order to get some insight into these effects, let us consider the idealised case of a rectangular gulf of length l and constant water depth h which communicates with a deep ocean at its open end. Additionally we assume that the gulf is sufficiently narrow for the Coriolis force to be neglected. For simplicity, the friction effects will also be omitted. The solution of eqs. (8.111) under the assumption that at the closed end of gulf the normal velocity is vanishing takes the form of a standing wave, i.e.:

$$\zeta = A_i \cos kx \cos \omega t, \tag{9.20}$$

where: A_i - amplitude at the gulf entrance $(x = l)$, $\quad \omega = \sqrt{ghk}$.
At the head of gulf $(x = 0)$, the amplitude $A = A_i / \cos(kl)$. Hence, if $\cos(kl) = 0$ or if $kl = \pi/2, 3\pi/2, \ldots (2n - 1)\pi/2$, $n = 1, 2 \ldots$, the resonance

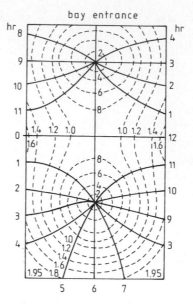

Figure 9.3: Kelvin wave in a bay. (From Bowden, 1983)

occurs. The first resonance mode is associated with the non–dimensional gulf length $kl = \pi/2$ or $l = L/4$, where L - length of the tidal wave. Thus:

$$l = \frac{1}{4}\sqrt{gh}\,T. \tag{9.21}$$

In practice the amplitude at the gulf head will not become infinite at resonance because the above simplifying assumptions are no longer valid. Nevertheless, the simple formula (9.21) may be used for the rough estimation of the period response for the natural gulf.

Let us apply the above results to the **Bay of Fundy** where the largest range of tide in the world is observed (spring tide ≈ 17.0 m at the head of the Bay). The length of Bay is about 250 km and its average depth is 70 m. Therefore eq. (9.21) yields that the resonance occurs at $T = 10.6$ hr which is close to the M_2 tidal period of 12.4 hr. A more accurate calculations show that the Bay of Fundy is not co–oscillating directly with the tides of the Atlantic Ocean. In fact, the Bay of Fundy and adjacent shallow Gulf of Maine form a system co–oscillating with ocean tides. Using this approach, Garrett (1972) estimated that the resonant period of such system is 13.3 ± 0.4 h.

In general, if the gulf is wider, the Coriolis effects are significant. The tidal wave entering into a gulf is reflected at the head of the gulf. As a result, the standing oscillation in a gulf produced by two Kelvin waves (entering and reflected) is observed. The distribution of amplitudes and phases over a gulf shows a remarkable pattern with the most characteristic point of zero tidal amplitude, known as an *amphidromic point* in tidal terminology (Fig. 9.3). The solid lines in this Figure are the *cotidal lines*, joining points at which high water occurs at the same time. The cotidal lines are radiated from the amphidromic points. The broken lines represent the *corange lines*, i.e. the lines of equal tidal range. In reality, the Kelvin waves can often be observed, but usually they are highly modified due to irregular bed topography, irregular shore geometry and the effect of bottom friction. For the prediction of the surface oscillations due to tides in such complicated areas, the numerical methods are more appropriate (Godin, 1972; Nekrasov, 1975; Marchuk and Kagan, 1977; Ramming and Kowalik, 1980). In the numerical approach, equations of motion and continuity are represented by a finite difference or finite elements techniques to a grid of points and the distribution of elevations $\bar{\zeta}$ and currents $\bar{U}$ is then computed.

9.4   Seiching

In previous section, the oscillations of the water surface in a gulfs and partially enclosed seas were forced by the ocean tides. However, any natural water basin, closed or partially open to a larger body of water will oscillate at its natural frequency if it is excited in some way. These oscillations are known as *seiches*. The most probable sources of the exciting mechanism are (Wilson, 1972):

  - passage of barometric fluctuations,
  - impact of wind gusts on the water surface,
  - release of pent–up water at a leeward shore through lapse of strong onshore winds,
  - flood discharge from rivers at one end of a bay, and
  - seismic oscillations of the earth during major earthquake.

Principal source of excitation remains meteorological. Thus, the coastal seiches are more often forced by direct action of wave energy at the mouth of a gulf or the entrance of a harbour. The oscillations produced by the imposition of an excitation of periodic nature can be both free and forced. The frequencies of the free oscillations (true seiches) are determined by the shape of the basin and the water depth. However, in bodies of water, such as gulfs, bays, estuaries and harbours, the excitations are produced by complete frequency spectrum of disturbances which are prevelant in the open sea.

The fundamental equations of motion of a water particle in the seiche oscillations can be derived by averaging the differential motion equations and mass equation over depth. Thus, the procedure similar to that developed in Chapter 8 yields:

$$\frac{\partial \zeta}{\partial t} + \frac{\partial}{\partial x_i} \int_{-h}^{\zeta} u_i \, dz = 0, \qquad i = 1, 2 \tag{9.22}$$

and

$$\frac{\partial}{\partial t} \int_{-h}^{\zeta} u_j \, dz + g(\zeta + h) \frac{\partial \zeta}{\partial x_j} = \frac{\tau_j^{(F)} - \tau_j^{(B)}}{\rho}. \tag{9.23}$$

In eq. (9.23) the nonlinear terms were neglected under the assumption that the motions are small. Let us define the volume of the flow per unit time across unit width of vertical cross section, measured between the surface and the bottom as:

$$q_j = \int_{-h}^{\zeta} u_j \, dz, \qquad j = 1, 2. \tag{9.24}$$

Eqs. (9.22) and (9.23) become:

$$\frac{\partial \zeta}{\partial t} + \frac{\partial q_j}{\partial x_j} = 0, \tag{9.25}$$

$$\frac{\partial q_j}{\partial t} + g(\zeta + h) \frac{\partial \zeta}{\partial x_j} = \frac{\tau_j^{(F)} - \tau_j^{(B)}}{\rho}. \tag{9.26}$$

For the bottom stress component $\tau_j^{(B)}$ we assume that they follow the quadratic law as functions of velocity (Wilson, 1972):

$$\frac{\tau_j^{(B)}}{\rho} = K' \, | \, u_j \, | \, u_j = \frac{K' \, | \, q_j \, |}{h^2} q_j \approx K_j \, q_j, \tag{9.27}$$

in which K_j are linearized component friction factors. Substituting (9.27) into (9.26) we get:

$$\frac{\partial q_j}{\partial t} + K_j \, q_j + g(\zeta + h) \frac{\partial \zeta}{\partial x_j} = \frac{\tau_j^{(F)}}{\rho}. \tag{9.28}$$

The right–hand sides of eqs. (9.28) represents the driving forces of surface wind stress and pressure fluctuations, which generate the fluid oscillations.

Consider now rectangular basin of length l and uniform depth h. If motion in the y direction can be neglected, the corresponding equations in the x - direction become:

$$\frac{\partial \zeta}{\partial t} + \frac{\partial q}{\partial x} = 0, \tag{9.29}$$

$$\frac{\partial q}{\partial t} + Kq + g(\zeta + h)\frac{\partial \zeta}{\partial x} = \frac{\tau^{(F)}}{\rho} = F_x(x,t), \tag{9.30}$$

or

$$\frac{\partial^2 \zeta}{\partial t^2} + K\frac{\partial \zeta}{\partial t} - g\frac{\partial}{\partial x}\left[(\zeta + h)\frac{\partial \zeta}{\partial x}\right] = -\frac{\partial F_x}{\partial x}, \tag{9.31}$$

$$\frac{\partial^2 q}{\partial t^2} + K\frac{\partial q}{\partial t} - g(\zeta + h)\frac{\partial^2 q}{\partial x^2} = \frac{\partial F_x}{\partial t}. \tag{9.32}$$

Usually, a surface ordinate ζ is much smaller than water depth h, i.e. $\zeta/h \ll 1.0$. Additionally we consider the simple case of the free oscillations in a basin, after the sudden withdrawal of the excitation force F_x. Therefore, eqs. (9.31) and (9.32) become:

$$\frac{\partial^2 \zeta}{\partial t^2} + K\frac{\partial \zeta}{\partial t} - C^2\frac{\partial^2 \zeta}{\partial x^2} = 0, \tag{9.33}$$

$$\frac{\partial^2 q}{\partial t^2} + K\frac{\partial q}{\partial t} - C^2\frac{\partial^2 q}{\partial x^2} = 0, \tag{9.34}$$

in which: $C^2 = gh$.
The solution for ζ or q takes the form:

$$\zeta(q) = X(x)T(t). \tag{9.35}$$

Thus:

$$\zeta \simeq a\,e^{-Kt/2}\cosh x\,\cos(\gamma t + \epsilon), \tag{9.36}$$

$$q \simeq \frac{a\gamma}{k}\,e^{-Kt/2}\sinh x\,\sin(\gamma t + \epsilon), \tag{9.37}$$

in which: a - amplitude of free oscillation and ϵ - phase of oscillation are depending on the original excitation:

$$\gamma = \omega\left(1 - \frac{K}{2\omega}\right)^{1/2},$$ (9.38)

and $\omega = kC$.

The value of wave number k is determined by the boundary conditions at the end of the vertical walls of the basin ($x = 0$ and $x = l$). Hence:

$$k_n l = n\pi, \qquad n = 1, 2, 3, \ldots$$ (9.39)

Eqs. (9.36) indicates that free oscillations in a basin are standing waves, decaying exponentially with time. If the damping coefficient K is small, $\gamma \approx \omega$, and the period of oscillation T_n can be approximated as:

$$T_n \approx \frac{2\pi}{k_n\sqrt{gh}} = \frac{2l}{n\sqrt{gh}}.$$ (9.40)

This formula is known as the Merian formula. The seiche would be unimodal if $l = L/2$, or bimodal if $l = L$, or $l = 3L/2$ and etc. For example, for the Lake Baikal, the Merian formula predicts $T_1 = 292\ min$ while the observed period is $T_1 = 278.2\ min$.

In the analysis given above, width and depth have been considered constant. However, for many natural basins, variable depth and width are more common features. Thus, let us consider the free oscillations in a narrow basin with variable depth, $b \equiv b(x)$. If friction K and driving force F_x are taken zero, eqs. (9.29) and (9.30) reduce to:

$$b\frac{\partial\zeta}{\partial t} + \frac{\partial q}{\partial x} = 0,$$ (9.41)

$$\frac{\partial q}{\partial t} + g(\zeta + h)b\frac{\partial\zeta}{\partial x} = 0.$$ (9.42)

Under the assumption that $\zeta \ll h(x)$, eqs. (9.41) and (9.42) can be presented in the form:

$$\frac{\partial^2\zeta}{\partial t^2} - \frac{g}{b}\frac{\partial}{\partial x}\left(bh\frac{\partial\zeta}{\partial x}\right) = 0,$$ (9.43)

$$\frac{\partial^2 q}{\partial t^2} - gh\left(\frac{\partial^2 q}{\partial x^2} - \frac{1}{b}\frac{\partial b}{\partial x}\frac{\partial q}{\partial x}\right) = 0.$$ (9.44)

Assuming the solution in the form:

$$\zeta = \zeta_0(x)\cos\omega t, \qquad q = q_0(x)\sin\omega t, \tag{9.45}$$

from eq. (9.43) we get:

$$\frac{1}{b}\frac{\partial}{\partial x}\left(bh\,\frac{\partial\zeta_0}{\partial x}\right) + \frac{\omega^2}{g}\,\zeta_0 = 0. \tag{9.46}$$

For the constant basin width, eq. (9.46) simplifies to:

$$\frac{\partial}{\partial x}\left(h\,\frac{\partial\zeta_0}{\partial x}\right) + \frac{\omega^2}{g}\,\zeta_0 = 0. \tag{9.47}$$

This result is in agreement with the formulas developed in Section 5 in Chapter 4 for long waves propagating over underwater channel with arbitrary water depth.

Wilson (1972) summarized several examples of the free periods of oscillation in lakes and in basins of simple geometric shape (constant width). For example, in parabolic basin with bottom profile equation $h(x) = h_0[1 - 4x^2/L^2]$ (origin of reference is taken in the still water surface in the middle of the basin), the fundamental period of free oscillation is $T_1 = 1.110\left[2l/\sqrt{gh_0}\right]$ and mode ratios T_n/T_1 for $n = 2,3,4$ are 0.577, 0.408 and 0.316, respectively.

The oscillations of bodies of water (bays, inlets, etc.) which are open on much larger water basins were mentioned in Section 2 in this Chapter. Wilson (1972) has pointed out that the modes of free oscillations are the odd modes at which some of equivalent closed basins, i.e. open–mouth basin and its mirror image about the mouth, would oscillate. Thus, the Merian formula (9.40) is still applicable provided that length l is replaced by $2l$ and n be restricted to the values $1, 3, 5, \ldots$, i.e.:

$$l = \frac{1}{4}\,s\,\sqrt{gh}\,T_n, \tag{9.48}$$

in which: $n = (s + 1)/2$.

If $n = 1$ ($s = 1$), eq. (9.48) is in complete agreement with eq. (9.21). Other examples of modes of free oscillations in semi-enclosed basins of simple geometric shape are summarized by Wilson (1972). For example, for the rectangular bay of length l and of triangular bottom profile $h(x) = h_1 x/l$ (origin of the reference is situated at the bay head, h_1–water depth at the bay entrance), the fundamental period of free oscillation is $T_1 = 2.618\left(2l\sqrt{gh_1}\right)$. Moreover, the ratios of the following modes T_n/T_1 are 0.435 ($n = 2$), 0.278 ($n = 3$) and 0.203 ($n = 4$).

For basins which are not of such simple form that their geometry can be expressed mathematically, more sophisticated methods should be used. For basins of constant depth but arbitrary plane form, numerical solutions have been obtained by Lee (1971) by the method of integral equations. The velocity potential $\phi(x, y)$ at any point in the bay and outside is determined in terms of a function $\phi'(x', y')$ at a boundary point (x', y') and a Green's function $G(x, y; x', y')$. The function $\phi'(x', y')$ satisfies the Helmholtz equation:

$$\frac{\partial^2 \phi'}{\partial x^2} + \frac{\partial^2 \phi'}{\partial y^2} + \left(\frac{\omega^2}{gh}\right) \phi' = 0, \tag{9.49}$$

and

$$\phi' = \int_s \left(\phi' \frac{\partial G}{\partial n} - \frac{\partial \phi'}{\partial n} G\right) ds, \tag{9.50}$$

in which s is a direction along the boundary and n is normal to the boundary. The resulting equations are applied to numerical computation. The published examples demonstrate a good agreement between theory and experiment. The extension of integral equations for varying depth can be developed by using the hybrid–element method (HEM). Then, the domain of motion is divided into two regions. The interior, comprising basin and its vicinity is discretized into finite elements, while the solution in the exterior is analytic and satisfies exactly the governing equation and the boundary condition at infinity (Mei, 1978).

The collected experimental data indicate that travelling pressure disturbances have been known to excite water oscillations. Let us consider the simple case of a positive pressure pulse moving with a constant velocity V over the water surface, along the longitudinal axis of the rectangular basin, considered above. If the pressure pulse can be represented in the form of Fourier series, the complete solution for ζ and q, for the forced oscillation are (Wilson, 1972):

$$\zeta = \sum_{r=1}^{\infty} \zeta_r, \tag{9.51}$$

and

$$q = \sum_{r=1}^{\infty} q_r, \tag{9.52}$$

in which:

$$\zeta_r = R \phi_r \, \mu \, \cos(k_r \, x) \cos (\sigma_r \, t - \epsilon) , \qquad (9.53)$$

$$q_r = R V \phi_r \, \mu \, \sin(k_r \, x) \sin (\sigma_r \, t - \epsilon) , \qquad (9.54)$$

where R, ϕ_r, μ, and ϵ are the functions of phase velocity C, pressure pulse velocity V and length l on which pressure pulse is distributed, frequency $\sigma_r = k_r V$, and wave number $k_r = r\pi/l$.

Eqs. (9.51) - (9.54) were used by Wilson (1972) for interpretation of influence of barometric fluctuations on the seiche type oscillations in some natural water basins.

9.5 Storm surges in coastal zone

The oscillation of water level over the coast line represents a vectorial sum of many constituents. One of them is the *storm surge*. A storm surge is defined as a disturbance of sea level due to meteorological factors resulting from wind and bottom stresses, atmospheric pressure changes, moving storm systems and bottom configuration. It should be noted that a surge may be either positive or negative. In the tidal area, the surge height is estimated by substracting of the predicted tide from the recorded level.

The storm surges may be expected to occur along coasts with relatively shallow waters which are affected by the passage of storms. In some areas storm surges have very disastrous effects. In North America surges frequently occur in the Gulf of Mexico and along the east coast of the United States of America, due to hurricanes. In Asia, the Bay of Bengal is particularly seriously affected and disastrous flooding has occurred on a number of occasions. In Europe, a comprehensive documentation of storm surges is related particularly to the North Sea and the Baltic Sea (Perry and Walker, 1977). For example, storm surge for January 31 - February 1, 1953, which caused serious flooding and loss of life in the Netherlands and eastern England, was one of the bigest storm surges observed in the North Sea. It was caused by the deep depression travelling eastwards the north of the British Isles and then south–eastwards into the North Sea, raising up the water level in the southern part of the North Sea (Bowden, 1983). The amplitude of storm surge on the eastern coast of England (Southend) reached about 2.75 m, while in some places at the Dutch coast it was greater than 3.95 m.

In the Baltic Sea, the highest storm surges are caused by the low pressure area which is travelling from the British Isles and the Scandinavian Peninsula south–eastwards into the South Baltic. The strong winds, associated with these cyclonic systems, generate high surges, of rather short duration. The

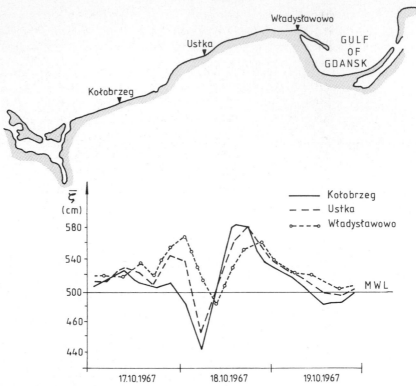

Figure 9.4: Surge heights recorded at some locations of the southern Baltic

maximum surge is moving along a coast of the South Baltic from west to east
(Majewski, 1983). In Fig. 9.4, surge heights at a number of location for the
storm in October, 1967, was shown. It should be noted that the tides in the
Baltic Sea are negligible small.

In general, the reduction in atmospheric pressure for one milibar is equivalent to about 1 cm of hydrostatic water head. Thus:

$$\Delta \bar{\zeta} = k \left(\bar{p}_a - p_a \right), \tag{9.55}$$

in which: $\Delta \bar{\zeta}$ - variation in water level, p_a - atmospheric pressure under
consideration, $\bar{p}_a$ - mean atmospheric pressure, and $k = (\rho g)^{-1}$. This is
so called *inverted barometer effect*.
If the atmospheric pressure disturbance p_0 moves with speed U, the surge will
move with it. Then, analysis given in previous Sections in this Chapter yields
the following momentum equation:

$$h \frac{\partial u}{\partial t} + gh \frac{\partial \bar{\zeta}}{\partial x} = \frac{h}{\rho} \frac{\partial p_0}{\partial x}. \tag{9.56}$$

In eq. (9.56), the pressure moving in x direction was assumed, i.e.:
$p_0 = f(Ut - x)$. Moreover, eq. (9.25) gives:

$$\frac{\partial \zeta}{\partial t} + h \frac{\partial u}{\partial x} = 0. \tag{9.57}$$

Under the assumption that wave–induced particle velocity is proportional to
the water surface displacement we get (Dean and Dalrymple, 1984):

$$u = \frac{U\bar{\zeta}}{h + \bar{\zeta}} \approx U \frac{\bar{\zeta}}{h}. \tag{9.58}$$

It will be convenient to assume the surface displacement $\bar{\zeta}$ in the form:

$$\bar{\zeta} = \bar{\zeta}(Ut - x). \tag{9.59}$$

Thus:

$$\frac{\partial \bar{\zeta}}{\partial t} = -U \frac{\partial \bar{\zeta}}{\partial x}. \tag{9.60}$$

After substitution of eqs. (9.57) - (9.60) into eq. (9.56) and integrating it, we
obtain:

$$\frac{\bar{\zeta}}{h} = \frac{p_0/\rho}{U^2 - gh}. \tag{9.61}$$

When the speed of the storm approaches the speed of the free waves, the
phenomenon of resonance should be considered. The complete resonance will
not occur, however, because of the frictional effects.

In general, the atmospheric pressure has a smaller effect on the rise of sea
level than the wind stresses at the surface, denoted by $\tau^{(F)}$. In specifying the
stress $\tau^{(F)}$ it is usually assumed that it acts in the direction of the wind and
that its magnitude is proportional to the square of the wind speed. Therefore:

$$\vec{\tau}^{(F)} = C_D \, \rho_i \left| \vec{V}_w \right| \vec{V}_w, \tag{9.62}$$

in which $\vec{V}_w$ is the wind speed measured at a given height (usually at 10 m
above the sea surface), ρ_a - density of the air and C_D is a drag coefficient.

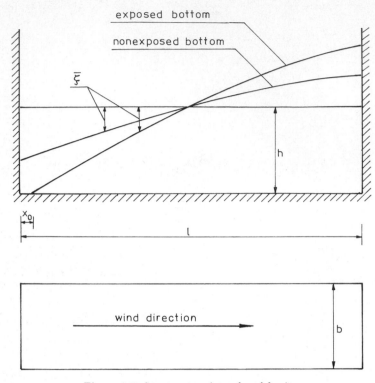

Figure 9.5: Storm surge in a closed basin

The value of C_D depends on the height at which velocity $\vec{V_w}$ is measured, the stability of the boundary layer of atmosphere and roughness of the sea surface. The detail analysis of the C_D coefficient for the various meteorological situations was given in Chapter 6.

Let us now consider the storm surge in the closed basin of the rectangular form and constant depth (Fig. 9.5). If the wind speed is constant and blowing along the channel axis, from eq. (8.74) we get:

$$\frac{d\bar{\zeta}}{dx} = \frac{\tau^{(F)} - \tau^{(B)}}{\rho g(h + \bar{\zeta})}. \tag{9.63}$$

Additionally, the following mass conservation conditions should be adopted:
- nonexposed bottom

$$\int_0^l (h + \bar{\zeta})dx = lh, \tag{9.64}$$

- exposed bottom

$$\int_{x_0}^{l} (h + \bar{\zeta})dx = lh. \tag{9.65}$$

Assuming that:

$$\frac{\tau^{(F)}}{\rho g} = k_1 \frac{V_w^2}{g} \quad \text{and} \quad \frac{\tau^{(B)}}{\tau^{(F)}} = k_2, \tag{9.66}$$

eq. (9.63) can be rewritten in the form:

$$\frac{d\bar{\zeta}}{dx} = \frac{k V_w^2}{g (\bar{\zeta} + h)}, \tag{9.67}$$

in which: $k = k_1(1 - k_2)$, $k_1 = 3.0 \cdot 10^{-6}$, and $k_2 = 0.1$. The nodal point, at which $\bar{\zeta} = 0$, is situated at the distance x_0 from the basin wall and:

$$\frac{x_0}{l} = 1 - \frac{\bar{\zeta}_{max}^2 + 2\,\bar{\zeta}_{max}\,h}{2k\,V_w^2\,l}. \tag{9.68}$$

The solutions of eqs. (9.67) - (9.68) in the form of function $\bar{\zeta}/h = f(x/l, k V_w^2 l/gh^2)$ are presented in Tables by Bretschneider (1967). In order to include some changes in water depth and basins width, it is useful to introduce so-called planform factor N, i.e.:

$$\frac{d\bar{\zeta}}{dx} = N(x)\,\frac{k V_w^2}{g\,\bar{\zeta} + h)}. \tag{9.69}$$

The factor $N(x)$ is a function of distance down the channel, changing depth, and changing width of the channel.

Let us apply this idea to the channel of constant sloping depth as:

$$h = h_0 - m\,x, \tag{9.70}$$

and of constant varying width:

$$b = b_0 - s\,x, \tag{9.71}$$

where m is the slope of the bottom and s is the slope of convergence of the sides. Under the assumption that $\bar{\zeta}/h \ll 1.0$, the factor N is (Bretschnei-

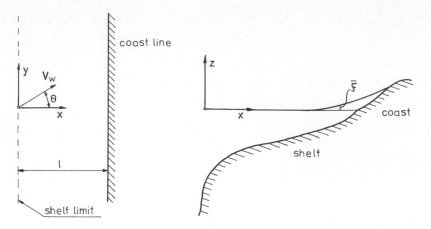

Figure 9.6: Coordinate system

der, 1967):

$$N = 2\left[\frac{1}{2} - \frac{7}{12}m + \frac{1}{8}m^2 - \frac{9}{320}m^3 + \frac{7}{160}m^4+\right.$$

$$\left. - s\left(\frac{1}{12} + \frac{1}{120}m^2 + \frac{1}{2120}m^4\right)\right]. \tag{9.72}$$

In more general case, when the quadratic or nonlinear effects are important, the numerical procedure should be used.

A number of analytical solutions can be obtained for areas of simple geometrical shape, e.g. a self–adjacent to a straight coast with constant or sloping bottom. Let us adopt a coordinate system normal to a coastline as in Fig. 9.6. The water depth h is linearly varying, i.e. $h(x) = h_0 (1 - x/l)$. Therefore, eq. (9.67) can be rewritten in the form:

$$(h + \bar{\zeta})\frac{d(\bar{\zeta} + h)}{dx} - (h + \bar{\zeta})\frac{dh}{dx} = \frac{k V_w^2}{g}, \tag{9.73}$$

in which: $dh/dx = -h_0/l$.
The separation of variables in eq. (9.73) gives:

$$-(h + \bar{\zeta})\left[\frac{h_0^2}{l}\left(\frac{h + \bar{\zeta}}{h_0} - \frac{k V_w^2 l}{g h_0^2}\right)\right]^{-1} d(\bar{\zeta} + h) = dx. \tag{9.74}$$

After solution we obtain:

$$l\left[\left(B - \frac{h+\bar{\zeta}}{h_0}\right) - B\ln\left(\frac{h+\bar{\zeta}}{h_0} - B\right)\right] = x + C, \tag{9.75}$$

in which: $B = kV_w^2 l / g h_0^2$.

The integration constant C is evaluated from the requirement that at $x = 0$, the set–up $\bar{\zeta}$ equals zero. Thus eq. (9.75) yields:

$$\frac{\bar{\zeta}(x)}{l} = \left(1 - \frac{h+\bar{\zeta}}{h_0}\right) - B\ln\left(\frac{\dfrac{h+\bar{\zeta}}{h_0} - B}{1 - B}\right). \tag{9.76}$$

A similar analysis for the constant water depth h_0 provides the following formula:

$$\frac{\bar{\zeta}}{h_0} = \sqrt{1 + 2B\frac{x}{l}} - 1.0. \tag{9.77}$$

Fig. 9.7 demonstrates the effect of the bottom slope on the storm surge. The sloping bottom generates a higher storm surge height. This effect is due to dependence of the surface slope on the local water depth (see eq. (9.67)).

The solutions (9.76) and (9.77) are appropriate for the small–scale systems, when the Coriolis forces can be neglected. The more general case of the wind blowing at an angle θ to the coast, requires the Coriolis force to compensate the hydrostatic gradient. In order to simplify the problem we assume that the onshore flow and the return flows are in balance. The wind system is uniform in the y direction. Under these conditions, the analytical solution, which also includes the shear stress at the sloping bottom, takes the form (Dean and Dalrymple, 1984):

$$\frac{x}{l^*} = \left(1 - \frac{h+\bar{\zeta}}{h_0}\right) - B^*\ln\left(\frac{\dfrac{h+\bar{\zeta}}{h_0} - B^*}{1 - B^*}\right), \tag{9.78}$$

in which: $B^* = kV_w^2 l^* / g h_0^2$ and:

$$l^* = l\left(1 - f\sqrt{\frac{8C_D\sin\theta}{f_c}\frac{V_w l}{g h_0}}\right)^{-1}, \tag{9.79}$$

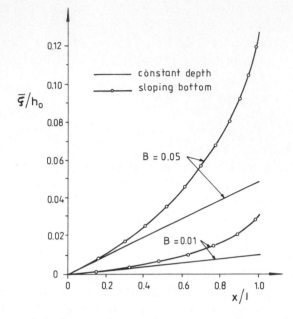

Figure 9.7: Influence of the bottom slope on the storm surge heights. (From Dean and Dalrymple, 1984)

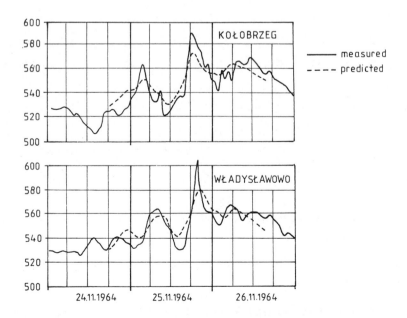

Figure 9.8: Comparison of predicted and measured storm surge heights for two locations at the coast of Southern Baltic (From Laska, 1980)

where: f - Coriolis coefficient, f_c - Darcy–Weisbach friction factor which was introduced for the bottom shear stress in y direction.

The solution (9.78) is exactly the same as the solution for surge without the Coriolis force (9.76) except that l is replaced by l^*. Thus, the Coriolis force modifies the bottom slope.

The methods of studying or forecasting storm surges for real sea basins now in use employ numerical models. The region under consideration is represented by a network of elementary areas which approximate to the actual coast line and which use the actual depths of water. Various finite difference or finite element models have been developed for the North Sea and the Baltic Sea. No attempt is made here to discuss the numerical formulation of the equations and their solutions. For information on these reference should be made to the publications by Voltsinger and Pyaskovskiy (1977), Heaps and Jones (1979), Laska (1980) and Ramming and Kowalik (1980). Fig. 9.8 demonstrates only the application of the numerical methods to predict the storm surge in some regions of the Southern Baltic Sea.

At the end it should be noted that the mean sea level itself is not constant. There are some long–terms changes of sea level due to various climatic reasons (Fuhrboter, 1986). However this problem lies outside the scope of this book.

9.6 References

Basco, D.R., 1982. Surf zone currents. Part 1. State of knowledge. US Army, CERC, Misc. Rep. 82, 243 pp.

Battjes, J.A., 1974. Computation of set–up, longshore currents, run–up and overtopping due to wind–generated waves. Comm. on Hydraulics. Delft Univ. Techn., Rep. 74–2, 244 pp.

Battjes, J.A. and Stive, M.J.F., 1985. Calibration and verification of a dissipation model for random breaking waves. *Jour. Geoph. Res.*, 90: 9159–9167.

Bowden, K.F., 1983. *Physical oceanography of coastal waters.* Ellis Horward Ltd., London, 302 pp.

Bretschneider, C.L., 1967. Storm surges. *Adv. in Hydriscience*, 4: 341–418.

Dean, R.G. and Dalrymple, R.A., 1984. *Water wave mechanics for engineers and scientists.* Prentice-Hall Inc., Englewood Cliffs, 353 pp.

Defant, A., 1961. *Physical oceanography.* Vol.II. Pergamon Press, 598 pp.

Druet, Cz. and Kowalik, Z., 1970. *Dynamika morza.* Wydawnictwo Morskie, Gdansk, 428 pp. (in Polish).

Fuhrboter, A., 1986. Veranderungen des sekularanstieges an der Deutschen Nordseekuste. *Wasser und Boden*, 9.

Garrett, C.J.R., 1972. Tidal resonance in the Bay of Fundy and Gulf of Maine. *Nature*, 238: 441–443.

Godin, G., 1972. *The analysis of tides.* Liverpool University Press, 264 pp.

Heaps, N.S. and Jones, J.E., 1979. Recent storm surges in the Irish Sea. In: J.C.J. Nihoul (Editor), *Marine forecasting.* Elsevier, Amsterdam, pp. 285–319.

Laska, M., 1980. On storm surge phenomena. Mathematical modelling of estuarine physics. *Lecture notes on coastal and estuarine physics.* Springer–Verlag, pp. 177–192.

Lee, J.J., 1971. Wave–induced oscillation in harbours of arbitrary geometry. *Jour. Fluid Mech.*, 45: 375–394.

Longuet–Higgins, M.S. and Stewart, R.W., 1964. Radiation stresses in water waves: a physical discussion, with applications. *Deep Sea Res.*, 11: 529–562.

Majewski, A., 1983. Maksymalne stany wody na polskim wybrzezu Baltyku i w delcie Wisly. *Inzynieria Morska*, 6: 443–444 (in Polish).

Marchuk, G.I. and Kagan, B.A., 1977. *Okeanskiye prilivy (matematicheskiye modeli i chislennyye eksperimenty).* Gidrometeoizdat, Leningrad, 266 pp. (in Russian).

Mei, C.C., 1978. Numerical methods in water wave diffraction and radiation. *Ann. Rev. Fluid Mech.*, 10: 393–416.

Nekrasov, A.V., 1975. *Prilivnyye volny v okraynykh moryakh.* Gidrometeoizdat, Leningrad, 285 pp. (in Russian).

Perry, A.H. and Walker, J.M., 1977. *The ocean–atmosphere system.* Longman Group Ltd., London, 267 pp.

Ramming, H.G. and Kowalik, Z., 1980. *Numerical modelling of marine hydrodynamics.* Elsevier Scientific Publishing Company, Amsterdam, 368 pp.

Voltsinger, N.E. and Pyaskovskiy, R.V., 1977. *Teoriya melkoy wody. Okeanologicheskiye zadachi i chislennyye metody.* Gidrometeoizdat, Leningrad, 206 pp. (in Russian).

Wilson, B.W., 1972. Seiches. *Adv. in Hydroscience*, 8: 1–94.

Author Index

Subject Index